Airpollution

AIR POLLUTION

FROM A LOCAL TO A GLOBAL PERSPECTIVE

Edited by Jes Fenger & Jens Christian Tjell

RSCPublishing

Polyteknisk
Forlag

Air Pollution – from a local to a global perspective
by Jes Fenger & Jens Christian Tjell (Eds.)

Published and distributed in Denmark, Norway and Sweden
by Polyteknisk Forlag, Denmark

Distributed in rest of the world by Royal Society of Chemistry, UK
Registered Charity Number 207890

© 2009 Polyteknisk Forlag
1. edition, 2009

Cover photo by Jens Landorph, Phy Grafisk
Drawings by Birte Brejl
Typesetting and cover by Phy Grafisk
Printed by InPrint

Printed in Denmark 2009

Polyteknisk Forlag ISBN-13   978-87-502-0967-6
Polyteknisk Forlag ISBN-10   87-502-0962-0
Royal Society of Chemistry ISBN-13 9781847558657
Royal Society of Chemistry ISBN-10 1847558658

Polyteknisk Forlag
Anker Engelunds Vej 1
DK-2800 Lyngby
Denmark

Phone: +45 7742 4328
Fax: +45 7742 4354
E-mail: forlag@polyteknisk.dk
Website: www.polyteknisk.dk

Royal Society of Chemistry
Thomas Graham House
Science Park
Milton Road
Cambridge CB4 0WF
UK

Phone: +44 (0) 1223 420066
Fax: +44 (0) 1223 426017
E-mail: sales@rsc.org
Website: www.rsc.org

Contents

Preface

Air pollution is a phenomenon that has increased both in geographical and temporal measure since the start of the civilisation. It is therefore a complex problem that covers many subjects. In this book they are treated by a series of experts, whose contributions are edited to form a comprehensive presentation.

The individual chapters have their own literature list that contains both direct references and recommended literature. The chapters can, in spite of a few cross-references, be read separately. This has led to some unavoidable repetitions – especially for national and international legislation that is treated briefly in the relevant chapters and repeated in more detail in a separate chapter at the end of the book.

The opinions of the different authors are their own responsibility and do not express any official attitudes of their respective institutions.

The book is primarily intended as a general presentation for university students at graduate level, but civil servants and employees in consulting companies can also use it. The interested layman can benefit from reading the more general chapters.

Many unnamed colleagues have contributed during the preparation of the text, but the editors and authors want especially to thank Birte Brejl for producing the figures. The publication has been financially supported by The COWI Foundation and the foundation R98.

Jes Fenger
National Environmental Research Institute, University of Aarhus, Denmark

Jens Chr. Tjell
Technical University of Denmark

The editors and authors

Mikael Skou Andersen

Department of Policy Analysis
National Environmental Research Institute, University of Aarhus
Grenåvej 12-14, DK 8410 Rønde, Denmark.
(Chapter 21)

Jørgen Brandt

Department of Atmospheric Environment
National Environmental Research Institute, University of Aarhus
Frederiksborgvej 399, DK 4000 Roskilde, Denmark.
(Chapter 14)

Anders Carlsen

National Board of Health
Medical Office of Health, Mid-Jutland
Lyseng Alle 1, DK 8270 Højbjerg, Denmark
(Chapter 15)

Jes Fenger

Department of Atmospheric Environment
National Environmental Research Institute, University of Aarhus
Frederiksborgvej 399, DK 4000 Roskilde, Denmark.
(Chapter 1, 2, 8, 16, 17, 20, editor)

Lise Marie Frohn

Department of Atmospheric Environment
National Environmental Research Institute, University of Aarhus
Frederiksborgvej 399, DK 4000 Roskilde, Denmark.
(Chapter 10)

Lars Kristian Gram

Force Technology
Park Alle 345, DK 2605, Brøndby, Denmark
(Chapter 11)

Gitte Brandt Hedegaard

Department of Atmospheric Environment
National Environmental Research Institute, University of Aarhus
Frederiksborgvej 399, DK 4000 Roskilde, Denmark
(Chapter 7)

Ole Hertel

Department of Atmospheric Environment
National Environmental Research Institute, University of Aarhus
Frederiksborgvej 399, DK 4000 Roskilde, Denmark
(Chapter 1, 8, 10, 14)

Jytte Boll Illerup

Department of Chemical and Biochemical Engineering
Technical University of Denmark, Building 229
2800 Kgs. Lyngby, Denmark
(Chapter 12)

Ib Johnsen

Department of Biology
University of Copenhagen
Øster Farimagsgade 2D, 2100 Copenhagen Ø, Denmark
(Chapter 18)

Matthew S. Johnson

Departmen of Chemistry
University of Copenhagen
Universitetsparken 5, DK 2100, Copenhagen Ø, Denmark
(Chapter 9)

Jan Erik Johnsson

Department of Chemical and Biochemical Engineering
Technical University of Denmark, Building 229
DK 2800 Kgs. Lyngby, Denmark
(Chapter 3, 4)

Ole John Nielsen

Copenhagen Center for Atmospheric Research
University of Copenhagen
Universitetsparken 5, DK 2100, Copenhagen Ø, Denmark
(Chapter 19)

Finn Palmgren

Department of Atmospheric Environment
National Environmental Research Institute, University of Aarhus
Frederiksborgvej 399, DK 4000 Roskilde, Denmark
(Chapter 13, 22)

Helge Ro-Poulsen

Department of Biology
University of Copenhagen
Øster Farimagsgade 2D, DK 2100 Copenhagen Ø, Denmark
(Chapter 18)

Jesper Schramm

Department of Mechanical Enginering
Technical University of Denmark, Building 403
DK 2800 Kgs Lyngby, Denmark
(Chapter 5)

Carsten Ambelas Skjøth

Department of Atmospheric Environment
National Environmental Research Institute, University of Aarhus
Frederiksborgvej 399, Roskilde DK 4000, Denmark
(Chapter 6, 8)

Jens Chr. Tjell

DTU Environment
Technical University of Denmark, Building 113
DK 2800 Kgs. Lyngby, Denmark
(Editor)

Practical information

The structure

This book is divided into 22 chapters in 6 groups. The first group gives the historical background and outlines the problem as seen in a general perspective. The second describes the various pollution sources and the technical means of emission control. Group 3 describes the dispersion, transformation and deposition of pollution in the atmosphere. Group 4 treats the various techniques, the various means of analyses and the mathematical modelling. Group 5 treats local, regional and global impacts – and to some extent the national and international laws and negotiations destined to reduce the pollution. Finally group 6 describes the legal background as well as economical, ecological and ethical aspects.

Figures, tables and formulas

Figures and other informative material are numbered separately for each chapter. As far as possible the text is illustrated with graphs. Direct measurements and calculations are shown with units on both axes; note that the scales may be shifted or be logarithmic. Illustrative curves are shown with only one or even no scales.

Figures showing air movements or dispersion are normally vertically expanded. Most air pollution phenomena take place below a few kilometres height, but may extend several thousand km horizontally.

Units

Basically the SI-system is employed, but not in all cases. The reason is first of all that to a large extent material is cited from other sources and that a recalculation may be misleading. In addition, quantities of widely different sizes appear, and it is reasonable to use units that give an immediate impression of the relative size. Note also that the same phenomenon can sometimes be described with units that may not be directly converted.

Concentrations of solids (particles) and gases in the atmosphere are often given in weight per volume as $\mu g \ m^{-3}$ or ppb (parts per billion). Larger concentrations are given in mg m^{-3} or ppm (parts per million). In air pollution gasconcentrations often refer to volume per volume, which is often indicated with a v (ppbv or ppmv).

It is only possible to convert from one system of units to the other at a given temperature and pressure. At 25 °C and 760 mm Hg some important transformations are thus:

Nitrogen dioxide, NO_2	1 ppbv = 1.882 µg m^{-3}
Nitrogen monoxide, NO	1 ppbv = 1.227 µg m^{-3}
Ozone, O_3	1 ppbv = 1.963 µg m^{-3}
Sulphur dioxide, SO_2	1 ppbv = 2.620 µg m^{-3}
Carbon monoxide, CO	1 ppbv = 1.146 µg m^{-3}

Emission is normally given as mass, but it can refer to many parameters: Amount of fuel, activity, time, area etc.

Deposition of compounds can be expressed as mass or equivalents in relation to area, precipitation etc.

The mass of a specific compound in relation to emission, deposition or concentration can be expressed both as the mass of molecules (e.g. CO_2) or the crucial element (e.g. C). The last is specifically used for cycles or mass balances of elements, where a given element can participate in more than one compound (e.g. CH_4 and CO_2). As an example 1 ton C corresponds to 3.66 ton CO_2.

The electromagnetic spectrum

Electromagnetic radiation originating from the sun is paramount to life on Earth, and the driving force in atmospheric photo-chemistry. The spectrum is divided into ultraviolet radiation, visible light and infrared radiation. The common names and the corresponding wavelength ranges are:

Common name	Abbreviation	Wavelength in nm
Ultraviolet C, short wave	UVC	100 - 280
Ultraviolet B, medium wave	UVB	280 - 315
Ultraviolet A, long wave	UVA	315 - 400
Visible light	VIS	400 – 700 (up to 750)
Near infrared	NIR, IR-A	750 - 1400
Short-wavelength infrared	SWIR, IR-B	1400 - 3000
Mid-Wavelength infrared	MWIR, IR-C	3000 – 8000
Long-wavelength infrared	LWIR, IR-C	8000-15,000

I

The atmosphere and its pollution

We all want to live in an environment that is healthy, safe and beautiful; but we also want a reasonable material standard of living. At the same time we feel an obligation to leave our descendants an Earth that is not completely plundered and uninhabitable. But these reasonable wishes are unfortunately often in conflict:

- Material standard of living is based on production that requires raw materials and energy.
- Application of lower grade materials or reuse of materials can cost more energy.
- The production of energy can take place in various ways; one may be pure, but involves a safety risk or requires raw materials that could be better used otherwise. Another may be more polluting, but is considered safer.
- Application of certain processes or materials gives more pollution than other processes, but will create work places or will be a benefit for the trade balance.
- Pollution reduction in one geographical area may lead to increase in production of waste products and thereby increased pollution in another area.
- Efforts and resources can be used to obtain the best location and design of factories, power stations or motor ways, but generally it would be nicer if they were not there at all.

Etc.

We want to consider all aspects, but even when the technological development can resolve some conflicts, it is necessary with compromises. One of the questions is how clean we want – and can have – our air. The question is political, and there is no definite answer. The problem can, however, only be discussed, if one knows what goes on in the real world. That is what this book is about.

This introductory part starts with a description of the scene: The atmosphere - as it would be without pollution. It is followed by a short historical overview of air pollution and its impacts.

1 The clean atmosphere

Jes Fenger and Ole Hertel

The Earth is surrounded by a layer of gases – the atmosphere, which is sustained by the gravity of the air masses. Thermo-molecular movements mean that the atmosphere has a certain thickness. The higher velocity a given molecule has, the higher it will go into space before its kinetic energy is transformed into potential energy and it starts to fall back towards Earth. The molecules follow Maxwell´s velocity distribution with decreasing number for increasing velocity. Therefore the density decreases towards zero with

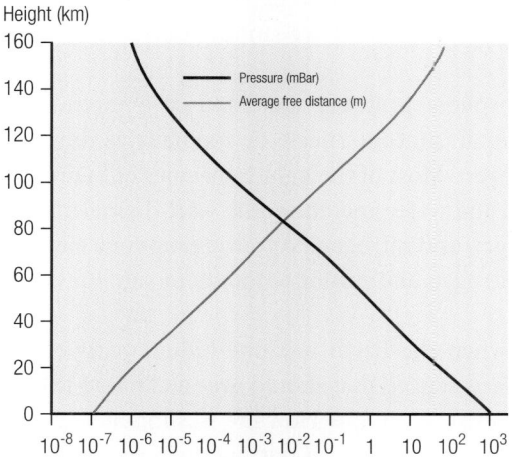

Figure 1.1 Pressure in the atmosphere rapidly drops with height. The average free molecular distance increases. Virtually all atmospheric mass lies below 100 km.

height. For a constant temperature the density will decrease exponentially and the pressure that is a consequence of weight of the mass of air lying over a given height therefore decreases rapidly (Figure 1.1).

About 90% of the air mass is found in the lowest 25 kilometres, and virtually all below 100 kilometres. Compared with the radius of the Earth, at about 6400 kilometres, this is a very thin layer. In this book all illustrations of the atmosphere and air movements over long distances are therefore greatly exaggerated in the vertical direction.

1.1 Origin and composition

Formation and change

Theories of how the Earth was formed are numerous throughout history. Thus in 1642 a certain John Lightfoot from Cambridge should have calculated that God created the world in the year 3928 before Christ – more precisely September the 17 at 0900 (presumably GMT). According to more modern theories the universe was formed in a global explosion – the so-called "big bang" – about 15 billion years ago. Larger amounts of matter were collected into galaxies some billion years later and the solar system was formed about 5 billion years ago. One possibility is that matter was gradually gathered as a rotating disk-shaped cloud, which later split up in a central sun and a series of planets. Different roles of nearby super nova explosions have also been discussed. According to these theories one should expect that the Earth and our next neighbours Mars and Venus would have similar atmospheres, but this is not the case anymore.

Originally the Earths atmosphere probably consisted of a mixture of hydrogen and helium with small concentrations of methane, ammonia, water, carbon dioxide and nitrogen. Most of the light molecules of hydrogen and helium escaped into space, and water and ammonia were dissociated by the solar radiation to form oxygen and nitrogen. Volcanic eruptions emitted more water, carbon dioxide, nitrogen and minor amounts of sulphur compounds.

About 4 billion years ago, when the Earth was one billion years old, a series of simple chemical reactions started that should eventually lead to formation of self-reproducing organisms – to life. How life was initiated at this planet is of minor importance in this context. Important, however, is it that these "live" molecules have developed into all species of plants and animals we know today. As a consequence of this development of life, and especially the appearance of green algae and plants to synthesize organic matter from carbon dioxide with release of oxygen, the atmosphere has gradually changed to its present composition with a high content of oxygen.

It was probably the initial low oxygen content in the early atmosphere that made the formation of life possible. When the first green plants a couple of billion years ago began to transform carbon dioxide to oxygen, it was a catastrophy for primitive organisms that had developed in an oxygen-poor atmosphere. These anaerobes are now forced to live secluded in the earth, in sediments and in various animal intestinal canals. But what is poison for them is now vital for higher life on the globe.

The stratosphere and the ozone layer

Up to about 100 km height the basic composition of the atmosphere is now fairly constant with a volumetric composition of about 21% oxygen, 78% nitrogen, 1% argon and varying minor concentrations of a series of gases of which many are normally considered pollutants. There is however one important exception: the ozone layer that was formed, when oxygen appeared in the atmosphere (Figure 1.2).

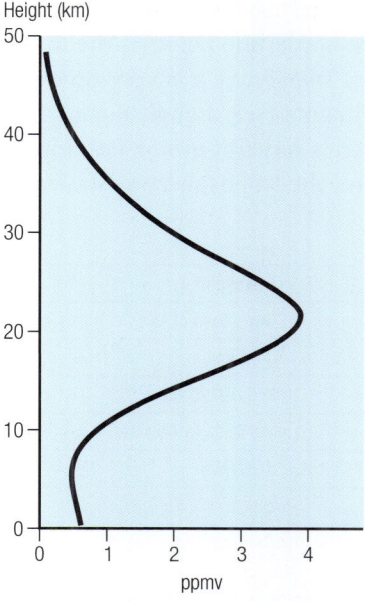

Figure 1.2 The ozone layer is a concentration of ozone with a maximum molar ratio at around 20 km height. The total amount of ozone in the atmosphere corresponds to a layer of about 3 mm at standard conditions (0 °C, 1 Atm.). This is defined as 300 Dobson units (DU).

Ultraviolet light dissociates oxygen molecules into oxygen atoms which combine with oxygen molecules and form ozone:

$$O_2 + h\nu \rightarrow O + O \tag{1.1}$$
$$O + O_2 + M \rightarrow O_3 + M \tag{1.2}$$

The reactions will be described in Chapter 9 in more detail, but it is obvious that the formation rate increases with UV-intensity which *increases* with height, and increases with di-oxygen concentration and concentration of M (a compound to take up excess energy) that *decreases* with height. The result is a maximum at a height of about 15 km over the poles and 30 km over Equator. This *ozone layer* protects the Earth from much of the short waved radiation, and the radiation absorbed by gases leads to localised heating by:

- absorption of solar UVC (<280 nm) radiation by O_2
- absorption of UVB (280-315 nm) and UVC radiation by O_3
- absorption of IR radiation from Earth by O_3

This creates an inversion that stabilises against vertical air movements, and thus forms the stratosphere (stratos, Greek for layer). It is thus misleading to say that the "ozone layer" is situated in the stratosphere. The ozone layer *forms* the stratosphere. The mixing over the borderline between the upper troposphere and the lower stratosphere – the *tropopause* – is very slow and normally not achieved by ordinary air pollutants (see though Figure 1.10 p. 32). Besides the basic composition there are a series of minor, but important differences between the stratosphere and the troposphere as shown in Table 1.1

	Stratosphere	Troposphere
Surface energy flux	None	Direct to/from earth
Height	10-50 km	0-10 km
Pressure	1-250 mbar	250-10123 mbar
Humidity (H_2O)	1 ppmv	1-3% (up to 7 vol %)
UV-light	>250 nm	>310 nm
Most important photooxidant	$O(^1D)$	OH
Number of compounds	Few hundreds	Thousands
Vertical mixing	Slow	Fast
Part of atmospheric mass	≈25%	≈75%

Table 1.1 Some key parameters to illustrate the difference between troposphere and stratosphere.

Higher layers

Above the stratosphere lies the *mesosphere*, where the temperature again decreases. At the *mesopause* the lowest temperature of the atmosphere is reached, and the temperature again increases in the *thermosphere*, where UV-radiation is absorbed by oxygen molecules forming free ions and electrons. Over the thermosphere, at above 700 km lies the *exosphere*, where an increasing number of ionised particles form the so-called Van Allen Belts. The atmosphere has of course no definite border to space, but it is assumed to extend to a height of 80,000 km – much more than the radius of the Earth. The lower part of the temperature profile is shown in Figure 1.3.

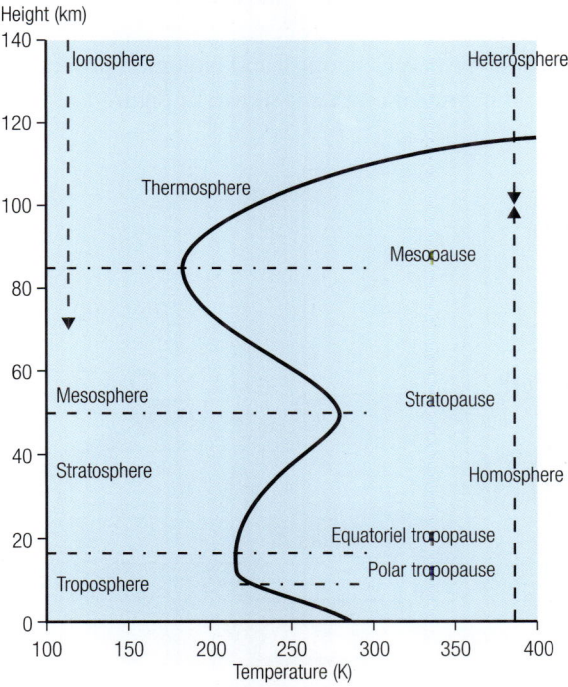

Figure 1.3 The atmosphere is divided into a series of layers, characterised by increasing or decreasing temperature. The pattern is similar all over the Earth, but the individual temperature maxima and minima lies higher over the Equator than over the poles. The figure only shows the lower part of the atmosphere.

1.2 Global atmospheric composition and circulations

Global circulations

The composition of the atmosphere is largely constant and a result of global cycles including emissions, possible transformations and reactions with other compounds and an eventual removal by deposition. The lifetimes of compounds in the atmosphere are from fractions of a second to several centuries depending upon the reactivity and the presence of various reactive compounds. The concentration is therefore a function of both the emission and the lifetime of the compound. If the emission increases, the concentration will increase until the deposition – or other removal processes – have increased enough to create a new equilibrium.

The global circulations include physical, chemical and biological processes. This is e.g. the case for the carbon circulation that is shown in Figure 1.4.

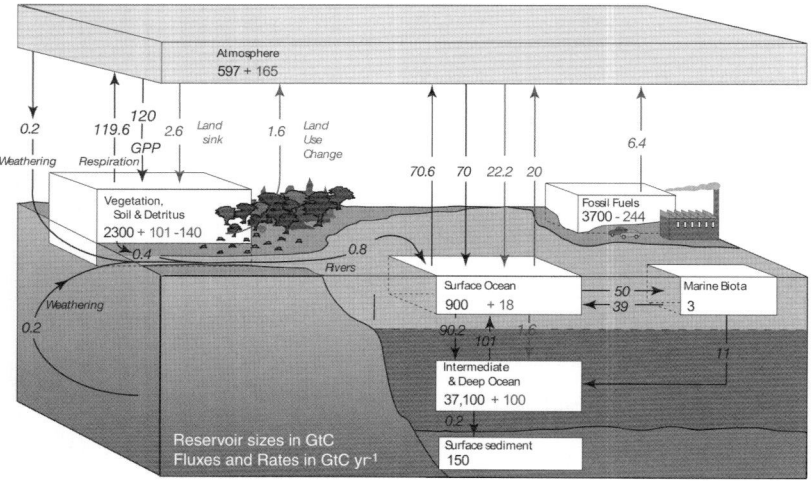

Figure 1.4 Schematic presentation of the global carbon cycle. The numbers in the boxes are reservoir sizes. The numbers at arrows are transport per year. All numbers are given in Gt C or Gt C yr^{-1}. Note that the human contribution in the form of emission of notably carbon dioxide from e.g. use of fossil fuels only constitute a modest part. Due to the long lifetime of carbon dioxide it will nevertheless lead to a substantial increase in atmospheric concentration (IPCC, 2007).

Average composition

The individual cycles, which are complicated in themselves, are more or less coupled in a not fully understood system. The oxygen cycle is thus coupled to the carbon cycle through, among other things, decomposition of organic material (e.g. plants) and weathering of minerals.

The resulting dynamical atmospheric concentrations are – for clean, dry air – as shown in Table 1.2. The basic components oxygen and nitrogen as well as the noble gases argon, neon, helium and krypton are present in nearly constant concentrations. For carbon dioxide the number is the yearly average, as the concentration varies a few percent due the yearly variation of the photosynthesis; this is overlaid by the steady increase due primarily to the use of fossil fuels (Chapter 20). Similar increases are seen for methane and nitrous oxide. For sulphur dioxide and other compounds, that also appears as pollutants in the troposphere, the variations are much larger. It is generally observed that the shorter the atmospheric lifetime the more variable is the concentration.

Compound	Concentration	Atmospheric Lifetime
Constant concentration		
Oxygen, O_2	20.95 vol.%	5000 years
Nitrogen, N_2	78.08 vol.%	1 mio. years
Argon, Ar	0.93 vol.%	Infinite
Neon, Ne	18.2 ppmv	Infinite
Helium, He	5.2 ppmv	Infinite
Krypton, Kr	1.1 ppmv	Infinite
Varying concentration		
Carbon dioxide, CO_2 (2005)	379 ppmv	120 years
Methane, CH_4 (2005)	1.75 ppmv	10 years
Hydrogen, H_2	0.5 ppmv	7 years
Ozone, O_3	40 ppbv	Min. – hours
Nitrous oxide, N_2O (2005)	319 ppbv	170 years
CFC, HCFC, HFC	1 ppbv	1 – 400 years
Highly varying concentration (polluted air, up to)		
Carbon monoxide	40 - 200 ppbv	60 days
Nitrogen dioxide, NO_2	0.01 - 0.3 ppmv	24 hours
Ammonia, NH_3	2 ppbv	10 hours
Sulphur dioxide, SO_2	0.02 - 1 ppbv	2 days
Hydrogen sulphide, H_2S	0.2 ppbv	5 days
Hydrocarbons	25 µg m^{-3}	Min. – days

Table 1.2 Vapour composition of the clean tropospheric air. In addition the air near the Earths surface contains up to about 7vol% water. Note the different units.

Besides the gaseous compounds the atmosphere contains solid material in the form of aerosols i.e. particles that are so small that they are staying for an extended time in the air. Even though these particles only constitute a very modest part of the atmosphere, they are important as condensation nuclei in formation of clouds. They also have important health impacts (Chapter 15) and act as carriers for long range transport of e.g. sulphur and nitrogen compounds (Chapter18). Their size varies over many decades as shown in Figure 13.6 p. 274.

Water vapour and clouds

A special role is played by the atmospheric water, which can exist in all three forms: vapour (water vapour), fluid (droplets, rain) and solid (ice and snow). The main component is water vapour, the concentration of which is strongly varying, but as a global average nearly 4% (by volume). Often the content is given as *relative humidity* defined as the ratio between the actual vapour

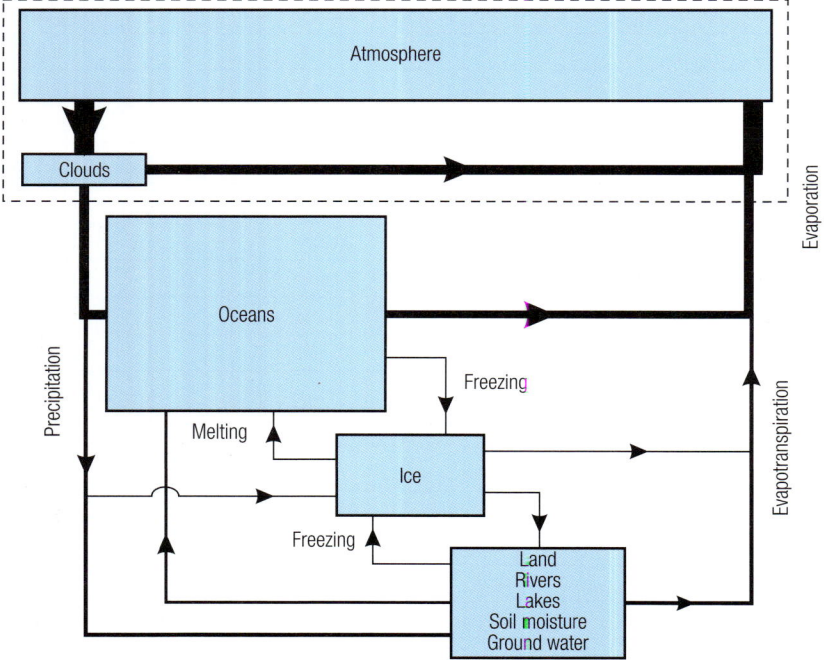

Figure 1.5 The global hydrological system. The atmosphere only contains 0.001% of the global water and the oceans 97.6%. The dominant exchange takes place between the atmosphere and the oceans with 86% of the evaporation and 78% of the precipitation. Only small changes in the transports can therefore drastically change the atmospheric humidity.

pressure and the vapour pressure of water at saturation at the given temperature. The atmospheric water is part of the hydrological cycle (Figure 1.5), and even if the atmosphere only contains 0.001% of the global water, the dominant transport of water between the oceans and the land masses takes place via the atmosphere.

The most important transport is by *evaporation* and *precipitation*. As water is the compound with a high evaporation heat (2501 Jg^{-1} at 0 °C) evaporation and condensation play an important role in the energy budget of the atmosphere (cf. Figure 1.7, p. 27).

Evaporation from a given surface depends upon several factors: the flux of energy, the relative humidity, wind movements and properties of the surface. If there is vegetation on the surface, it gives a further possibility for water transport to the atmosphere as evaporation through the stomata of the plants (Chapter 18). This so-called transpiration can for dense vegetation constitute more than half of the total flux. Since the two phenomena are simultaneous, it can be difficult to distinguish between them, and they are often treated collectively as evapo-transpiration. Generally the evapo-transpiration is larger over oceans than over land with a maximum over the tropical oceans.

Condensation in the free atmosphere does not necessarily lead to precipitation, but stops at formation of clouds, which are divided into three main types:

- *Stratus*. Layered clouds that are formed by large-scale vertical movements of the order of cm s^{-1}. The typical lifetime is one day over the oceans.
- *Cumulus*. Bubble-formed clouds that are formed locally, typically as a consequence of heating of the earth surface. The rising velocity is of the order of m s^{-1}. Lifetime typically one hour.
- *Cirrus*. Tread-formed thin clouds that are formed at large heights (above 5 km) and consist of ice-particles.

On average half of the Earths surface is covered by clouds that play an important role for the global heat balance, both by reflecting short-waved radiation from the Sun back to space and radiating long-waved radiation back to Earth. The total global impact of clouds is not fully understood, but it is assumed to be a cooling factor.

There is a large variation in the precipitation over the Earth where a series of factors act on different scales. Generally there is relatively high precipitation in areas where large air masses collide and are forced upwards – i.e. around the Equator and at middle high latitudes. Conversely there is relatively little precipitation in the subtropics and near the poles where the air masses move downwards.

The landmasses play a decisive role as most of the atmospheric water comes from the oceans. The interior of the continents will generally have less precipitation than the coastal areas, and if the wind blows across larger mountain areas the forced upwards movement will give adiabatic cooling and thus precipitation. A typical example is the high precipitation on the Norwegian west coast. The significance of these atmospheric movements for air pollution is treated in Chapter 8.

1.3 The Sun and the Earth

The sun and its radiation

It is the Suns radiation that, in connection with the rotation of the Earth, drives the movements of air and water. The Sun is a large gaseous globe essen-

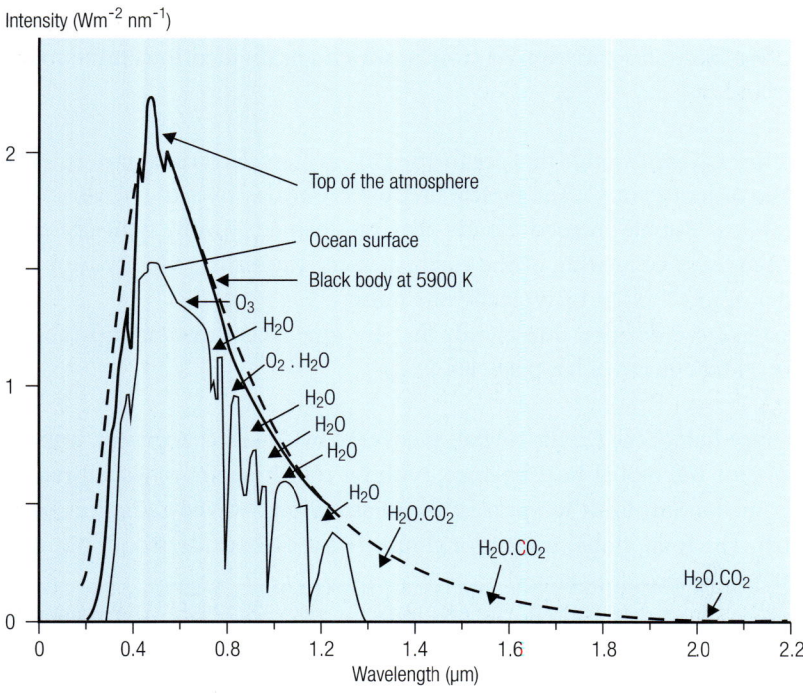

Figure 1.6 The Solar spectra at the top of the atmosphere and at the surface of the Earth after partial absorption in the atmosphere (fully drawn curves). For comparison is also shown the radiation from a black body at 5900 K.

tially composed of hydrogen and helium and kept warm by internal nuclear fusion processes. The Sun does not have a definite surface, but usually the outer 1000 km is called the photosphere and is conceived as the surface. This is what we see of the Sun, since it does not permit radiation from lower layers to penetrate. The radius of the photosphere is nearly 0.7 million km.

The solar spectrum – i.e. the distribution of the emitted radiation – can with good approximation be conceived as radiation from a black body with a temperature of about 5900 K (Figure 1.6), although there is less UV-radiation than would have been expected from a black body.

The energy balance of the Earth

The Earth moves around the Sun in an elliptic orbit with the Sun situated in one of the focal points. The distance from Earth to Sun is thus not constant and neither is the energy reaching the Earth. On average the energy reaching a surface perpendicular to the radiation is 1367 W m^{-2}, the so-called Solar constant. The surface of the atmosphere as such receives on the average ¼ of this, or 342 W m^{-2}, with large differences between the Equator and the poles. At the surface of the Earth the incoming radiation is marked by a series of absorption bands from various compounds in the atmosphere, notably water vapour and carbon dioxide (Figure 1.6).

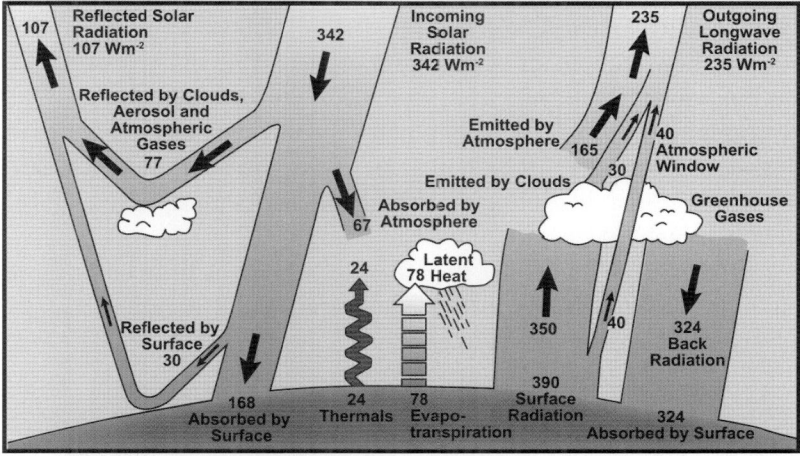

Figure 1.7 The energy balance of the Earth. The incoming solar radiation is partly reflected or absorbed in the atmosphere, but a fair amount reaches the Earths surface and heats it up. The surface radiates energy to the atmosphere and receives nearly the same energy back from greenhouse gases. Further energy is transported from the Earth to the atmosphere by air convection and evaporation of water. (After IPCC, 2007).

The solar radiation, the atmosphere and the Earths surface constitute a system that determines the temperature. In-coming solar radiation is of fairly short wavelengths passes through the atmosphere and heats up the Earths surface. This energy must be radiated back to space, but since the Earth is much colder than the Sun, it happens at a longer wavelength (IR), where the atmosphere absorbs better due to presence of the so-called greenhouse gases.

A greenhouse gas (GHG) is able to absorb (near-) IR radiation at specific wavelength(s) causing low-energy bending vibrations in the gas molecules. Thus only a gaseous molecule with three or more atoms is able to act as a GHG, like H_2O, CO_2, O_3, N_2O and e.g. CFCs. Molecules like O_2 and N_2 can not be GHGs.

The energy absorbed in the atmosphere is re-irradiated in all directions as IR radiation – some towards the Earth. This heats up the Earth, until the incoming and outgoing energy fluxes are balanced.

The energy flux back to space approximates the Stefan-Boltzmanns law:

$$J = \sigma T^4 \ (W\ m^{-2})$$

Where $\sigma = 5.67 \cdot 10^{-8}\ W\ m^{-2}\,K^{-4}$ (Stefan's constant) and T the temperature in K. For more details see: http://en.wikipedia.org/wiki/Stefan-Boltzmann_law

The present Earth surface temperature can quite accurately be calculated to approximately 288 K or 15 °C, when taking into account the albedo's of the surfaces and other confounding factors. This so-called natural greenhouse effect is increasing the Earths average temperature to about 33 °C above what it would have been without the warming atmosphere. The natural greenhouse effect is thus an important prerequisite for life in the form we know it.

The present day problem is that human activities emit vast quantities of compounds and particles that both increase and decrease the natural greenhouse effect, but mainly leading to a much debated anthropogenic global warming (Chapter 20).

As it appears from Figure 1.7 the radiation scheme is in practice more complex with energy absorption in the atmosphere and energy exchanges due to reflections, convection, evaporation and precipitation.

1.4 Forces acting on the air

Nearly all phenomena concerning air pollution take place in the lowest layer of the atmosphere, the troposphere. In this region the air contains varying amounts of water vapour, which is often transformed to fluid (water) or solid phase (ice and snow). Large amounts of energy are connected to these transformations, and it is the interplay between them and the incoming solar radiation that gives rise to the pressure differences that lead to wind. But as soon as the wind is blowing it acts on the pressure differences, modified by the rotation of the Earth.

Horizontal movements

The air movement in the horizontal plane is essentially determined by three factors: Pressure differences, Coriolis force and friction. The pressure force (F_p) is proportional to the pressure gradient (dP/dx), but with opposite direction, and is inversely proportional to the air density (ρ):

$$F_p = 1/\rho \; dP/dx \qquad\qquad\qquad (1.3)$$

The Coriolis force (F_c) is a virtual impact due to the rotation of the Earth and it is perpendicular to the movement and proportional to the velocity (v). It further increases with the latitude (φ):

$$F_c = 2v \; \omega \; \sin\varphi \qquad\qquad\qquad (1.4)$$

where ω is the angular velocity of the Earth. On the Northern hemisphere the Coriolis force will give a deflection to the right and on the Southern hemisphere to the left.

   Initially an air mass will move from a high pressure towards a low pressure i.e. perpendicular to the isobars (lines with same pressure). As the air moves towards low pressure regions it will be influenced by the Coriolis force. The air accelerates in the pressure field and the Coriolis force deflects the air mass until the two forces compensate each other, and eventually the air moves parallel to the isobars (Figure 1.8). This is the geostrophic wind with the velocity:

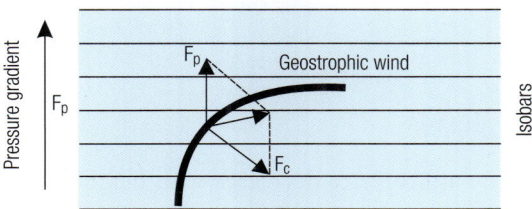

Figure 1.8 The geostrophic wind is a result of the air being deflected by the Coriolis force and moving along the isobars (here shown for the northern hemisphere).

$$V_g = 1/\rho \ dP/dx \ 1/2v \ \omega \ \sin\varphi \qquad\qquad (1.5)$$

Near the surface the moving air will be influenced by a friction force opposite the velocity. It means that the air velocity decreases towards the ground and rotates in direction of being perpendicular to the isobars (Figure 1.9). The effect depends upon the roughness of the surface and will over oceans typically give a 10-20° rotation and a 40% loss in velocity. Over land, where the friction is larger, the wind will be rotated 25-35°, and the velocity reduced about 60%.

The layer where the surface roughness is important for the air movements is called the atmospheric boundary layer (ABL). The ABL varies in depth, but is generally confined to the lowest 2-3 km of the atmosphere. Some phenomena even take place in the lowest few hundred meters.

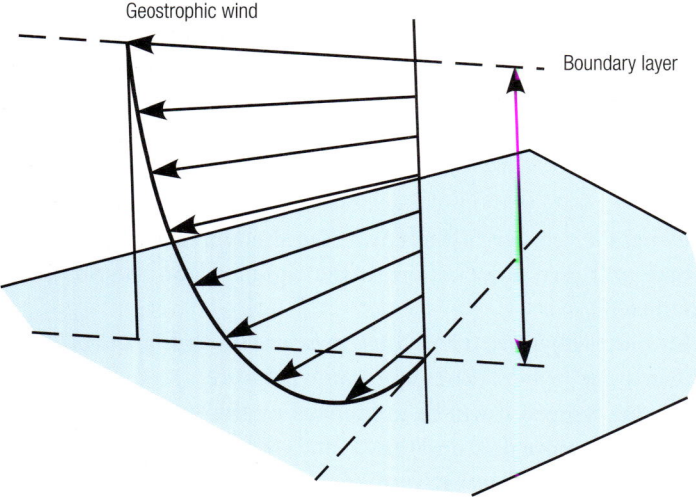

Figure 1.9 The interplay between pressure gradients, the Coriolis force and the friction against the earth surface mean that the wind velocity decrease towards the earth and simultaneously turn left on the northern hemisphere.

Vertical movements and stability

Generally the temperature of the atmosphere decreases with height in the troposphere (Figure 1.3), but locally different temperature profiles appear and are determining for vertical air movements.

If an air mass for some reason moves upwards, it will expand, due to decreasing pressure, and its temperature will fall. This temperature fall is generally adiabatic (without gain or loss of heat to the surroundings). In an unsaturated atmosphere the dry adiabatic temperature decrease is about 1 °C per 100 m. If the air is saturated and water condenses, the condensation heat is released and the cooling is reduced. This saturated adiabatic temperature decrease is on the average 6.5 °C per km.

This has importance for the vertical stability of the atmosphere. If the temperature decrease is less than 1° per 100 m – or even is negative, an ascending air mass will always be colder than the surroundings and a descending air always warmer. Vertical movements will therefore be prevented, and the air is stable. If, on the other hand, the temperature falls more than 1 °C per 100 m, an air mass that accidentally moves upwards will allways be warmer than its surroundings and will thus continue its movement. If it moves downwards, it will always be colder than the surroundings and will likewise continue the movement. The air is therefore unstable.

An unstable situation can appear during summer when the Sun heats the soil surface and thus the lower air. The air expands and ascends and colder air sinks down. The phenomenon is called *convection* and results in a strong mixing of the air. If conversely the soil surface is cooled e.g. by radiation to a clear night sky, the mixing is prevented. The phenomenon where the temperature increases with height is called *inversion*. If the inversion extends down to the soil surface it is called *ground inversion*. The stratosphere is as mentioned earlier a very stable global inversion.

These inversions are radiation induced phenomena. As described in Section 1.5 inversions can also be induced by horizontal air movements.

1.5 Air movements on different scales

The energy, the Earth receives, is not evenly distributed and it is on the average much larger near the Equator than at the poles. This difference is, however, partly compensated by general circulations in the atmosphere and in the oceans.

The general atmospheric circulation

In principle heated air ascends over the Equator, moves towards the poles and returns at low height to the Equator. Therefore the two hemispheres are practically separated and the air pollution that is emitted in the heavily populated Northern hemisphere largely stays there. In practice, however, the

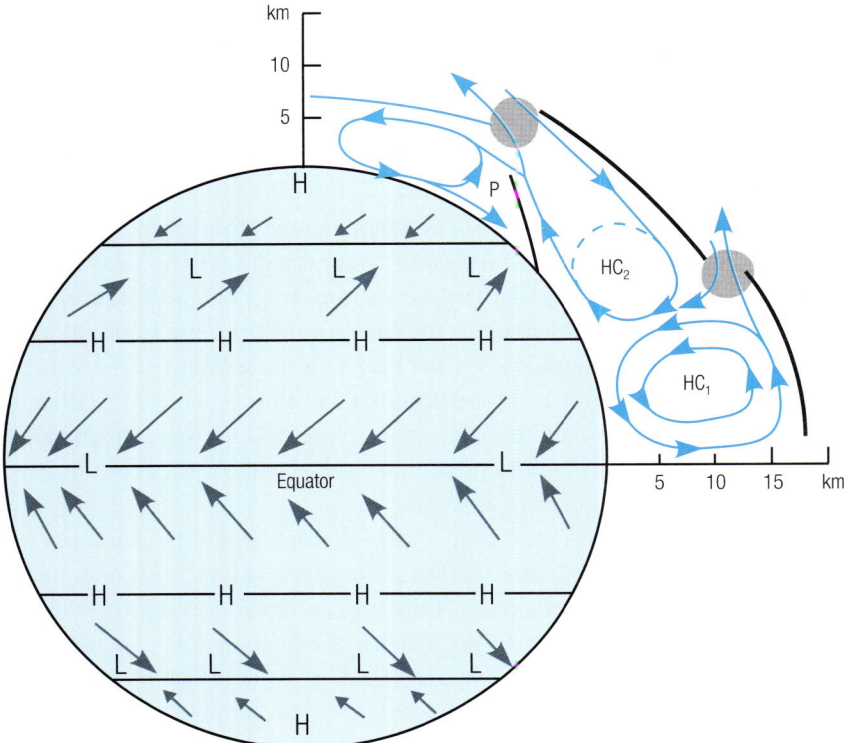

Figure 1.10 Global atmospheric circulations. There are three "cells" on each hemisphere, where the air performs a spiral movement. The air ascends around the Equator and sinks again around 30° latitude where high pressures appears and where we find the great deserts, e.g. Sahara on the Northern and Kalahari on the Southern hemisphere. From these high pressure regions the trade winds blow along the surface towards the Equator. Similar winds blow against the two low pressure regions around 60°. Note also that there are holes in the tropopause near the so-called jetstreams, where there is a possibility of exchange between the troposphere and the stratosphere.

Earths rotation and the Coriolis force split up the movement in circulation cells (Figure 1.10).

When the warm air moves from the Equator towards the poles it will be deflected by the Coriolis force towards east. Around 30-40° latitude the deflection is complete and air accumulates. This forms the permanent high pressure belt on the North African latitude. Driven by this high pressure air moves near the earth both north and south. On the northern hemisphere the south-going air and on the southern hemisphere the north-going air are forced towards west and two circulation cells (*Hadley cells*) are formed, where the air moves around the Earth in spirals.

Correspondingly, so-called *Ferrel cells* are formed between 30° and 60° latitude when air from the subtropics moves towards the poles and thus collide with the cold polar air at the so-called polar front. Finally *polar cells* are formed, when relatively warm air sinks down near the poles.

This three-cell model roughly describes the observed predominant wind patterns on the Earth. Of special interest is the north-going air from the sub-tropical high pressure that is bending towards east and becomes the predominant westerly wind in temperate Europe. It has great importance for the distribution of transboundary air pollution where countries in the west generally send pollution towards countries in the east.

The polar front that separates the temperate westerly wind from the polar easterly wind is heavily tilted against north, due to the difference in temperature, velocity and humidity. This is e.g. the reason for the unstable weather in northern Europe.

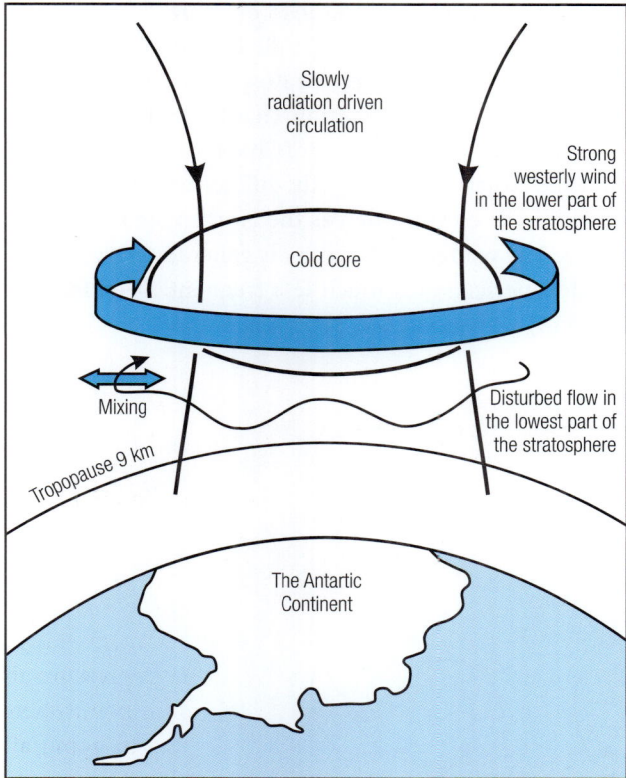

Figure 1.11 In the stratosphere over the South Pole the air circulates with velocities up to 100 ms[-1]. This so-called polar vortex keeps the air masses together and may be a contributing factor to ozone depletion (Chapter 19).

Jet streams and polar vortex

Above the boundary layer the geostrophic wind can change direction and strength due to changes in pressure and temperature. The general temperature decrease from south towards north (on the northern hemisphere) thus gives a rotation towards east with increasing height. This creates strong winds, the so-called jet-streams, where the circulation cells collide in the stratosphere (Figure 1.10). The subtropical jet-stream runs at latitudes between 20 and 35° at a height of 10-12 km. It is strongest during the winter where it can have velocities above 65 ms^{-1}. The polar jet-stream that runs between 35 and 65°, is fairly irregular and has a velocity of about 25 ms^{-1}. It results in normally shorter flying times towards east than towards west. The jet-streams give a hole in the tropopause and thus permit some mixing between the two atmospheric layers.

A similar circulation arises around the South Pole when the air is cooled and sinks during winter. This results in a strong westerly wind in the lower part of the stratosphere with velocities of 100 ms^{-1} or more during spring (autumn on the northern hemisphere). This so-called *polar vortex* (Figure 1.11) develops a kernel of very cold air with stratospheric clouds. In July to September it is in practice a gigantic reaction chamber which plays an important role in the depletion of the ozone layer (Chapter 19).

The situation around the North Pole is slightly different. The South Pole is a continent surrounded by an ocean, whereas the North Pole is an ocean surrounded by continents. The winter stratosphere is generally warmer over the North Pole and stratospheric clouds much less frequent. Therefore the ozone depletion (Chapter19) will also be less pronounced.

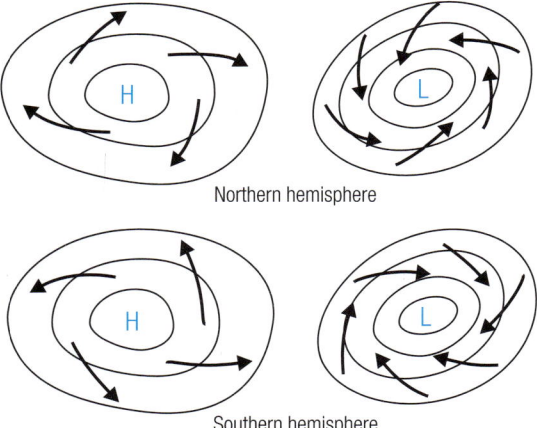

Northern hemisphere

Southern hemisphere

Figure 1.12 On the northern hemisphere the air will move outwards clockwise from a high pressure and inwards anticlockwise towards a low pressure. On the southern hemisphere it is opposite.

High and low pressures

If an air mass over a larger area is heated it will ascend and a low pressure will be created that is filled by air moving in horizontally. Due to the Coriolis force the air will be deflected and (on the Northern hemisphere) move in a spiral against the clock. In the centre of the area, where the air ascends, there will often be condensation and precipitation. The phenomenon is called a cyclone. Correspondingly an anticyclone with downwards air movements in the centre and a clear sky around a high pressure is formed (Figure 1.12). These phenomena are large scale wind systems ("synoptic") – typically 1000 km in size – and must not be confused with waterspouts and similar local phenomena.

In the subtropics partly permanent anticyclones can appear in connection with high pressure, whereas wandering cyclones and anticyclones are much more frequent in the temperate regions. Their lifetime is generally a few weeks, and they move with an average velocity of 800 km d^{-1} towards east. If an anticyclone stops, a so-called stagnation appears. It can result in air pollution events because of the overlying inversion and the generally weak winds.

Front systems and inversions

When two air masses with different temperature, humidity, and maybe also pressure, collide, a so-called front is created. Within this front the meteoro-

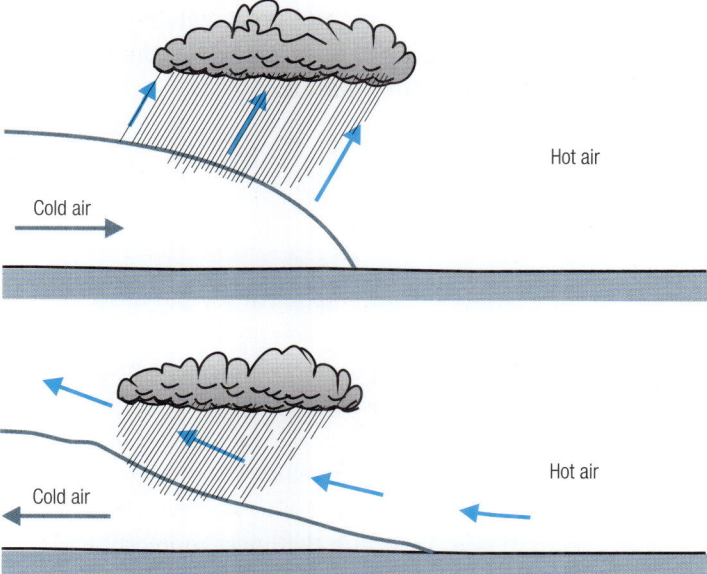

Figure 1.13 Formation of a cold front (upper curve) and warm front (lower curve). In both cases it will result in precipitation, and it will normally appear after the cold front and before the warm front.

logical parameters change rapidly with location and height. One example is the polar front. If it is the warm air that moves, it is called a warm front. If it is the cold air, it is called a cold front. In both cases warm air will be situated above the cold air and an inversion is created. Often front systems are accompanied by cloud formation and precipitation when hot air is forced upwards and cooled (Figure 1.13).

The inversion can also appear when warm air is forced over a mountain ridge and meets cold air on the other side.

Meso- and micrometeorology

Mesometeorology comprises weather systems from 1000 km down to less than 100 km; typical examples are thunderstorms, sea- and land breezes, tornados, mountain and valley winds and katabolic winds. In many cases they are due to large differences in heights of the landscape. In flat countries micrometeorological phenomena are quite important for dispersion of pollution, and air movements down to the scale of individual streets are studied. There is no clear distinction between meso- and micrometeorology, but it is a common feature that the time scale is short and the air movement is near the ground so that terrain, buildings and vegetation become important.

Many regions have characteristic wind systems, which can be caused by special topographic features. In mountainous regions ascending air over warm slopes may result in cloud formation and precipitation. During the night the process is reversed, and the cold air is situated in the valley and results in a stability that can give pollution episodes.

1.6 Ocean streams

The air movements in the atmosphere are important for the movement of water in the oceans. The trade winds in the tropics and the westerly winds in north and south thus drive the whirls that on the northern hemisphere run clockwise and on the southern hemisphere anticlockwise.

Most important is the global thermo-haline cycle that in the North Atlantic is driven by oppositely directed forces: Differences in temperature drive a surface stream towards north. An input of fresh melt water (with low density) in the north, and an evaporation in the south that gives more salt and thus high density water acting the other way. At present the first process dominates and creates a circulation as shown in Figure 1.14 with warm surface water running up through the Atlantic. As shown in Figure 1.10 a predominant

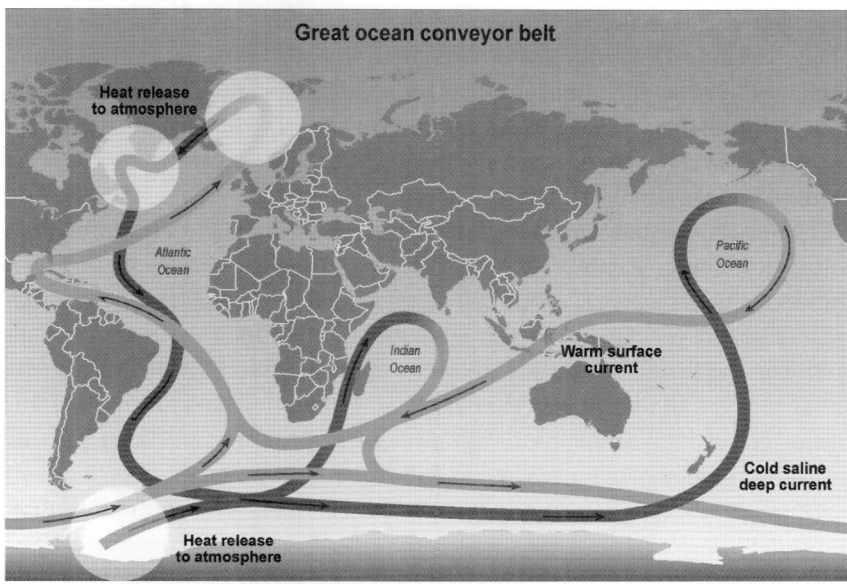

Figure 1.14 The present thermo-haline circulation. Of special interest for Northern Europe is the warm surface stream up through the Atlantic that returns as cold bottom stream (IPCC, 2001).

westerly wind that gives North Europe a warmer and more humid climate than at similar latitudes in North America is present here..

A similar phenomenon is known as the so-called El Niño that is the most important source of year to year variations in the global climate, although not all parts of the Earth is influenced. In principle it is a shift between the atmospheric movements over the sea between Indonesia and the area around the Easter Islands. Normally the trade wind blows along the Equator towards West from Peru to Indonesia and results in a corresponding water movement. When the water arrives near Indonesia the lower air masses become humid and warm. Every 6-10 years, however, the direction of the wind cycle changes with the result of an oppositely directed water movement and large changes in precipitation pattern. As the water transports nutrients, the fish life will also be influenced and fish catches are normally drastically reduced. At the same time large amounts of carbon dioxide is released, partly because less is bound in plankton, partly because of temperature increase.

1.7 Weather and climate

The condition of the atmosphere and its variations are described by temperature, pressure, wind forces and precipitation. The short term situation is denoted *weather*, whereas the general condition over a longer period (typically 30 years) is denoted *climate*.

The climate is not constant, but oscillates with various frequencies. The energy emission from the Sun is nearly constant, but the way the Earth moves around the Sun results in varying amounts energy being received. Over a year there is a variation of about 6% – most in January and least in July. The form of the solar orbit varies with a period of 100,000 years, the tilt of the axis of rotation with a period of 40,000 years and the orientation of the axis of rotation in space with a period of 21,000 years. These periods (so-called Milankovitch cycles) can be found in variations of the climate of the Earth – especially the appearance of recent ice ages. Also continental drift appears to have had an impact on climate by influencing the macroscale air movements.

The climate belts

Generally the temperature decreases from the Equator towards the poles, because the energy input per area unit falls with the cosine of the latitude. The temperature differences are, as mentioned, diminished by the general circulation, where warm air moves towards the poles and returns as cold. Although there are no sharp borders, the Earth can be divided in 7 climatic zones:

- The tropical zone, where the coldest month is never under 15 °C, and where frost never occurs. The vegetation is typically rain forest and savanna
- Two subtropical zones, where strong or long-lasting frost never occurs, and where the warmest mean monthly temperature is at least 21 °C
- Two temperate zones (45-65° latitude), where the warmest monthly mean temperature is at least 10 °C. The natural vegetation is deciduous or grass steppe with coniferous trees near polar regions
- Two polar zones with tundra and glaciers

The zones are more pronounced on the *northern* hemisphere which has largest land areas. There are slight differences in distribution between summer and winter, because the Earth moves in an elliptic orbit around the Sun. Further the Antarctica is colder than the Arctic regions because it is a continent surrounded by oceans, whereas the Arctic is an ocean surrounded by continents.

1.8 Literature

Graedel, T.A. and Crutzen, P.J. (1995): *Atmosphere, Climate, and Change.* Scientific American Library, New York.

Greadel, T.E. and Crutzen, P.J. (1995): *Atmospheric Change – An Earth System Perspective.* W.H. freeman and Company, New York.

IPCC (Intergovernmental Panel of Climate Change) **(2001):** *Climate Change 2001* – Three large working group reports and a Synthesis Report. Cambridge University Press. Cambridge. This and more detailed reports can be downloaded from the internet at: www.ipcc.ch

IPCC (Intergovernmental Panel of Climate Change) **(2007):** *Climate Change 2007* – Three large working group reports and a Synthesis Report. Cambridge University Press. Cambridge. This and more detailed reports can be downloaded from the internet at: www.ipcc.ch

2   The short history of air pollution

Jes Fenger

All life influences its surroundings. That is a natural, acceptable process, as long as the impacts are not greater than nature can still be in balance or the development proceeds so slowly, that a running adaptation is possible. For air the classical example has been the green vegetation, with a photosynthesis that over billions of years has changed the concentration of oxygen from practically nothing to presently 21%v. At the same time the sedimentation of carbonates as chalk in waters has reduced the concentration of carbon dioxide that was once dominating, to presently (so far) below 0.04%. Such processes have led the English philosopher and scientist J.E. Lovelock to propose the so-called *Gaia Theory* (Lovelock, 1979), according to which life on Earth is a system that regulates its own conditions and possibilities of development.

That, however, is not the case for the increasing human impact anymore. The world population has roughly increased exponentially from an estimated 10 million 10,000 years before Christ, when agriculture and urbanisation started. World wars, plaques and economic disasters have only had a marginal impact. The population is now above 6 billion, and it will in all probability be about 10 billion before it may level off at the end of this century. Half of the population lives in urban areas, and the urbanisation is increasingly caused by mechanisation of farming and opportunities in new industries and public services. The car park is increasing at an even greater rate (Figure 2.1). The world has now more than 70 urban areas with a population above 3 million and about 20 with more than 10 million. The largest part is situated in the developing countries.

Growth, index 1950=100

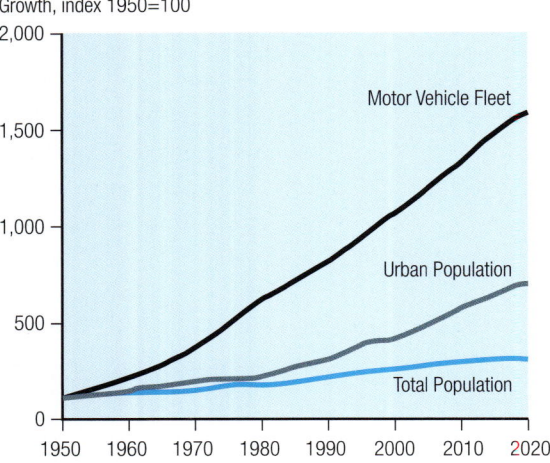

Figure 2.1 Growth in global population, urbanisation and car numbers since the middle of last century – and the expected development (UNEP, WHO, 1992).

The global use of energy has increased even faster than the population and has doubled since the beginning of the 1970's (Figure 2.2). Globally about 85% is produced by fossil fuels resulting in air pollution with oxides of sulphur, nitrogen and carbon and a series of other compounds in minor amounts. Similarly, the increasing food production has led to pollution with ammonia, methane and nitrous oxide. In addition, a fast growing industry often pollutes with high amounts of exotic compounds.

Mtoe y^{-1}

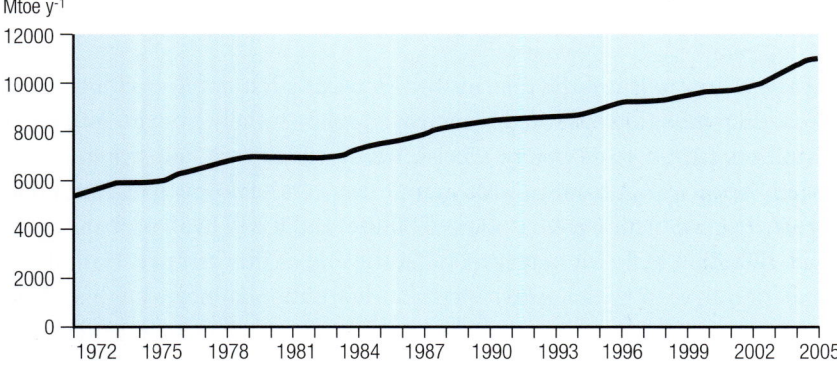

Figure 2.2 Global total primary energy supplies since 1971. Mtoe y^{-1} (IEA, 2005).

The human domination over nature was summed up in a review article in Nature some years ago (Vitousek et al. 1997):

> *Human alteration of Earth is substantial and growing. Between one–third and one-half of the land surface has been transformed by human action; the carbon dioxide concentration in the atmosphere has increased by nearly 30 percent since the beginning of the industrial revolution; more atmospheric nitrogen is fixed by humanity than by all natural sources combined; more than half of all accessible surface fresh water is put to use by humanity; and about one-quarter of the bird species on Earth have been driven to extinction. By these and other standards, it is clear that we live on a human-dominated planet.*

This has led some to the provocative comparison of humanity with a cancer which in its unlimited growth threatens the organism that is the foundation of its own existence.

2.1 Early observations

Air pollution is sometimes perceived as a modern phenomenon resulting from an uncontrolled technological development. This is, however, a truth with severe reservations. In 1866 the Danish author Hans Christian Andersen is in Paris and writes in his dairy on April the 13th, how he looks at some

> *- apparently dead trees. They could, as I, not stand the Paris atmosphere, and when they could not get away they have died.*

In the same year the Norwegian author Henrik Ibsen in his poetic drama "Brand" (Fire) lets the main person exclaim:

> *More evil times, more evil visions*
> *Illuminate the dark midnight of the future!*
> *A heavy pall of coal-smoke falls black*
> *across the countryside, it's British blackness*
> *besmirching all the fresh green growth,*
> *and stifling all the fair new shoots!*
> *It swirls low, mixed with poisons,*
> *robbing our fields of sun and daylight,*
> *sifting down like rains of ash*
> *upon that doomed and ancient city.*

Five years earlier (in 1861) the English scientist Tyndall had shown that
changes in atmospheric concentration of carbon dioxide could have global
climatic effects. But some of the problems facing us today were known much
earlier. Indeed, the literature is abundant with descriptions of the bad air
in cities. Seneca writes in year 61 about the smoke and smell of cooking in
ancient Rome:

> As soon as I had got out of the heavy air of Rome and from the stink of the smoky
> chimneys thereof, which being stirred, poured forth whatever pestilential vapours
> and soot they had enclosed in them, I felt an alteration of my disposition.

In the Odyssey from about 700 years before Christ the homecoming Odys-
seus asks his son Telemachus to hide the weapons from the hall before the
ultimate fight against his mother's suitors.

> - and when the suitors miss them and ask you questions, put them off with a
> winning story: "I stowed them away, clear of the smoke. A far cry from the arms
> Odysseus left when he went to Troy, fire-damaged equipment, black with reeking
> fumes".

And maybe the sixth Egyptian plague, which is described in Exodus from
about 1000 years before Christ, is one of the first examples of industrial pollution:

> And the Lord said unto Moses and unto Aaron, take to you handfuls of ashes of
> the furnace and let Moses sprinkle it towards the heaven in the sight of Pharaoh.
>
> And it shall become small dust in all the land of Egypt, and shall be a boil breaking
> forth with blains upon man, and upon beast, throughout all the land of Egypt.

All over the world urban air pollution was a nuissance for millennia and several half-hearted attempts of mitigation were performed. London was since
the end of the Middle Age the prototype of a polluted city (Brimblecombe,
1987). Already around the year 1300 a commission was set up to investigate
the air pollution from use of coal in London. Apparently, however, its regulations were largely ignored. An often-cited story about an offender, who in
1307 should have been tortured, hanged and/or beheaded, has never been
verified in primary sources.

In the middle of the 17th century there is a sharp increase in the occurrence of rickets, which could have a connection with the reduction of sunlight due to the nearly constant fog during winter. At the same time a strong

degradation of buildings and works of art is noticed. In 1661 John Evelyn thus writes:

> *This is that pernicious Smoake which fullyes all her Glory, superindusing a sooty Crust or Furr upon all that it lights, spoiling the moveables, tarnishing the Plate, Gildings and Furniture, and corroding the very Iron-bars and hardest stones with those piercing and acrimonious Spirits which accompany its Sulphure; and executing more in one year, than exposed to the pure Aer of the Country it could effect in some hundreds.*

The visible "smog" (a contraction of the words "smoke" and "fog") arising mainly from heating was not the only problem. Also the primitive industry had its impact, e.g. by production of alkali salts as sodium sulphate used in production of glass. As a by-product hydrochloric acid was formed and discharged directly into the air.

In 1863 the Parliament passed the so-called "Alkali Act" and R.A. Smith was appointed the first "General Inspector of Alkali Works for the Government". Besides his administrative work Smith was actively engaged in scientific investigations that culminated in the publication of the now classical book "Air and Rain: The Beginnings of a Chemical Climatology" from 1872, where he for the first time used the term "acid rain".

The problems were thus not ignored, but the attitude to them was ambivalent. It is tempting to say that until quite recently air pollution was considered a symbol of wealth and prosperity – words, that earlier had a more unconditionally positive sound than today. Examples of this can be found in

Figure 2.3 Poster for a British exhibition in Copenhagen 1932 (Ib Andersen) and an earlier Jugend style advertisement for cars (Gerda Wegener).

advertisements prior to the Second World War (Figure 2.3): Smoking stacks and cars that pass in a cloud of dust. Hardly images anyone would like to cultivate today!

2.2 A growing awareness

The first European air pollution episode to influence the development of pollution policy was possibly the Meuse Valley disaster in Belgium December 1930. The pollutants and mechanisms in the heavily industrialised valley have never been fully identified, but 63 people died and a larger number fell ill (Firket, 1931). It was, however, not until after the Second World War and only in industrialised countries that the efforts had real effect. In London a series of incidents, notably a disaster in 1952 with thousands of surplus death (Figure 2.4) was necessary to give the impetus to the Clean Air Act in 1956. London is now reasonable clean, at least with respect to the sulphur pollution from domestic heating that formed the main problem in 1952.

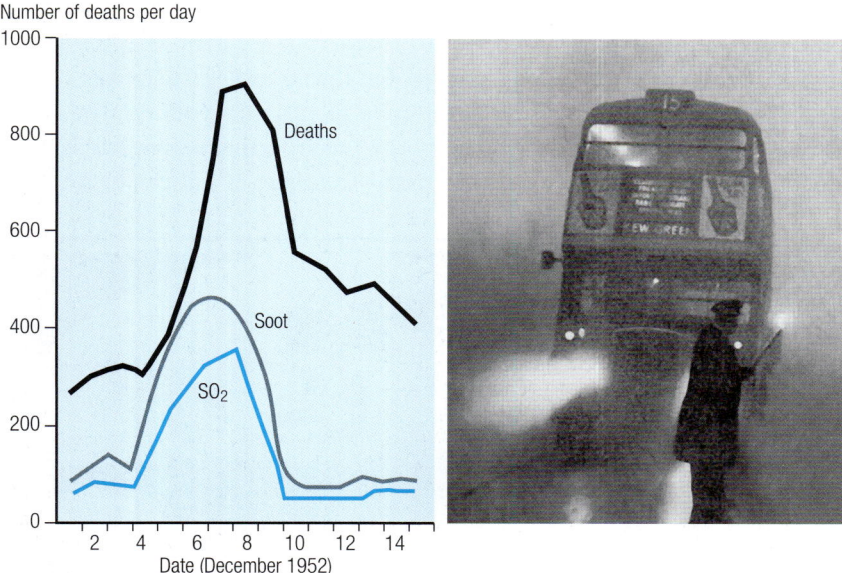

Figure 2.4 The London smog episode in 1952 caused a premature death of several thousand people. The peaks of sulphur dioxide and soot are about 2 mg m^{-3} (Wilkins, 1954). The episode was the result of poor dispersion conditions in combination with very high sulphur emissions from domestic heating within the city area. This disaster in London gave the final impetus to the clean air act in 1956. The figure to the left shows the measured facts. The photo to the right gives an impression of the situation during the episode.

SO$_2$ concentration (µg m^{-3})

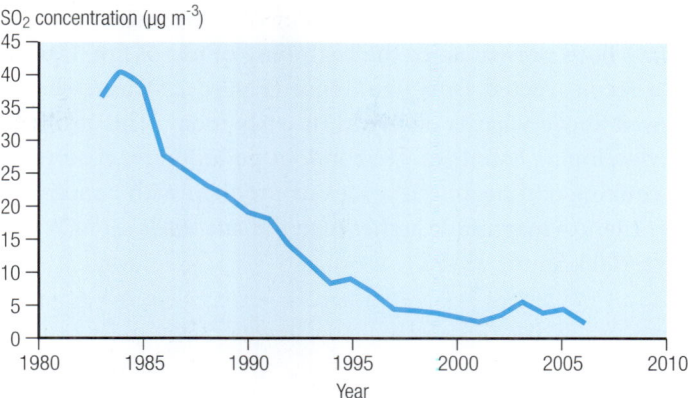

Figure 2.5 Recent SO$_2$ concentrations in Copenhagen. Yearly averages composed from measurements from different stations.

Since then the SO$_2$ levels in the industrial countries have decreased, especially since the 1980s after international agreements to reduce emissions. Figure 2.5 shows how the present day concentration in Copenhagen is only 1/10 of the level in the early 1980's and well below present guidelines.

Figure 2.6 Indoors in a reconstructed Iron Age house at the Research Centre, Lejre. The open fire in the house gives an air pollution that is much higher than in the centre of Copenhagen. Photo: Climate-X-experiment, Research Centre, Lejre, Denmark.

A special type of air pollution is indoor pollution from heating and cooking. This must have been present since the beginning of use of fire. Danish experiments in a reconstructed Iron Age house (Figure 2.6) have shown pollution levels well above what is observed in cities today. This problem still exists in the developing countries. Here indoor pollution from biomass combustion for cooking and heating is a serious problem with concentrations of particles orders of magnitude higher than the safe levels set in WHO guidelines (WHO, 2000; Smith, 1988).

2.3 Photochemical air pollution

In the middle of the 1940s a new type of pollution had been observed in Los Angeles, USA. It gave a brown colour to the air, a reduction of sight, damages to vegetation, and for humans respiratory troubles and smarting of the eyes. Not until 5-10 years later it was understood that it was the result of photochemical processes. Here hydrocarbons react with nitrogen oxides under the influence of sunlight and form ozone and other oxidants. A significant source of the reacting compounds is car traffic. It was therefore not a coincidence that the phenomenon was first observed in Los Angeles. A contribut-

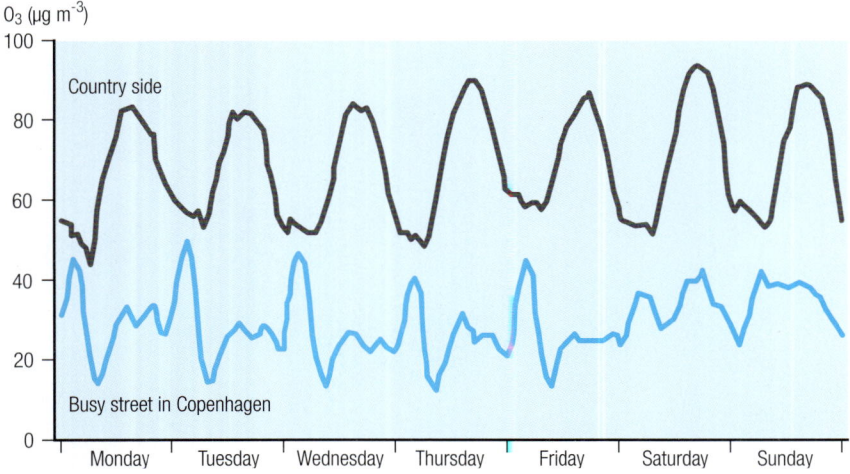

Figure 2.7 In North European cities photochemical air pollution is different from the phenomenon in Southern Europe or The United States. The figure shows average weekly variation of the ozone level in Copenhagen, Denmark. It appears that the levels are highest during weekends, when the traffic emission of NO is lowest. The ozone levels in a nearby rural area are shown for comparison. Here the ozone concentration is generally higher. (Based on data from The Copenhagen Environment Control).

ing factor was the warm and sunny climate, and that the city is situated in a valley that keeps the reactants trapped together.

After the Second World War photochemical air pollution was also observed in Europe, but in more northern cities it had a different form (Figure 2.7). Here the ozone concentration is generally low with high traffic and high with low traffic. The reason is that the observed ozone is largely due to long range transport from Central Europe and the decrease in ozone level is caused by reactions with the primary compound NO emitted from cars (Chapter 9).

All the reactions with ozone and other photochemical oxidants can be described with mathematical models and may involve hundreds of different compounds.

With catalytic converters on cars the photochemical problem is also on its way to a solution. Now the emerging problem is small particles, especially from diesel cars. It appears that the smaller the particles the more dangerous they are. Attempts are now made to measure concentrations of particles down to 1 µm or less. So far it is has not been quite clear whether the particles as such are harmful or the impact is due to compounds attached to them.

2.4 Lead in petrol

The urban air pollution with lead (Pb) increased slowly until the middle of the 1960's due to increasing industrial applications, but then a drastic growth began with the increasing use of lead additives in petrol to increase the octane number (Nriagu, 1990). In spite of severe warnings the use continued to the late 1970s, when restrictions were imposed.

In the industrialised countries lead in petrol is now practically phased out. Ironically, part of the argumentation against lead was that it poisoned catalytic converters that were introduced at about the same time. After lead was banned the octane number of petrol was maintained by addition of benzene, which is carcinogenic. That was then substituted by other hydrocarbons, e.g. MTBE that may contaminate the ground water. This is now again substituted by other less harmful hydrocarbons. Although lead is thus not a problem anymore in the industrialised world, it is still a major air pollutant in urban areas in many developing countries.

2.5 Long range transport

Until the beginning of the 1970's it was the general perception that air pollution was by and large a local problem that could be solved with cleaner fuels for domestic heating and high stacks on power plants. High stacks of course lead to less influence in the nearby area and the pollution is thus diluted before it reaches the ground. But it was a short-sighted solution. In 1977 the Organisation for Economic Co-operation and Development (OECD) reported on a five-year investigation into the mechanisms and magnitude of long-range transport of air pollutants, and in 1982 the Swedish EPA "Naturvårdsverket" published its report on "Acidification today and tomorrow" (Aniansson, 1982) which spelled out the problem:

> This circumstance is of vital importance, because within regions the size of Western Europe, activities that involve the emission of sulphur may cause damage in neighbouring countries. Adjacent countries intervene in each others economies through the effect of atmospheric pollutants, so knowledge of the atmosphere as a recipient of pollutants, and of its function in the transport and spreading of these substances, are obviously of vital importance. The fact that atmospheric pollutants are spread over great distances means that countries affect each other.

> That the fish have died in thousands of lakes we have known for many years. But not until recently have we been able to establish that drinking water from springs and wells may, in consequence of acidification, contain sufficient amounts of toxic heavy metals to be a threat to health. That forest trees on acidified land may begin to show slower growth is so far only a suspicion – it will be at least another two decades before we know for certain.

A significant factor in the understanding of this problem and its later investigation was the appearance of large electronic computers in the early 1970 that permitted a mathematical treatment of the dispersion on a large scale.

Later investigations of ice cores revealed that long- range transport of other compounds had taken place for centuries. Thus lead from production in Rome more than 2000 years earlier was found in Greenland ice (Hong et al., 1994).

2.6 Global impacts

In the mid 1970s passengers and crew on highflying aircrafts operating between New York and Tokyo complained of headache and respiratory trouble. That was the same problems as met in photochemical episodes. It was not so strange, as they had flown through the ozone layer. This could be solved by filtering the air in the aircraft, but the most important problem was the flight emissions of nitrogen oxides which might deplete the ozone layer.

The first signs of an increase to global impact scale came in 1982, when English scientists actually measured depletion of the ozone layer that protects humans as well as nature against harmful UV-radiation. The measurements were performed in the Antarctic with an earth-based so-called Dobson spectrometer. In fact this "ozone hole", as it was later to be called, had been registered a couple of years before by the American satellite "Nimbus", but all ozone measurements below a certain value had been discarded in the data treatment as being false (Gribbin, 1988).

The cause of the ozone depletion appeared, however, not to be emissions from high-going aircrafts that did not after all became as abundant as expected, but, quite unexpected, emission of the industrial Chloro-Fluoro-Carbon-compounds (CFC's). These compounds were originally considered an environmental asset, because they have extremely high chemical stability and thus are completely non-toxic. Exactly that was the reason why they could be mixed up into the stratosphere and, after being decomposed by UV-radiation, attack the ozone (Chapter 19). They are now under de-phasing, and the problem is thus in principle solved.

At the same time another rather more serious global problem appeared: global climate change due to increased greenhouse effect (Weart, 1997). Climate change had, as mentioned earlier, been discussed since the middle of the 19th century. It was, however, not until after an international conference "The Greenhouse Effect, Climate Change and Ecosystems" in 1985 (Bolin et al. 1986), that the frightening possibilities of impacts of greenhouse gases for the coming centuries became apparent. In spite of some disagreements this is now generally considered *the* coming air pollution problem.

Climate change is a much more controversial problem than ozone depletion. The causes of ozone depletion are fairly well understood and technical solutions are available, although it may take some decades before the human impact has faded out. Climate change has impacts that are to some extent so far only demonstrated in model calculations. The impacts are different all over the world and not always negative, and most important: mitigation in the form of emission reductions touches practically all human enterprises.

2.7 Sources and emissions

Emission inventories

Air pollution problems are the result of emissions from various types of sources with highly different characteristics concerning release height and temporal variability. Most important is pollution from use of fossil fuels. These so-called primary pollutants can arise in various ways:

- As a product of the combustion, where it can be the formation of carbon dioxide
- As impurities or additives to the fuel. Typical examples are sulphur in oil and lead in petrol
- During the combustion that can be incomplete or lead to formation of new compounds. Typical examples are carbon monoxide, nitrogen oxides and hydrocarbons

Other sources, typical industrial, may contribute with the same compounds, only in different mixing ratios, e.g. more hydrocarbons (solvents). Some industrial pollution is purely human in origin. Agriculture yields ammonia, methane and nitrous oxide that are also naturally emitted, but human activities have increased the emissions manifold.

Secondary pollutants are compounds that are formed in the atmosphere in reactions between primary pollutants. The most important is ozone formed from hydrocarbons and nitrogen oxide.

Constant emission from a stack placed in a flat terrain is the type of source that is simplest to characterise. However, such sources are rare. Emissions

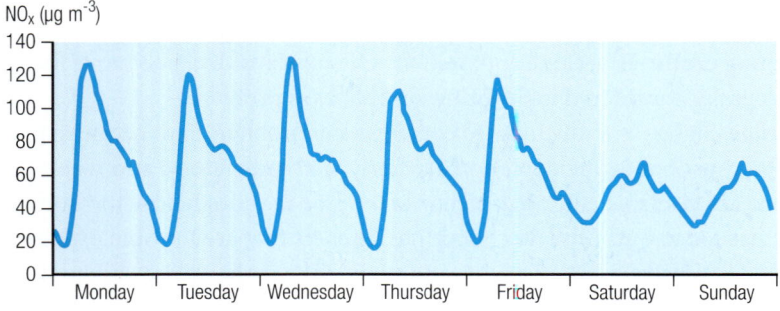

Figure 2.8 Weekly variation in NO_x concentrations in Copenhagen with peaks in the morning rush hour and lower values during he weekend.

from energy and heat production depend upon the needs of the market, and therefore they typically follow a seasonal variation with maximum emissions during winter, whereas emissions from traffic to a larger extent follow a diurnal variation. This results in corresponding variations in the pollution levels. Figure 2.8 shows the NO_x levels during a week in Copenhagen. The levels are highest during weekdays, when the traffic is dense. The levels are also highest in the mornings with rush hour and poor dispersion conditions.

A proper estimate of the emissions is crucial in air pollution modelling, where the obtained results will never be better than the quality of the information provided about the emissions. Therefore all air pollution studies are based on more or less detailed emission inventories, and much effort is currently being put into improving the quality and degree of detail in these inventories (Chapter 12). The resolution in emission inventories can cover scales in time from seconds to years, in space from individual point sources to global emissions and from a single to hundreds of compounds. Also the conditions of release can be indicated as e.g. a stack height and a flue gas temperature. The results take the form of emission from point sources, line sources or area sources. Often emission inventories have an aggregated form as e.g. the total emission from cars in a street or the total emission from a whole country. Also aggregation of compounds with similar effects are performed e.g. by use of efficiency factors. Typical examples are the use of ozone depletion potentials (ODP) aiming at the destruction of the ozone layer (Chapter 19) or global warming potentials (GWP, Chapter 20) to describe the increasing greenhouse effect.

The time resolution is an important point. Generally, the shorter the geographical distance the shorter the time step necessary. If the calculation shall support urban regulations with limit values, a one-hour resolution or better may be necessary. Long-range calculations with a geographical scale of 1000

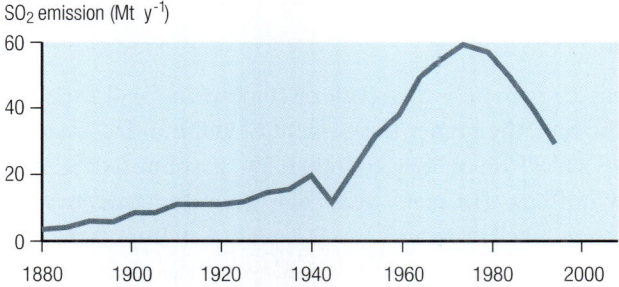

SO_2 emission (Mt y^{-1})

Figure 2.9 European emissions of SO_2 since 1850. There has been a drastic increase after the Second World War until the middle of the 1970's, when the importance of long-range transport became apparent and mitigation was initiated. Such inventories can be used to demonstrate general trends, but are not sufficiently detailed for more accurate studies (After Mylona, 1993).

km and aiming at e.g. acidification can (or must) do with yearly averages.

Also the type of final impact may determine the time resolution. Compounds with local immediate impact on the respiratory system, as e.g. nitrogen dioxide require a much shorter time resolution than e.g. compounds with long term carcinogenic effects or material damage. Generally a high resolution in space also requires a high resolution in time. A general trend can best be illustrated with low resolution, which is not suited for dispersion modelling (Figure 2.9).

The compounds

To date nearly 3000 different anthropogenic air pollutants have been identified, most of them are organic (including organo-metals). Combustion sources, especially motor vehicles, emit about 500 different compounds. However, only for about 200 of the pollutants have the impacts been investigated, the ambient concentrations are determined for a much smaller number, and time series for even fewer.

The pollutants can be divided into two groups (Wiederkehr, Yoon, 1998): the traditional *major air pollutants* (MAP, comprising sulphur dioxide, nitrogen dioxide, carbon monoxide and ozone) and *hazardous air pollutants* (HAP, comprising chemical, physical and biological agents of different types). The HAPs are generally present in the atmosphere in much smaller concentrations than the MAPs, and they often appear more localised (typically in urban areas or near industries), but they are – due to their high specific harmful action – nevertheless toxic or hazardous. Both in basic investigations and in abatement strategies HAPs are difficult to manage, not only because of the low concentrations, but also because they are in many cases poorly identified.

Major air pollutants

Sulphur dioxide, SO_2 in air is mainly the result of sulphur in fossil fuels. In general, the heavier the fuel, the higher the content of sulphur. Un-cleaned coal may contain up to a few % S; oil may contain 0.5%, gasoline 0.05%, and natural gas practically nothing. For many years SO_2 from domestic heating was the classical pollutant in urban areas. Later SO_2 played an important role in transboundary air pollution. In the industrialised world the problem is to some extent solved, by use of purified fuels and desulphurising systems in the exhaust. Sulphur dioxide concentrations in modern cities is often well below the air pollution guidelines, but long range transport with acidification is still a problem.

Particles have many sources and a large variety of sizes, shapes and compo-sition. Previously, soot from incomplete combustion was most important. Today the interest is centred on small particles from car exhaust, which have an important health impact. It is not quite clear whether the impact is due to the particles as such or to compounds attached to them. The emission can be reduced by changing the combustion process or with various types of fil-ters, but it may for a foreseeable future remain a significant urban air quality problem (European Commission, 2000).

Nitrogen oxides, NO and NO_2 is formed from the free nitrogen (N_2) in com-bustion air at high temperatures in combustion processes. It can also originate from nitrogen content in the fuel. To a large extent the emission thus depends upon the combustion conditions. In general the main part (typically 90 to 95%) is emitted in the form of nitrogen monoxide NO that is subsequently oxidised by ozone in the atmosphere to nitrogen dioxide, NO_2. In emission inventories the sum of NO and NO_2 called NO_x is normally indicated. If the emission is expressed in terms of weight, it is assumed that all the nitrogen is present in the form of NO_2. The emission can be limited by changing the combustion conditions or by using catalytic converters. In spite of the intro-duction of catalytic converters, the levels in cities can still be too high due to the increasing traffic volume. Nitrogen oxides have further impacts on nature as a toxic gas, or as fertiliser leading to eutrophication.

Nitrous oxide, N_2O is mainly formed by transformation of nitrogen com-pounds in the soil, and the use of fertilisers is therefore an important factor. There are further small contributions from various industrial processes, e.g. the production of nylon. A small amount comes from cars, where the use of catalytic converters can actually increase the emission. The compound is not toxic in itself, but it is an important greenhouse gas. The global concentration has increased about 15% since the 19th century.

Ammonia, NH_3. A few percent of ammonia in air come from traffic, but the absolute dominant source is agriculture with the use of animal manure as fertiliser and other agricultural practices. Here the emission can be reduced by careful storage and effective distribution. Filters can reduce emission from stables. The smell of ammonia and especially related compounds is a nui-sance. Ammonia can harm vegetation and promote eutrophication.

Carbon monoxide, CO is mainly formed by incomplete combustion of fossil fuels. Traffic is the main source. The emission can be reduced by use of cata-lytic converters and the concentration reduced to below harmful levels.

Carbon dioxide, CO_2 is the end product in combustion of practically all fuels. It is in itself harmless in present concentrations and can even act as a promoter for vegetation growth. The rising global atmospheric concentration, however, increases the greenhouse effect and thus changes the global climate. Many attempts have been made to limit the global emissions (Chapter 20), but on a global level they have so far had very limited effect. So far the atmospheric concentration has increased about 30% since the start of the industrialisation in the 19th century.

Methane, CH_4 is mainly formed in anaerobic decomposition of organic material, both in the soil and in the gastric system of especially ruminants. An important human source is wet rice paddies. It can also arise from accidental releases from natural gas systems. It is in itself harmless in present concentrations, but it plays a role in photochemical air pollution. It is also an important greenhouse gas. Emissions from refuse dumps are therefore sometimes collected and used as fuel. The global concentration is more than doubled since the 19th century.

Photochemical oxidants, including ozone, are secondary pollutants i.e. they are not emitted from sources, but formed in the atmosphere from primary pollutants. The oxidants have impacts on the human respiratory system and harm vegetation. Ozone in the troposphere is also a greenhouse gas. In Europe it appears that the ozone concentration in the troposhere has doubled since the first measurements were made around 1900 (Volz, Kley, 1988).

Hazardous air pollutants

Heavy metals include first of all lead, mainly from petrol additives. By and large it is phased out in the industrialised countries, but it is still important in some developing countries. Further there is a large range of metals and organo-metals from industrial sources.

Non Methane Volatile Organic Compounds (NMVOCs) have a series of different sources: industrial processes, use of solvents, evaporation of liquid fuels, incomplete combustion etc. They can be precursors of photochemical pollutants or can be toxic themselves. Reduction in emissions is carried out by many means. Most important is the use of catalytic converters in cars.

Benzene has had a special interest as a possible substitute for lead in petrol to increase the octane number. It is now phased out, and its importance is declining.

Dioxins are formed by combustion of refuse containing chlorine compounds. Of 210 possible compounds especially 12 (the dirty dozen) are toxic.

Chloro-Fluoro-Carbons, CFCs form a family of man-made compounds. They are very stable and non-toxic and have had a series of applications as propellants in spray cans, in refrigeration systems, in fire extinguishers etc. For many years they were considered an environmental asset. They have, however, proved to deplete the ozone layer and to be effective greenhouse gases. The use is therefore phased out. They are substituted by related compounds such as HFCs and PFCs.

Natural sources and emissions

With the exception of some industrial hydrocarbons and CFCs the pollutants are ordinary chemical compounds, which also exist in the natural, "unpolluted" atmosphere. Whether a compound is a pollutant or not is therefore to a large extent a question of concentration. It can also be ambiguous to establish whether a given emission is human. E.g. changes in the use of agricultural or forest areas can change the emissions from natural systems.

On a global scale it is often natural emissions from vegetation, volcanoes, forest fires or sea spray that dominate the content of trace compounds in the atmosphere. It may often be important to determine those emissions as e.g. natural hydrocarbons that play an important role in photochemical air pollution. Thus in Europe about one third of the volatile hydrocarbons arise from forests and agriculture; they are typically in the form of isoprene or terpenes. Particles from natural sources, e.g. sea spray, can constitute a large proportion of the total atmospheric content of particles, but they are not particularly poisonous, and they cannot be limited by mitigation.

2.8 Impacts of air pollution

Impacts can have various forms:

Health. Immediate impacts are determined by short-term exposure where regulation is attempted by setting short term limit values (hours). Typical examples are respiratory symptoms. Long-term impacts are e.g. due to toxic heavy metals or carcinogenic compounds. These are determined by average exposure over long time and attempted regulated by setting limit values for long term average exposure (months to years).

Materials. Impact on materials is a long-term process, and in principle long-term average values of air pollution should be sufficient to assess the effect. On the other hand, however, the processes depend upon other varying parameters – first of all temperature and humidity.

Vegetation. Vegetation can be damaged by direct impacts that are short-term e.g. photochemical air pollution, but also by long term impacts as e.g. acidification. Some air pollutants as e.g. nitrogen compounds are fertilisers, and their impacts are in principle beneficial. Some natural ecosystems are, however, adapted to low levels of nutrients and over-fertilising (eutrophication) can disturb the equilibrium and lead to loss of biodiversity.

Global impacts. Some compounds have lifetimes of a year up to centuries in the atmosphere. They will be more or less homogeneously distributed in the entire atmosphere. This is the case for CFCs that attack the ozone layer and for CO_2 that is the most important human greenhouse gas.

Air pollution is of course not the only environmental threat. But sometimes the impacts are similar or interacting. A typical example is the forest die back in Denmark in the 1970's and 1980's.

Exposure to e.g. heavy metals can proceed through various channels. Thus lead can be deposited on vegetables and later eaten by humans or by animals that are later eaten by humans. Lead can also be a content or addition to foodstuffs. Thus it has been argued that lead poisoning through food and wine was one of reasons for the collapse on the Roman Empire. Such effects must be taken into account when air quality standards, guidelines and limit values are determined.

2.9 The general development

Income and pollution levels in cities

In general the urban pollution in the industrialised world has gone through a development with increasing wealth *and* pollution until the standard of living was so high that environmental protection was found reasonably affordable (Figure 2.10). Also in cities in economies in transitions this development is on its way. However, large cities in the developing world are still facing serious pollution problems often as a result of tremendous unregulated traffic loads and poorly regulated industries in rapidly growing urban areas. In

Asia, Latin America and Africa the urbanisation has been accompanied by the proliferation of slums and squatter settlements.

In some cases polluting industries transferred from industrialised countries with stricter environmental legislation and higher wages has aggravated the situation.

Air pollution concentration

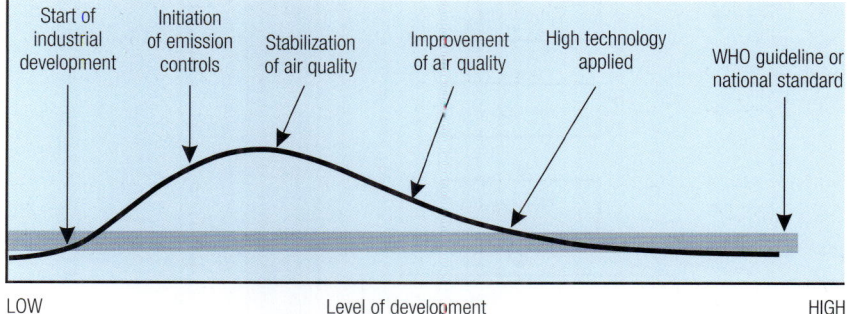

Figure 2.10 Development of pollution with per capita income in a large city. The curve only shows a general tendency, with marked deviations. Thus in the early 1980's Kuwait had at the same time some of the highest incomes and highest pollution levels. (After Mage et al. 1996).

In line with this typical development The World Commission on Environment and Development in its report "Our common future" (1987) conceives technological development and rising standard of living as a prerequisite for environmental improvement. Or as Bertolt Brecht has put it:"*Erst kommt das Fressen, dann kommt die Moral*" – in modern form "*First development and later pollution control*".

The pollution chain

The air pollution problem can be perceived as a chain of events (Figure 2.11). It starts with a technical or political *decision* e.g. to produce energy by burning coal. Since this fuel is not clean and the combustion processes can form new compounds, this leads to *formation* of various pollutants that are *released* to and *dispersed* in the atmosphere, often with simultaneous *transformations*. The pollutants are *deposited* by various processes on the ground, on materials or vegetation, or in the human respiratory system. Here it can have unwanted *impacts*. Some compounds as e.g. CO_2 can have impacts directly in the atmosphere. Other compounds may take a round in an ecological system, before they reach the target. This may e.g. be the case for heavy metals in foodstuffs. Finally the impacts may lead to a societal *response* in terms of regulations based on limit values. This response can in principle be applied at any link of the chain, from different means of production, over flue gas cleaning and dispersion, to various protective measures.

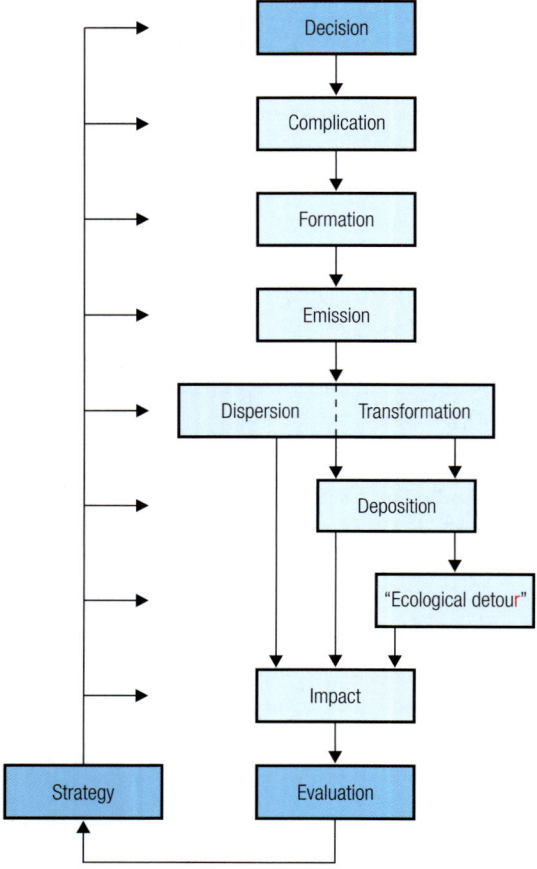

Figure 2.11 Air pollution phenomena can be perceived as a chain of events from a political or technical decision to a final impact. Mitigation can be applied at any stage.

In the late 1980s this awareness led to development of the so-called DPSIR concept (Driving force, Pressures, State, Impact, and Response) by the Dutch institute RIVM. It describes the connections between social conditions and global as well as regional and/or local environmental impacts.

Mathematical models can in principle describe all steps in such a system, and the ultimate goal is a combination of the individual models in a comprehensive system for scientific investigations and decision support (Chapter 14). There are examples of such coupled model systems that have been developed and currently are in use. Caution must be taken, however, when the results from these systems are interpreted. Often, the included sub-models have been highly simplified in order to reduce calculation time and make the system operational for fast scenario studies. The results may therefore in some cases not reflect the governing processes in sufficient detail. These considerations call for careful validation of the applied tools. Vast improvements have, however, been obtained in recent years due to improved understanding of processes and tremendous increase in computer resources being available.

The need for an integrated assessment

Cost effective management of air pollution requires knowledge and understanding of processes of very different nature. For all processes there are mathematical models. The ultimate goal is to combine them in a comprehensive system covering a description from emission to impacts and their economic consequences. This is not an easy task, because air pollution is a phenomenon

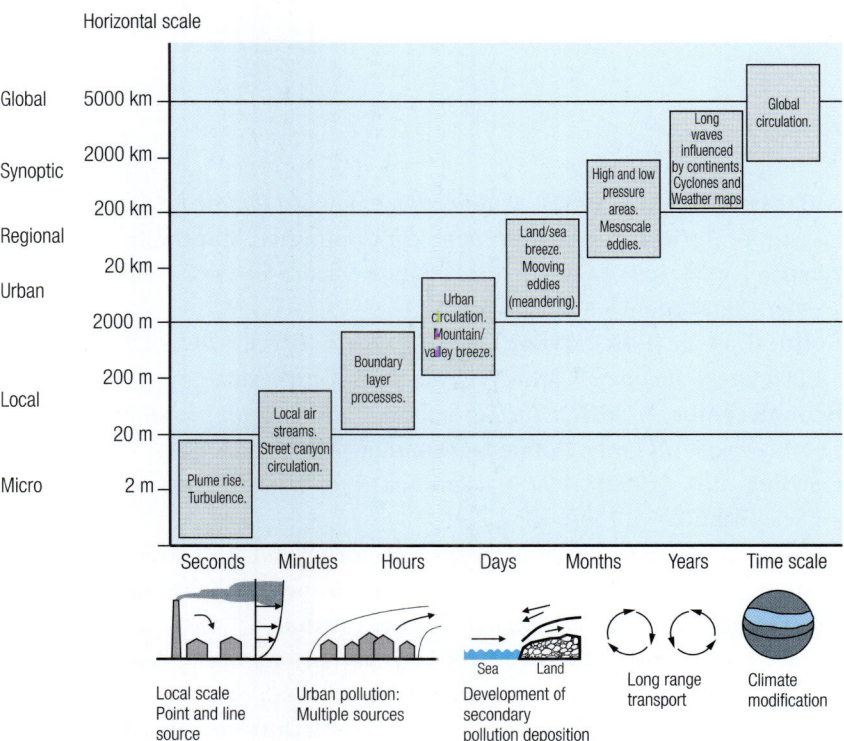

Figure 2.12 Time and spatial scales of atmospheric processes governing air pollution.

that is played on widely varying scales (Figure 2.12) from local "here and now" events to phenomena that takes centuries and cover the entire Earth.

In recent time various attempts have been made. In international negotiations so-called *The Rains Model* has played an important role. The model is based on linear determinations of pollution levels from a fixed matrix and it can calculate the economy in emission reductions in the entire Europe, and e.g. define where the mitigation efforts should most economically be applied. The culmination so far in this work is the so-called Gothenburg Protocol, which was signed in 1999. It is both a "multi-pollutant" and a "multi-effect" protocol aiming at reducing acidification, eutrophication and damages from tropospheric ozone. The strategy is a comprehensive effort against trans-boundary pollution with sulphur dioxide, nitrogen oxide, hydrocarbons and ammonia. The protocol operates with fixed targets for improvement of the state of the environment in all the areas considered. The emission reductions, however, are carried out where it is most cost-effective, i.e. where the benefits to the affected areas are largest, and where actions so far have been taken at least towards emission control and further efforts therefore will be relatively cheap.

2.10 Literature

Aniansson, B. (ed.) (1982): *Acidification today and tomorrow.* Swedish Ministry of Agriculture, Environment ´82 Committee, Stockholm.

Austin, J. et al. (eds.) (2002): *Air Pollution Science for the 21ˢᵗ Century.* Elsevier, Amsterdam.

Bolin, B. et al. (Eds.) (1986): *The Greenhouse Effect, Climatic Change and Ecosystems.* Scope 29. John Wiley & Sons, Chichester.

Brimblecombe, P. (1987): *The Big Smoke. A history of air pollution in London since medieval times.* Routledge, London.

Cowling, E.B. (1982): *Acid precipitation in historical perspective.* Environ. Sci. Technol. *16*, 110A-123A.

EMEP (Co-operative programme for monitoring and evaluation of the long-range transmission of air pollutants in Europe) publishes yearly reports through the Norwegian Meteorological Institute.

Fenger, J. et al. (eds.) (1998): *Urban Air Pollution – European Aspects.* Kluwer Academic Publishers. Dordrecht.

Firket, M. (1931): *Sur les causes des accidents survenus dans la valee de la meuse lors des brouillard de Decembre 1930.* Bulletin de L´academie Royale de Medicine de Belgique 11,126.

Graedel, T.A. and Crutzen, P.J. (1995): *Atmosphere, Climate, and Change.* Scientific American Library, New York.

Gribbin, J. (1988): *The Hole in the Sky.* Corgi Books.

Hong, S. et al. (1994): *Greenland Ice Evidence of Hemispheric Lead Pollution Two Millennia Ago By Greek and Roman Civilizations.* Science *265*, 1841-1843.

IEA (International Energy Agency) **(2005):** Key world energy statistics.

IPCC (Intergovernmental Panel of Climate Change) **(2001):** *Climate Change 2001* – Synthesis Report. Cambridge University Press. Cambridge. This and more detailed reports can be downloaded from the internet at: www.ipcc.ch

IPCC (Intergovernmental Panel of Climate Change) **(2007):** Climate Change 2007 – Synthesis Report. Cambridge University Press. Cambridge. This and more detailed reports can be downloaded from the internet at: www.ipcc.ch

Kowalok, M.E. (1993): *Research Lessons from Acid Rain, Ozone Depletion, and Global Warming – Common Threads.* Environment *35.6*, 13-20, 35-38.

Lovelock, J. (1979): *Gaia. A new look at life on Earth.* Oxford University Press, Oxford.

Mage,D. et al. (1996): *Urban Air pollution in Megacities of the World.* Atmospheric Environment *30*, 681-686.

Mylona, S. (1993): *Trends of sulphur dioxide emissions, air concentrations and depositions of sulphur in Europe since 1880.* EMEP/MSC.W Report 2/93. Norsk Meteorologisk Institut, Oslo. 35 pp + appendix. (Later published in Tellus (1996), *48B*, 662-689).

Nriagu, J.O. (1983): *Lead and lead poisoning in antiquity.* Wiley, New York.

Nriagu, J.O. (1990): *The rise and fall of leaded gasoline.* The Science of the Total Environment, *92*, 13-28.

Smith, K.R., (1988): *Air pollution. Assessing total exposure in developing countries.* Environment *30*, 16-20, 28-30, 33-35.

The World Commission on Environment and Development (1987): *Our common future.* Oxford University Press, Oxford.

UNEP, WHO (1992): *Urban Air Pollution in Megacities of the World.* Blackwell, Oxford.

Vitousek, P.M. et al. (1997): *Human Domination of Earth´s Ecosystems* Science *277*, 494-499.

Volz, A. and Kley, D. (1988): *Evaluation of the Montsouris series of ozone measurements made in the nineteenth century.* Nature *332*, 240-242.

Weart, S.R. (1997): *The discovery of the risk of global warming.* Physics Today. January 1997. 34-40.

WHO (World Health Organisation) **(2000):** *Air Quality Guidelines for Europe. Second Edition.* WHO Regional Publications, European Series, No.91.

Wiederkehr, P. and Yoon, S.-J. (1998): *Air Quality Indicators.* In: Fenger, J. et al. (Eds.) Urban Air Pollution. European Aspects. Kluwer academic Publishers. Dordrecht, pp. 403-418.

Wilkens, E.T. (1954): *Air pollution aspects of the London fog of December 1952.* Quarterly Journal of the Royal Meteorological Society, *80,* 267-271.

II

Sources and control

Air pollution arises from many different sources which emit a spectrum of compounds. Most come from the application of fossil fuels for energy production i.e. coal, oil and gas products, and it is to a large extent the same compounds that are emitted – only in varying proportions. The different fuels and combustion processes are therefore first described in general terms – and in connection with this the principles of emission control (Chapter 3).

There are, however, many different sources of air pollution. They can be divided into two groups, stationary sources (Chapter 4) and mobile sources (Chapter 5). Stationary sources are typically power plants, industrial plants, district heating and individual heating systems in houses, as opposed to mobile sources, typically cars, ships and aircraft. Agriculture (Chapter 6) is a special type of sources, where the border to natural sources (Chapter 7) is less well defined. A survey of sources and emissions is given later in Chapter 12.

3 Pollution Control Principles

Jan Erik Johnsson

The major sources of air pollutants are transport, energy conversion, industrial production processes and agriculture. Therefore combustion of solid, liquid and gaseous fuels is very important for the emission of air pollutants. Most fuels contain carbon and hydrogen, and there may be a content of oxygen, sulphur, nitrogen, water and mineral substances depending on the fuel type. The mineral part of the fuel is forming ash, while the remaining constituents are converted to gaseous components. An overall reaction scheme for complete combustion of a fuel on an ash and water free basis can be written as follows:

$$C_aH_bO_cS_dN_e + (a + b/4 + d + xe/2 - c/2)\ O_2 \rightarrow$$

$$a\ CO_2 + b/2\ H_2O + d\ SO_2 + xe\ NO + (1-x)e/2\ N_2 \tag{3.1}$$

The content of carbon, hydrogen and sulphur is oxidized to CO_2, H_2O and SO_2, and if ash is present in the flue gas a small part of the SO_2 may be captured by the ash. In most stationary sources an almost complete combustion to CO_2 and H_2O is usually obtained, but smaller amounts of CO, CH_4 and NMVOC (Non Methane Volatile Organic Compounds) may be emitted (Flagan and Seinfeld, 1988; Turns, 1996; Warnatz et al., 1996). In the internal combustion engine used in mobile sources a larger part of the carbon content leaves the combustion chamber as CO, CH_4 and NMVOC.

Only a fraction, x, of the nitrogen content in the fuel is converted to NO, the major part forms N_2. The degree of conversion to NO depends on the design and operation of the combustor and can be lowered by low-NO_x techniques as described in Chapter 4. At high combustion temperatures NO may be formed from O_2 and N_2 in the combustion air (thermal NO_x), even if the fuel is without a nitrogen content. Some trace metals like Hg, Cd and Pb may be emitted as aerosols or partly vaporized dependent on the mineral content in the fuel and the flue gas cleaning techniques. The conclusion is that the emissions from combustion depend on the fuel and plant type (Flagan and Seinfeld, 1988), as described in Chapter 12 Emission inventories. A substantial part of the emissions from production processes and agriculture is due to the energy consumption, but a variety of emissions are related directly to the processes. Examples of process related emissions are solvents and other organic compounds from chemical production, solvents from printing and ammonia from agriculture.

In this chapter the emphasis will be on different methods to reduce the emission of air pollutants to the atmosphere. First minimizing the formation and emission of air pollutants using the principles of Clean Technology and Life Cycle Assessment (LCA) will be described, and then the methods for control of particulate and gaseous emissions from stationary sources will be addressed in Chapter 4, and from mobile sources in Chapter 5.

3.1 Clean Technology

Obviously the best way to avoid the emission of pollutants to the environment would be to carry out all production processes and other human activities without the formation and emission of any pollutants. That is a complete recycling of all materials in the complex ecosystem encompassing the whole world. In theory it is possible to completely recycle and re-use the materi-

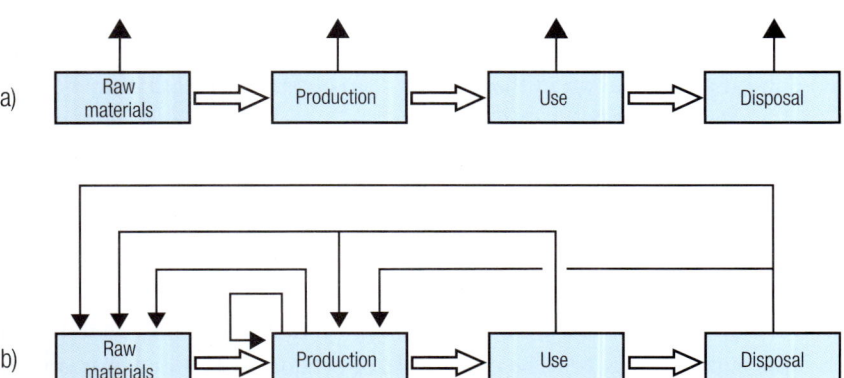

Figure 3.1 a. Traditional production process with emissions to the atmosphere. b. Clean technology, recirculation of waste materials.

als we are using, but our state of technology is not sufficiently developed. Furthermore the energy consumption for 100% recycling of material may be very high, giving rise to high emissions from energy conversion processes, and so the total emissions may increase. However, it is possible to come a long way in reducing the emission of pollutants to the environment from production, use and disposal of products by proper planning of the production processes. Figure 3.1a shows a production process, where all steps from the processing of raw material to the disposal of the used product will give rise to emissions to the environment.

Traditionally focus has been on the production process, which will in many cases be a major source of emissions. The use of new or different processes or technologies with lower emissions has been termed "clean technology". In many cases it has been possible to develop technologies with very low emissions. In principle, this can be done in different ways by:

- Substitution of raw materials giving rise to emissions in the production process
- Substitution of products, e.g. change the product so that polluting materials are not needed
- Choice and optimisation of production processes e.g. by altering the process to require minimal amounts of polluting material
- Use, re-use and recycling of waste streams and by-products, e.g. by enclosing the process and recycle material so polluting compounds will not be emitted to the atmosphere.

This way of preventing the formation or reducing the emission of pollutants will often be a better solution than gas cleaning, and the use of clean technology should always be considered, when a new production or other activity with a potential for emissions to the environment is planned. In existing processes it may be very costly and in many cases the technology for zero emission has not yet been developed, and gas cleaning and dispersion in the atmosphere will still be important methods to improve the air quality.

The use of clean technology has often focused on the production process, while emissions from the production of raw materials, use of the produced items and final disposal was not considered. However, there is a potential for emission of pollutants from all steps in the total life cycle of the product from the processing of raw materials to the final disposal. The ultimate way to reduce emissions would be to recycle all waste and close the total process completely or to use all waste and by-products in other production processes, as illustrated in Figure 3.1b. The use of alternative raw materials to reduce emissions from the production process might increase the emission from the processing of raw materials, and thus lowering the emissions from the whole process shown in Figure 3.1 may be much more difficult than to introduce clean technology in the production process alone. The way in which the environmental impact of the complete process is addressed in a formal way is by using the Life Cycle Assessment (LCA), a family of methods for looking at materials, services, products, processes and technologies over their entire life, as described later.

To illustrate the concept of clean technology a few examples from the field of energy use and conversion will be given. In the emission inventories in Chapter 12 it can be seen that transport and residential heating are very important sources of air pollution in cities. For NO_x, transport is the major source in all cities. For SO_2, individual heating of houses has always been a major contributor to the concentration levels measured in cities at geographical locations where heating is necessary.

Previously the major energy resource for heating purposes was solid fuels, especially coal in larger installations and so-called smokeless fuels like coke in individual houses. Air pollution with dust, smoke and odorous compounds were major problems in cities as described in Chapter 2. In the period 1950-1970 solid fuels were gradually exchanged with various oil products, residual oil for larger installations and gas oil was used for individual heating and in many other small installations for heating or industrial purposes. From one air pollution point of view this was a clean technology, because the emissions of particulate matter decreased dramatically. From another point of view, that is the emission of SO_2, it was not a clean technology, because the crude oil came mainly from the Middle East where the sulphur content of the oil is high. As a consequence the emissions of SO_2 increased with increasing energy consumption in Europe. After the energy crisis in 1972-73 the energy consumption in many European countries levelled off, and the interest for different clean technologies started. Over the last 30 years many of these clean technologies have been implemented, although to a varying degree in the many European countries, depending on local circumstances and needs. The description of the clean technology options for heating will be given under the headings mentioned above starting with the choice of raw material.

Substitution of raw material

When liquid or gaseous fossil fuels are burned, the sulphur content in the fuel is converted completely into SO_2 and emitted with the flue gas. Therefore one way of reducing the emission of SO_2 has been to use oil and gas with low sulphur content.

Over the years oils with relatively high sulphur content have been substituted with oils with lower sulphur content. This change has been forced by legislation. There are two ways of obtaining oil with low sulphur content. One way is to make a more effective desulphurisation at the oil refinery. In modern refineries the sulphur compounds removed are converted to sulphur, which is then sold and used as raw material for production of sulphuric acid. The other way is to use a crude oil with lower sulphur content. The North Sea oil is relatively low in sulphur, and therefore this option has been possible in several countries.

Substitution of oil with natural gas decreases the emission of SO_2 from the heating system to close to zero. The natural gas may have a content of H_2S when it comes from the gas field, but the H_2S is removed at a gas treatment plant, converted to sulphur, and used for sulphuric acid production. The emission factors for SO_2 for commonly used fuels are shown in Chapter 12.

The oil may of course be substituted with other energy sources like straw and wood, and the emission of SO_2 will be in the same range as for gas oil.

However, if climate changes and emissions of greenhouse gases are considered, the conclusion is different. A change in fuel from coal to natural gas can lower the emission of CO_2 by about 40%, but a change to renewable fuels like wood and straw can in practice eliminate the emission of CO_2, since the CO_2 emitted from combustion of wood and straw will be fixed by photosynthesis in the next growing season.

Substitution of products

The purpose of indoor heating is to obtain an appropriate temperature. When it is steady, the heat input is equal to the heat loss, and therefore this temperature can be obtained either by a large heat input or a low heat loss. In Europe the interest in insulation increased dramatically after the energy crisis in 1972-73. Older houses were insulated and stricter standards for insulation of new houses were enforced during the next years. The purpose was to reduce energy consumption and in many countries to reduce the dependence on oil import. There was a very short pay back time on insulation of existing houses because of the high energy prices, and as a result insulation was done to a great extent and the energy consumption for heating purposes was reduced substantially. As a side effect the emission of SO_2, NO_x and other air pollutants was reduced and the air quality improved.

In the beginning of the 21st century fuel prices began to increase and now there are two main driving forces for lowering the energy consumption: economy and reduction of emission of greenhouse gases and other air pollutants.

Choice and optimisation of production processes

In many cities district heating is used for heating purposes. Hot water is produced in a central unit and distributed to the individual houses. The use of district heating usually will improve the air quality locally, because the emission is moved from many small and low chimneys to one large and high chimney, which may be located outside the city. But the benefit is even larger, if the hot water is produced by cogeneration or combined heat and power production.

The most important method of electricity production in Europe is coal combustion, although some countries get all or some of their electricity from hydropower, nuclear power or combustion of gas, oil and bio-fuels. Figure 3.2 shows a modern coal fired power plant located at the sea or at a river.

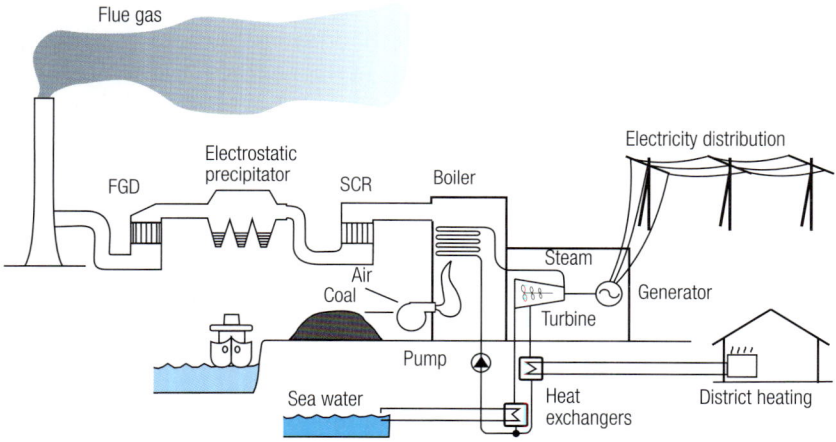

Figure 3.2 Modern power plant for combined heat and power production.

Coal is transported to the plant, crushed to a fine powder and blown into the combustion chamber with air. The heat of combustion is transferred to water flowing in tubes along the walls of the combustion chamber. The water evaporates and steam at high pressure and high temperature is formed. The steam expands in a steam turbine and a generator produces electricity. After expansion the steam is condensed with cooling water from the sea or from a cooling tower and the water is pumped back to the combustion chamber. In a modern power plant producing electricity with high efficiency more than 40% of the energy content of the fuel is lost with the cooling water, almost 50% may be converted to electricity and the balance, about 6% of the energy input, is heat loss, mostly with the hot flue gas.

If the steam is taken from the turbine before it is completely expanded, its temperature is high enough to heat water for district heating, and it is condensed by cooling with return water from the district heating system. This is the so-called cogeneration or combined heat and power production. The Carnot cycle efficiency for electricity production decreases and about 40-43% of the energy content of the fuel can be converted to electricity. But about 50% will be converted to district heating, and nothing will be lost with cooling water. The total energy efficiency of the combined heat and power production is about 94%. As a result there is a large saving in fuel consumption compared to production of electricity and hot water for district heating in separate plants.

As seen in Figure 3.2 flue gas from the combustion chamber is cleaned for NO_x, fly ash and SO_2 before leaving via the chimney. These flue gas cleaning processes will be described in Chapter 4. District heating with combined heat and power production results in a major improvement of the air quality. The fuel consumption is much lower than for individual heating or conventional district heating and the flue gas can be cleaned before entering the atmosphere from a high stack. Combined heat and power production in other types of plants is also possible, and many municipal waste incinerators, gas fired plants and gas turbines with combined heat and power production are being built.

The use of clean technology options described above and the general trend towards building higher stacks have been the major reasons for the very substantial improvement in the air quality with regard to SO_2 which has been observed in many European cities over the last 25 years. Details about the air quality are given in Chapter 13.

3.2 Life Cycle Assessment (LCA)

"Life cycle assessment" (LCA) is an objective process to evaluate the environmental burdens associated with a product, process, or activity. This is done by identifying and quantifying energy and materials usage and environmental releases, and to assess the impact of those energy and materials uses and releases on the environment, and further to evaluate and implement opportunities leading to environmental improvements. The assessment includes the entire life cycle of the product, process or activity, encompassing extracting and processing raw materials; manufacturing, transportation, and distribution; use, re-use, maintenance; recycling; and final disposal." (Graedel and Allenby, 1995). Using LCA it is possible to ensure that a proposed clean technology will not increase emissions from another step in the life cycle and eventually increase the total emissions. It is very useful for assessment of the environmental impact when comparing different technologies, production processes and products (Wenzel et al., 1997; GaBi , 2004).

Some perspective on LCA is provided by an example comparing the use of fuel to power gasoline and electric automobiles. Providing gasoline involves crude oil production, transporting crude oil, refining the oil, transporting and delivering gasoline and use of the gasoline in the vehicle. For an electricity powered automobile, the energy comes mostly from fossil fuel combustion at large stationary power generation facilities, but in some countries electricity is produced mainly from hydropower and/or nuclear power. In this way an LCA for vehicles will include emissions from stationary sources.

The LCA inventory analysis considers the emissions from each type of operation and compares the total emission from the whole process with criteria relevant for the specific emissions. An important consequence is that most emissions attributable to the gasoline powered vehicle occur where it is being operated, often in the cities, while most emissions attributable to the electric vehicle occur at power plants which may be placed in rural areas or at least have high stacks. Hence for pollutants with a short lifetime the population exposure is greater from the gasoline powered vehicles. Also control technologies are better applied and monitored at a few point sources than over a lot of small emission sources.

The scope of the analysis in this case determines the conclusion: If emissions from the actual vehicle are all that is considered, the electric vehicle obviously is superior. If emission of greenhouse gases is of concern from

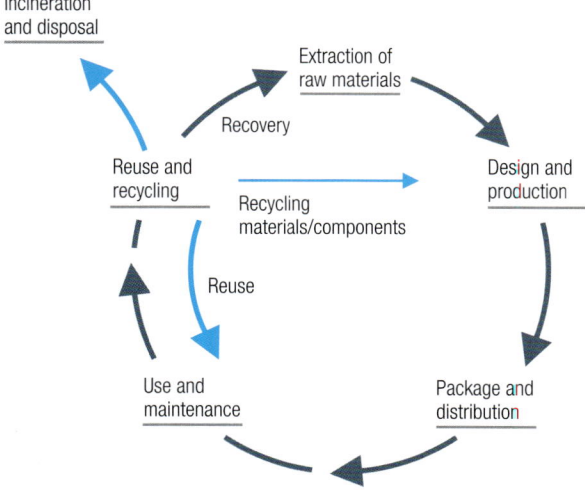

Figure 3.3. The principles of Life Cycle Assessment, LCA (Remmer and Münster, 2003).

a global warming standpoint, a different conclusion might be the result, although estimates indicate that the CO_2 emission will be lowered slightly, even if the electricity is produced on coal fired power stations, (Horstman and Jørgensen, 1997). Another complication is that the state of flue gas cleaning on the power stations is different from country to country and some countries use mostly hydropower, and therefore the analysis is valid only for the specific local conditions it is made for.

This example was actually only part of a LCA. A comprehensive LCA would consider in addition the emissions and environmental impacts of the vehicles themselves; their materials, their manufacture and their eventual recycling. And also the energy consumption and the consumption of resources are included in the LCA. Figure 3.3 illustrates the whole life cycle of a product. The consumption of energy and resources and the environmental impact of all steps in the life cycle must be considered in the LCA.

To perform a proper LCA is not an easy task, and the result will depend on many factors. A large effort has been put into the development of standard methods for LCA (ISO 14040-14043, 1997-2000) and software (GaBi, 2004). LCA is an excellent method in product development and also when choosing between different new clean technologies or air pollution control techniques taking the total environmental impact into account.

3.3 Air pollution control

As described above, even if clean technology may be the best way to lower emissions and improve air quality in the long run, control of air pollution by gas cleaning has been, and still is, a very important method to reduce emissions to the atmosphere.

There are two different ways to lower the emission of air pollutants from a given source:

- Removal of air pollutants
- Conversion of pollutants

In the first case a waste product or by-product will emerge and must be taken care of. The solution of an air pollution problem must not give rise to water pollution or a waste problem. In the second case the pollutant is converted to other chemical compounds, and the environmental impact of the compounds must be considered.

Flue gas desulphurisation on power plants is an example of the first case.

Sulphur dioxide can be removed from the flue gas in a scrubber system using limestone or chalk to react with SO_2 and the product will be gypsum:

$$SO_2(g) + \tfrac{1}{2}O_2(g) + CaCO_3(s) + 2H_2O(l) \rightarrow$$

$$CaSO_4, 2H_2O(s)(gypsum) + CO_2(g) \qquad\qquad (3.2)$$

In this case gypsum of a saleable quality is produced and may be sold for the production of wallboard or cement. However, an emission of CO_2 will also take place, but is minor compared to the emission of CO_2 from the combustion of coal or oil in the power plant. In this case a useful product is the result of the gas cleaning process and the process is a kind of clean technology, because exploration of natural resources of gypsum is avoided.

Conversion of NO_x ($NO + NO_2$) in the flue gas from power plants is an example on the second type of gas cleaning methods. In this case ammonia is injected into the flue gas and reacts with NO_x over a catalyst. The overall reaction for reduction of NO is:

$$4\,NO + 4\,NH_3 + O_2 \rightarrow 4\,N_2 + 6\,H_2O \qquad\qquad (3.3)$$

In this case nitrogen and water are the only products, and no harmful compounds have to be taken into consideration. More details about the flue gas cleaning methods will be given in Chapter 4.

Another example of gas cleaning by conversion of the pollutants is the use of three way catalysts on cars with spark ignition petrol engines. The active materials on the catalyst are noble metals, and the catalyst is placed in the exhaust system. The catalyst simultaneously removes CO, hydrocarbons and NO_x, hence the name. More details about this method are given in Chapter 5.

3.4 Summing up

To reduce the harmful effects of air pollutants in an efficient and economic way is not an easy task. There are several ways of doing this, and they may be grouped as follows:

• Environmental regulations and administrative procedures
• Clean technology
• Gas cleaning
• Dispersion in the atmosphere.

In the next chapters about stationary sources (4), mobile sources (5) and agriculture (6) emphasis will be on gas cleaning and clean technology.

3.5 Literature

Flagan, R.C. and Seinfeld, J.N. (1988): *Fundamentals of Air Pollution Engineering,* Prentice Hall, New Jersey,.

GaBi (2004): IKP, University of Stuttgart.

Graedel, T.E. and Allenby, B. R. (1995): *Industrial Ecology,* Prentice Hall, Englewood Cliffs, New Jersey.

Horstman, J. and Jørgensen, K. (1997): *The future for electric vehicles in Denmark,* Orientering fra Miljøstyrelsen, Nr. 1 (in Danish),.

ISO (1997): ISO 14040: *Life Cycle Assessment – Principles and Framework,* ISO.

ISO (1998): ISO 14041: *Life Cycle Assessment – Goal and scope definition and inventory analysis,* ISO.

ISO (1998): ISO 14042: *Life Cycle Assessment – Life cycle impact assessment,* ISO.

ISO (2000): ISO 14043: *Life Cycle Assessment – Life cycle interpretation,* ISO,

Remmer, A. and Münster, M. (2003): *An introduction to Life-cycle thinking and management.* Environmental News 68. The Danish Ministry of Environment.

Turns, S.R. (1996): *An Introduction to Combustion,* McGraw-Hill Inc., Singapore,.

Warnatz, J. et al. (1996): *Combustion,* Springer-Verlag, Berlin Heidelberg,.

Wenzel, H. et al. (1997): *Environmental Assessment of Products – Methodology, Tools and Case Studies in Product Development,* Chapman and Hall, London.

4 Control techniques for stationary sources

Jan Erik Johnsson

This chapter falls in two parts, the first dealing with particulate removal and the second with treatment of gaseous compounds. At last some examples of important gas cleaning methods in power production will be described. In the text the term "gas cleaning efficiency, η" will be used. The definition is straightforward:

$$\eta = \frac{\text{mass of the pollutant removed by the cleaning device}}{\text{mass of the pollutant entering the cleaning device}} \qquad (4.1)$$

4.1 Primary particulates

As it will be discussed later, a large part of the fine particulates in the atmosphere are secondary particles. Nonetheless, the control of primary particles is a major part of air pollution control. Although primary particles are generally larger than secondary particles, many primary particles are small enough to be respirable and are thus of health concern. The most important types of control devices are (Johnsson, 1998; Wark et al., 1998; Nevers, 1995; Heinsohn et al., 1999):

- Gravity settlers
- Cyclones
- Scrubbers
- Bag filters
- Electrostatic precipitators

The choice of dust collector will depend on many parameters; the most important are the volumetric gas flow, the particle concentration, the particle size distribution and the physical and chemical properties of the particles. The efficiency of a dust collector depends on the aerodynamic properties of the particles, and the terminal velocity is used to characterise these properties. This important property will be shortly described before the different gas cleaning methods are treated.

Terminal velocity

The "terminal gravitational settling velocity" is an important characteristic of particles suspended in gas or liquid. At steady state, i.e. zero acceleration, three forces act on the suspended particle: the gravitational force, the drag force and the buoyancy. From a force balance the terminal velocity can be calculated. In Stokes' flow regime, i.e. for particles in air in the size range 5 to 50 μm, the terminal velocity, V_t, can be written as given in Equation 4.2:

$$V_t = \frac{g \cdot d_p^2(\rho_p - \rho_g)}{18\mu} \tag{4.2}$$

g: acceleration of gravity, d_p: particle size, ρ_p: particle density, ρ_g: gas density, μ: gas viscosity. The equation can be used as an approximation in the particle range from 1 to 100 μm (Nevers, 1995). This size range is very important for control of particulates, because the majority of the particles to be removed will be in that interval. Table 4.1 shows the terminal velocity of spherical particles in air at ambient conditions.

Particle size, μm	0.01	0.1	1.0	10	100	1000
Terminal velocity, ms^{-1}	$1.3 \cdot 10^{-7}$	$1.8 \cdot 10^{-6}$	$6 \cdot 10^{-5}$	$6 \cdot 10^{-3}$	0.5	7

Table 4.1 Approximate terminal velocities of spherical particles in air at ambient conditions (298 K and 101.3 kPa). The specific gravity of the particles is assumed 2000 kg m^{-3}.

Particle sizes with a large terminal velocity are easy to remove from the gas stream, and as seen in Equation 4.2 the terminal velocity is proportional to the square of the particle size. The equation can be used with little error in the range from 1 to 100 μm. This means that small particles are the most difficult to remove by inertial forces.

Gravity Settlers

A gravity settler is simply a chamber through which the contaminated gas passes slowly, allowing time for the particles to settle by gravity. It is an old style unsophisticated device, but it is simple to construct, requires little maintenance, and has some use in industries treating very dirty gases, e.g. some smelters and metallurgical processes. In Stokes' flow regime the terminal velocity depends on the square of the particle diameter, and the terminal velocity is low for small particles. As a consequence fine particles will pass the gravity settler and only large particles with large settling velocity will be captured. The collection efficiency is low for particles with sizes below 40 to 50 μm and respirable dust is not removed. In general this method is not useful to solve an air pollution problem, but it may be used to pre-clean a gas with a heavy dust load before the final cleaning device.

Cyclones

A cyclone is a centrifugal separator, where the centrifugal force is used to move the particles, and because the centrifugal force is more powerful than the gravity force, the particle velocity is much larger than the terminal velocity and smaller particles can be removed from the contaminated gas. A cyclone consists of a vertical cylindrical body and a conical bottom. The gas enters through a rectangular inlet arranged tangentially to the cylindrical part of the cyclone, so that the entering gas flows along the circumference of the cylindrical body, not radially inwards, see Figure 4.1.

The gas spirals along the outer part of the cylindrical body with a downward component, then turns and spirals upward, and the clean gas leaves through the central outlet at the top of the cyclone. During the outer spiral of the gas the particles are driven to the wall by centrifugal forces. They collect at the wall, attach to each other, and form larger agglomerates that slide down the wall and collect in the dust hopper in the bottom of the conical part. The particles are removed from the cyclone through a rotary valve to avoid any gas leakage in the bottom.

The centrifugal acceleration is larger than the acceleration of gravity, and so the particle velocity is larger than the terminal velocity, but depends on particle size and density and gas viscosity and density as given in equation 4.2. Cyclones are low cost and cheap to maintain, are useful for gas cleaning when the particles are relatively large, but they are not efficient for fine particles. As a rule of thumb, they have a good efficiency for particle sizes above 10 μm. A better efficiency is obtained with many small cyclones in parallel, a so-called multi-tube cyclone. In this case the centrifugal force is larger because the cyclone diameter is smaller, and the travelling distance for

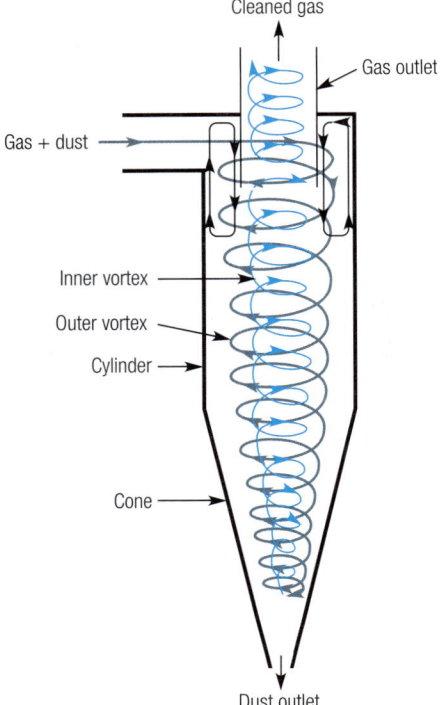

Cleaned gas

Gas outlet

Gas + dust

Inner vortex

Outer vortex

Cylinder

Cone

Dust outlet

Figure 4.1. Cyclone, the arrows indicate the gas flow pattern. (Adapted from Wark et al., 1998).

the particles to the wall is short. Consequently the efficiency of the cyclone becomes better.

The efficiency of a dust cleaning device is often characterised by the particle cut size diameter i.e. the particle size removed with an efficiency of >50%. Figure 4.2 shows the efficiency of different types of cyclones. It is seen that the efficiency is very dependent on the particle size, and the cut size varies between 2 and 13 μm for different cyclone designs and operating conditions.

It is difficult to design a cyclone from theory, because the flow conditions in the cyclone are complicated, but by combination of theory and experience it is possible to formulate an equation to calculate the cut size diameter $d_{p,50}$. For the standard cyclone shown in Figure 4.1 the result is:

$$d_{p,50} = 0.27 \sqrt{\frac{D_c \mu m}{V_c \cdot (\rho_p - \rho_g)}} \qquad\qquad (4.3)$$

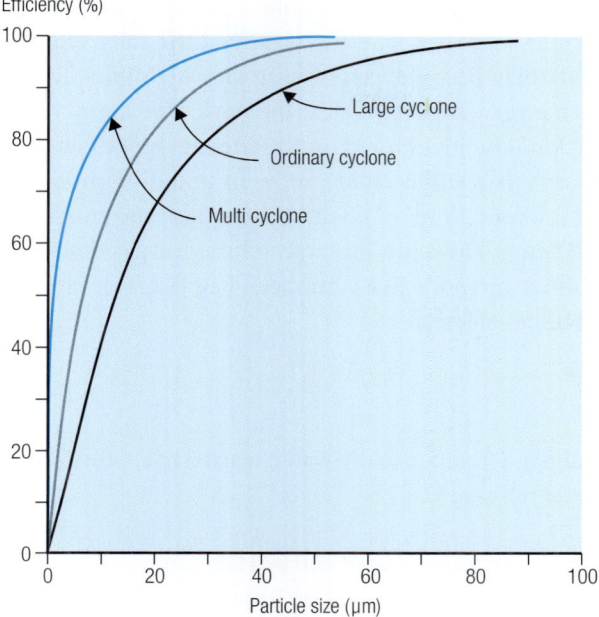

Figure 4.2. Fractional collection efficiency as a function of particle size for several types of cyclones. (Adapted from Wark et al. 1998).

In equation 4.3 μ is the gas viscosity, D_c is the cyclone diameter, V_c is the inlet gas velocity and ρ_p and ρ_g are the particle and gas density, respectively. It is seen, that a high inlet velocity and a large particle density lowers the cut size diameter and so increases the efficiency, while a high gas viscosity and a large cyclone diameter has the opposite effect.

The particles to be removed never have exactly the same size, and in design of a dust cleaning device the particle size distribution must be taken into consideration. For a given cyclone a curve like the ones given in Figure 4.2 must be known to calculate the efficiency for all particle sizes in the gas, and then the total efficiency can be calculated by summation. If this grade efficiency curve is not available it can be calculated from the cut size diameter, and in this way different designs can be compared.

$$\eta = \frac{(d_p / d_{p,50})^2}{1 + (d_p / d_{p,50})^2} \tag{4.4}$$

More details about cyclone design are given by Nevers (1995), Wark et al. (1998) and Heinsohn et al. (1999). As seen in Equation 4.3 the efficiency of a cyclone can be improved by increasing the inlet velocity. However, the pressure drop of the cyclone increases with the square of the inlet velocity, and so the operating costs may become excessive at very high inlet velocities. The cyclone has been a very important cleaning device for many years. It has a simple construction, low investment and moderate operating costs and so has been widely used. It is still useful to solve air pollution problems for small gas flows (below about 10 m^3 s^{-1}), with coarse dust and moderate demands on cleaning efficiency. However, for large volumetric gas flows and more efficient cleaning other methods like scrubbers, bag filters or electrostatic precipitators must be considered.

Scrubbers

In a wet collector or scrubber a liquid, usually water, is used to capture particles. The principle of wet scrubbing is:

- Formation of liquid droplets with a size of 100 to 1000 μm
- Contact between gas and liquid and transfer of particles from the gas phase to the liquid phase
- Separation of droplets from the gas phase

Fine particles, both liquid and solid, ranging from 0.1 to 100 μm, can be effectively removed from a gas stream by scrubbers. The actual mechanism of particle removal may be impaction, direct interception or diffusion as illustrated in Figure 4.3. Particles are carried along at approximately the same velocity as the gas. Owing to its extreme lightness the gas moves in streamlines around any object in its path. However, particles with a much higher density than gas resist changes in motion. The larger the particle, the less tendency it has to change direction because of its inertia. Inertial impaction, shown in Figure 4.3a, is associated with the relatively larger particles that travel on collision course with the drops. Inertia keeps them on the path, even though the gas and the smaller particles tend to diverge and pass around the drop or another interceptor.

In direct interception, illustrated in Figure 4.3b, some of the smaller particles, even though they tend to follow the streamlines, may contact the interceptor at the point of closest approach. This occurs because the streamlines tend to converge as the gas passes around the element, and the particle size is greater then the distance between the streamline and the element. Finally, collection by diffusion is shown in Figure 4.3c. In this case, very small particles (usually less than 1 μm) impinge upon the drop as a result of random molecular (Brownian) motion or diffusion. Impaction is the most important

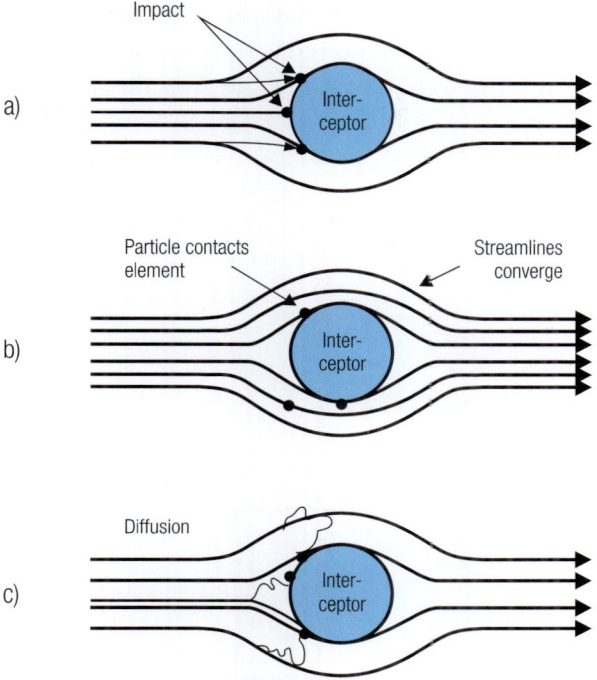

Figure 4.3 Mechanisms of particle removal. a) Inert al impaction; b) Interception; c) diffusion. (Adapted from Wark et al., 1998).

removal mechanism in scrubbers. Consequently a large relative velocity difference between the gas and the droplets is favourable and particulates with a large terminal velocity are easy to separate.

The size and concentration of droplets is important for the removal efficiency. Droplet sizes between 100 and 1000 μm are optimal for particulate collection. If the droplets are smaller than 100 μm, they tend to follow the gas and the relative velocity difference between particle and droplets approaches zero. If the droplets are larger than about 1000 μm their concentration will be small, and the chance of collision between a particle and a droplet becomes smaller. When the particulates have been captured by the droplets, the problem has changed from removing fine particulates from a gas stream to separation of much larger droplets, an easier task. Many different arrangements of basic equipment are available, but only spray scrubbers will be described here.

The spray tower or spray scrubber is one of the simplest devices for wet collection of particulate. It is an empty vessel, where liquid droplets are produced by suitable nozzles located across the gas flow passage. The polluted gas flows counter-currently, co-currently or in cross-flow to the liquid, and the particles collide with the droplets which will in turn settle by gravity to the bottom of the scrubber. The most common construction is a circular or rectangular tower with at least two levels of spray nozzles over the cross section (Figure 4.4).

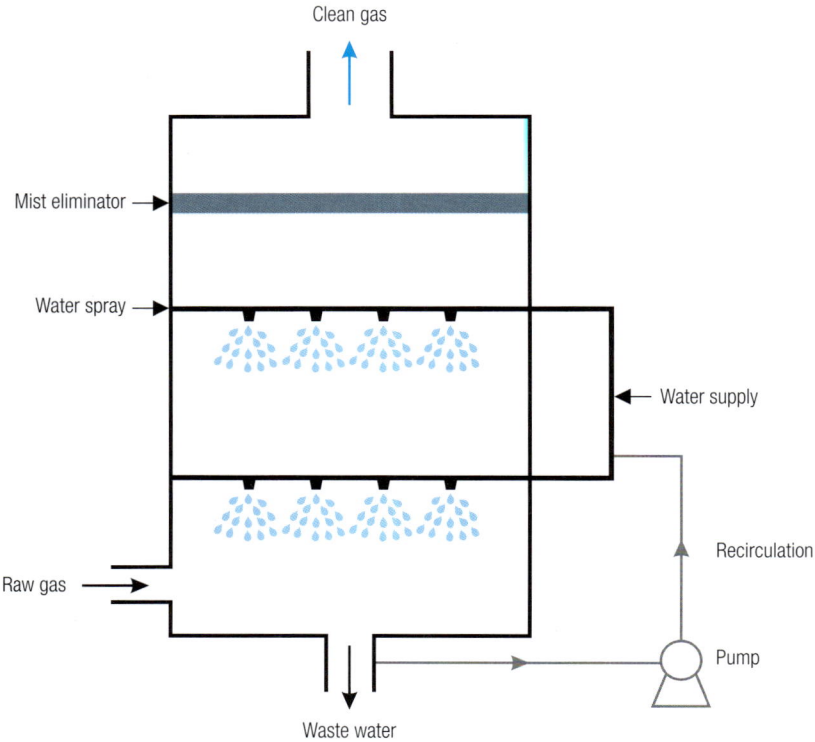

Figure 4.4 Schematic spray scrubber for particle removal.

The gas enters close to the bottom, and the gas velocity is low enough to allow the droplets to settle. For a 500 μm droplet the terminal velocity is about 2 m s^{-1} and the gas velocity is usually below 2 to 3 m s^{-1}. To avoid carry-over of small droplets there is a mist eliminator in the top of the scrubber. The liquid is recycled from the bottom of the scrubber to the spray nozzles to reduce water consumption and the amount of wastewater. The liquid load on the tower is typically 0.3-1.2 kg water per m^3 of gas and the pressure drop is relatively low, about 250-500 Pa. The efficiency is good for particle sizes above 10 μm, and the spray scrubber is superior to the cyclone. The use of packing can increase the efficiency but at the same time the pressure drop and so the operating costs increase. Polymer packing, e.g. tellerettes as shown later in Figure 4.12b, is common in scrubbers for particle removal. If very high efficiency is needed, venturi scrubbers may be used, but the pressure drop is very high because of the high gas velocity obtained when the gas passes a narrow slit in the venturi throat. In general, the scrubber is a relatively simple and cheap construction, and it has been used widely for solving air pollution problems.

The scrubber has some advantages compared to dry methods:

- Particles and gases may be removed simultaneously
- There is no risk of dust explosion and no dust problem by handling the material separated from the gas
- The method is useful for humid gases and sticky material
- Corrosive particles and droplets may be removed and neutralised
- The equipment is compact compared to bag filters and electrostatic precipitators

But also some disadvantages:

- The particles must be separated from the scrubbing liquid as a sludge or maybe by crystallisation if they are soluble
- Wastewater must be cleaned before discharge
- There is water consumption due to evaporation, and the gas becomes saturated with water
- There may be problems with corrosion
- There may be problems with freezing in cold weather on outdoor plants.

A comparison to other methods can be seen in Table 4.2.

Bag filters

Bag filters are also called fabric filters or in some cases just filters. Filtration is one of the oldest and most widely used methods of separating particles from a gas, and the efficiency of particle removal is very high for all particle sizes. The total efficiency for a bag filter may be above 99.9%. A filter generally is any porous structure composed of granular or fibrous material or a membrane that tends to retain the particulate as the carrier gas passes through the voids or the fine holes of the filter material. It is exactly the same principle used in a conventional household vacuum cleaner.

The filter material can be of any kind compatible with the gas and particulate, and especially the temperature of the gas is important for the choice of filter material. Textile, glass, polymers, metals and ceramics may be used for filter material, depending on process conditions. The cheapest materials are cotton and wool, but they are not applicable at temperatures above 80 to 90 °C, and they are not resistant to acids and bases. Many polymers can stand 200 °C, but at higher temperatures metals or ceramics may be necessary (Theodore and Buonicore, 1994).

Fabric filters are usually formed into cylindrical tubes and hung in rows to provide large surface areas for gas passage. A typical size of the filter bags in an industrial filter is a diameter of 150 mm and a length of 4 to 8 m, but smaller filters are also used (Figure 4.5).

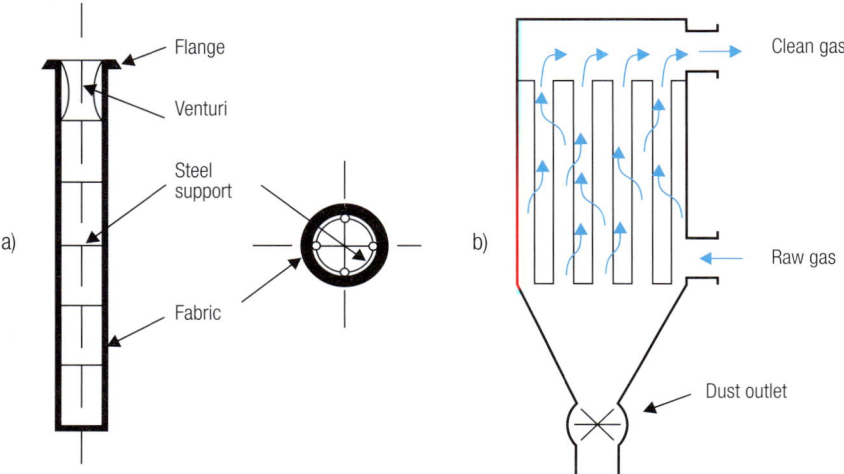

Figure 4.5. Bag filter. a) A single bag with steel support. b) Chamber with several parallel rows of bags.

For application at low flow rates the filter material may be plates or cylinders or any other convenient configuration.

Woven or felted filter material is very open in the structure and usually has a porosity of 80%. The distance between the fibres is larger than the particles to be removed and the principle of particle removal is not sieving. The mechanisms for separation of the particles from the gas using a fibrous filter material are impaction, interception and diffusion as illustrated in Figure 4.3. For larger particles the inertia forces are important and the particles hit the fibres when the gas flows around them. For the small particle in the range 0.001-1μm diffusion is important, and for small particles electrostatic effects may play a role also. The gas flow is almost perpendicular to the fibres and Figure 4.3 shows the cross section of a single fibre. The particles penetrate into the fibrous material, and as filtration goes on, a dust cake builds up on the surface of the filter, and this cake improves the filtration efficiency. Another type of filter material is based on membranes with very fine holes, and in this case the mechanism of particle separation is sieving, and the filter cake collects on top of the filter material.

When the filter cake grows, the pressure drop increases, and at some point the filter cake must be removed. This has to be done on-line in an automatic way, and the most common practice for this operation is illustrated in Figure 4.6. The filter bags are supported by a metal frame and the gas to be cleaned passes from the outside of the bags to the clean inner side. The filter cake builds up on the outside and is removed by a short pulse jet of compressed air from the top of each filter bag. The filter material expands a little during pulsing and the dust cake breaks off, falls to the bottom and is removed automatically. One bag or one row of bags may be cleaned at a time depending on filter size and dust loading. Figure 4.6 shows a bag filter for a large installation. It is important to notice, that removal of the dust from the bags takes place on stream, while the polluted gas is cleaned on the other bags in the filter.

The important issue in design of a bag filter is a matter of estimating the total filter material area. The size of the filter depends mainly on the volumetric gas flow and the acceptable pressure drop. The particle concentration is not important for the filter size, but important for the frequency of pulse cleaning the bags. The capacity of a bag filter is often specified as the

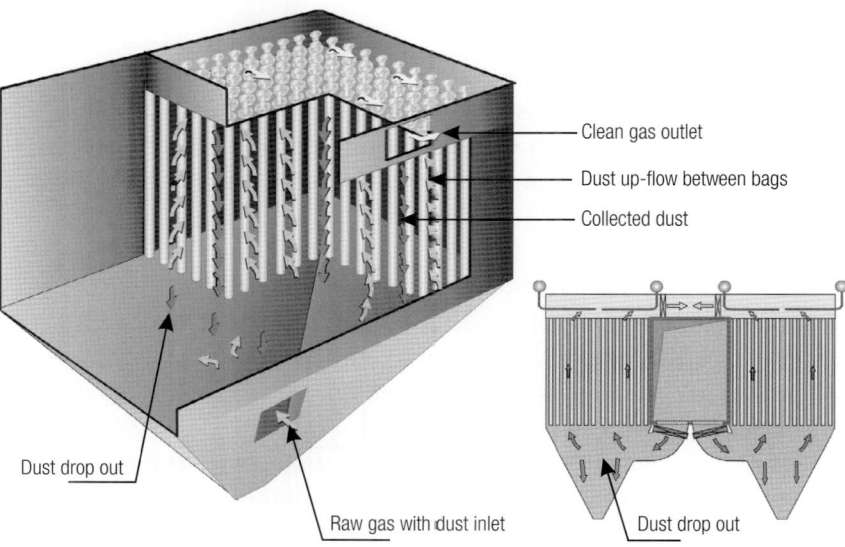

Figure 4.6 Industrial bag filter with jet pulse cleaning. (Courtesy of FLSmidth Airtech)

volumetric flow rate per filter area, the air to cloth ratio, and a typical value is 0.015-0.02 m s^{-1} for a pulse jet filter, but it can vary in the range 0.005-0.04 m s^{-1}. This number is actually the gas velocity perpendicular to the filter surface.

During filtration a dust cake builds up on the filter surface, and the pressure drop increases, the variation is typically between 750 and 2000 Pa, but the pressure drop can be kept low by frequent cleaning of the bags. However, for bags of felted or woven material, the efficiency of the bag decreases upon cleaning, and improves when the dust cake builds up. This means that frequent cleaning of the bags lowers the efficiency, because part of the filtration takes place in the dust cake on the bags. The bags may be of the membrane type, and in this case the efficiency is highly independent of the thickness of the dust cake as the separation of gas and particles takes place at the filter surface. Membrane bags could be made of Teflon, they are expensive and have a higher pressure drop than woven fabrics.

There are many advantages of using a bag filter for dust cleaning:
• The efficiency is high for all particle sizes
• Many possibilities for choice of filter material and filter construction
• Can be made in all sizes covering volumetric gas flows from very low, 0.05 m^3 s^{-1}, to very large 500 m^3 s^{-1}
• Moderate pressure drop and energy consumption
• Different types of particles can be removed in the same installation

But also some disadvantages:
• A large space is required; the bag filter is a voluminous construction
• There may be a risk of dust explosion for certain materials
• Hygroscopic materials my give rise to problems when cleaning the bags
• Humid gases must have a temperature above the dew point to avoid condensation and problems with cleaning of the bags

Fabric filters have a very good efficiency also for small particle sizes, more than 99% of 0.5 μm particles are removed, and even particles as small as 0.01 μm may be removed to a great extent by the diffusion mechanisms. This is very important from an air pollution point of view, because respirable dust is removed very efficiently with fabric filters. Fabric filters are versatile and can be used for many different purposes, e.g. from very small to very large volumes of gas, and they have widespread use in industrial production processes.

Electrostatic precipitators

Particle collection by electrostatic precipitation is based on movement of electrically charged particles in an electrical field. This method is most convenient for large installation because of the large investment in electrical equipment. The electric force in an electrostatic precipitator is strong compared to the centrifugal force of a cyclone and the collection efficiency for small particles is much better. Particles in the range 0.05-200 μm can be removed with high efficiency, and a total efficiency above 99% is possible. The pressure drop of an electrostatic precipitator is usually in the range of 250-500 Pa, i.e. lower than for cyclones, bag filters and efficient scrubbers.

The principle of an electrostatic precipitator is that the particles are charged and move in an electric field. An electrostatic precipitator consists of a large number of emission and collecting electrodes. The emission electrodes are wires and the collection electrodes are plates. Figure 4.7a shows schematically the wires used for emission electrodes and the plates used for collection electrodes.

The wires are charged with 20 to 100 kV below ground potential and the plates are grounded. The gas flows horizontally between the plates, and the gas molecules are ionised by the corona from the emission electrodes. There will be a strong electrical field between the wires and the plates.

The principles of the electrostatic precipitator may be described as follows:

- The dust particles are charged
- The dust particles move because of the electric field
- The dust particles are collected on the collecting electrodes
- The collected dust is removed from the collecting electrodes

The emission electrodes may be metallic wires with many points like barbed wire. Around the points the electrical field is strong and a corona generation takes place with a emission of electrons that ionize the gas molecules. The negatively charged gas ions are repulsed by the negative emission electrode and during their movement they hit the particles, which then become negatively charged. The charge is proportional to the square of the particle size, and large particles have a very high charge.

The negatively charged particles move in the electric field, and the migration velocity can be found from a force balance, setting the drag force on the particle equal to the electrostatic force:

$$w = \frac{k(D) \cdot E^2 \cdot d_p}{\mu} \hspace{4cm} (4.5)$$

w is the migration velocity, k(D) is a constant depending on the dielectric constant of the particle, E is the electric field strength, d_p is the particle size and μ is the gas viscosity. Equation 4.5 shows that large particles have higher velocities and are easy to remove, and it is obvious that the electric field strength is determining the migration velocity and so the removal efficiency. The negatively charged particles are repulsed by the negative emission electrode and move towards the collecting electrode.

When the particles hit the collecting electrode they loose their charge. A dust layer builds up on the collecting electrodes and adhesive, cohesive and electrical forces prevent re-entrainment of the particles into the gas stream. One property of the dust layer, which is extremely important in precipitator operation, is the dust electrical resistivity. When the dust resistivity is too low (high conductivity) there is a rapid movement of charge from the deposited dust to the collector plate. In this case insufficient electrostatic forces remain on the collected dust particles to keep them together. Re-entrainment back into the gas stream frequently results and collection efficiency suffers. On the other end of the scale, a high resistivity dust layer (low conductivity) acts as an insulator and a part of the voltage drop occurs over the dust layer and only part of the total corona power is available to ionise the gas and move the charged particles to the collection electrode. As a rule of thumb the resistivity should be within given limits (10^4 to 10^{10} ohm · cm (Wark et al., 1998), 10^7 to $2 \cdot 10^{10}$ ohm · cm (Nevers, 1995), 10^4 to 10^{11} ohm · cm (Soud, 1995). In extreme cases when the dust has a very high resistivity there may be ionisation or corona generation on the collecting electrodes. In this case the particle loose their charge before reaching the electrode and the filter efficiency suffers. The resistivity depends on the particle properties and the gas conditions. Water and other polar compounds adsorb to the particle surface and lower the resistivity at gas temperatures below about 150 °C

The dust layer is removed regularly with a rapping system vibrating the collecting electrodes. The dust falls to the hopper in the bottom of the precipitator.

An electrostatic precipitator for industrial use has a horizontal flow of gas as shown in Figure 4.7b.

The gas is distributed evenly across the cross section of the precipitator and flows parallel to the collecting electrodes. The emission electrodes are suspended from the roof of the precipitator and fixed to a frame in the bottom. The distance between the wires and the plates are typically 0.15-0.4 m.

A typical size of a precipitator at a power plant could be height · width · length = 16 · 30 · 20 m³ but the size depends on the flow rate and the required efficiency. If the flow rate is very high, say 500 m³ s⁻¹ parallel units may be used, and if very high efficiency is needed more sections are put in series.

Design of an electrostatic precipitator is very dependent on experience.

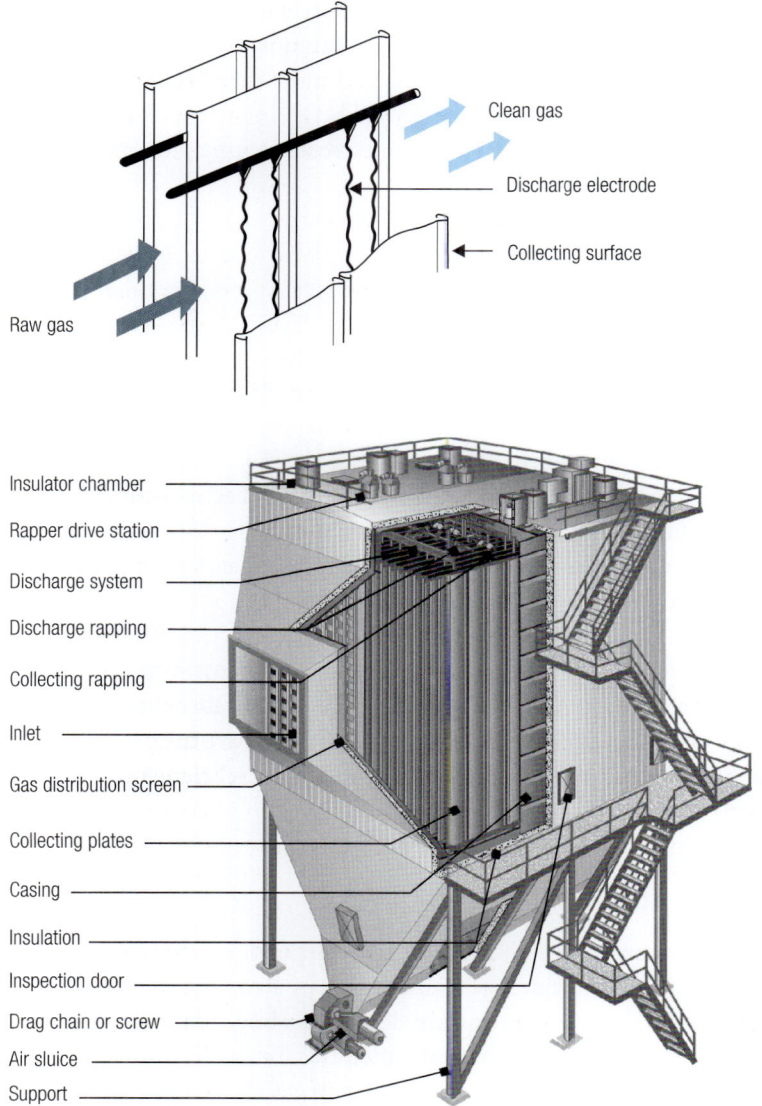

Figure 4.7 a) Principles of electrostatic precipitation. b) An example of electrostatic precipitation. (Courtesy of FLSmidth Airtech).

The classical equation used in design is the Deutsch-Anderson equation:

$$\eta = 1 - \exp\left(-\frac{A \cdot w}{Q}\right) \tag{4.6}$$

A is the total area of the collecting electrodes (both sides of the plates), w is the particle migration velocity and Q is the volumetric flow rate of the gas. The migration velocity depends on the particle size as seen in Equation 4.5, but in general it is difficult to estimate the migration velocity theoretically. There are other design equations available and more details can be found in (Nevers, 1995).

There are many advantages of using electrostatic precipitators for dust cleaning:

- High efficiency for all particle sizes
- Can be used for very large installations, 500 m³ s⁻¹ gas can be cleaned in one unit
- Low pressure drop and low energy consumption
- Can be used at high gas temperatures of 400-500 °C

But there also disadvantages:
- The properties of the particles are important for the removal efficiency and the dust resistivity must be within certain limits
- Too expensive for small scale installation, say less than 5-10 m³ s⁻¹
- Large investment

Electrostatic precipitators are used mostly at power plants, waste incinerators, cement factories and other large scale processes, because the investment is high. The collection efficiency is very good, for new plants it is usually more than 99.5%, and also respirable dust with a particle size below 10 μm is removed effectively.

Summing up – primary particulates

The choice of gas cleaning method for particles depends on many parameters as mentioned in the introduction. Table 4.2 gives a comparison of the different methods of air pollution control described above, and the table may be used as a first guideline to choose the methods for further consideration. More information about design and operation of gas cleaning equipment for particles is given by Theodore and Buonicore (1994), Nevers (1995), Wark et al. (1998), Heinsohn et al. (1999) and Soud (1995).

	Gravity settlers	Cyclones	Scrubbers	Bag filters	Electrostatic precipitators
Particle sizes with a high removal efficiency	> 50-100 μm	> 5-10 μm	> 2-5 μm (>0.3 μm for Venturi scrubbers)	All particle sizes	All particle sizes
Temperature limit	Depends on the material of construction (> 1000 °C is possible)	Depends on the material of construction (> 1000 °C is possible)	Water consumption increases with temperature due to evaporation	200-250 °C (higher with filters of metals or ceramic material)	400-500 °C
Influence of water content	Condensation must be avoided	Condensation must be avoided	No influence	Condensation must be avoided	The efficiency depends on the water content
Pressure drop, Pa	50-130	1000-2000	500-1300 (Venturi scrubbers 2500-125000)	1000-1500	50-130
Other comments	Can be used as a pre-separator but not for final air pollution control	Useful for coarse dust	Sludge and waste water must be taken care of	The optimal choice if a very high efficiency is needed	Not to be used for a gas flow below 5-10 m³ s⁻¹ because of high investment

Table 4.2 Comparison of methods for control of particulate air pollutants.

4.2 Gases and odorous compounds

Many different methods are used for removing gaseous air pollutants, and the choice of method depends very much on the type of compound to be removed. The main processes are:

- Oxidation (thermal or catalytic)
- absorption
- adsorption
- biological methods (bio-filters or bio-scrubbers)

These are described by Johnsson (1998), Nevers (1995), Wark et al. (1998) and Heinsohn et al. (1999).

The choice of method depends on many parameters, the most important are the volumetric gas flow and the properties of the pollutants: are they combustible, are they water soluble, are they acidic or basic, are they present in high or low concentrations and last but not least, are they valuable and regeneration for reuse in the production process is a must.

Oxidation

Combustible air pollutants like hydrocarbons, solvents and odorous substances may be oxidised to H_2O and CO_2, as shown by the net reaction for hydrocarbons in Equation 4.7.

$$C_mH_n + (m + n/4)O_2 \rightarrow n/2\ H_2O + mCO_2 \qquad (4.7)$$

The reaction mechanism is complicated and many intermediates are formed and oxidized before we have the final products H_2O and CO_2. An important intermediate compound is CO, and the overall reactions may be written as Equations 4.8 and 4.9:

$$C_mH_n + (m/2 + n/4)O_2 \rightarrow n/2\ H_2O + mCO \qquad (4.8)$$

$$CO + 1/2O_2 \rightarrow CO_2 \qquad (4.9)$$

In practice the oxidation of CO may be the critical design parameter.

If the pollutants contain S, Cl or N secondary air pollutants like SO_2, HCl and NO_x may be formed, and thermal NO_x may be formed from O_2 and N_2 in the air in any case, if the flame temperature is above 1300 °C.

If the concentration of combustibles is above the lower explosion limit the gas will burn if ignited, and in this case we have a fuel and not an air pollution problem. Table 4.3 shows the upper and lower explosion limit for selected compounds.

Compound	Lower explosion limit vol%	Upper explosion limit vol%
Ethylene	2.7	34.0
Carbon monoxide	12.5	74.0
Methane	4.6	14.2
Methanol	6.4	37.0
Propane	2.4	8.5
Toluene	1.2	7.0
Hydrogen	4.0	76.0

Table 4.3 Upper and lower explosion limit for selected compounds. (1 bar, ambient temperature and ignition energy 10 J). (Brauer and Varma 1981). Courtesy of Chemviron.

A mixture of air and combustible gases can explode if the concentration is between the explosion limits. In this interval the mixture can burn if ignited. The explosion limits depend on the chemical compound, the composition of the mixture, the temperature, the pressure and the ignition temperature or energy. If the concentration is outside the explosion limits the gas will not burn when ignited. If the concentration is above the explosion limit, the mixture may be diluted with air, as is the case with other gaseous fuels in combustion processes. If the concentration is below the lower explosion limit we must either heat the gas to the temperature necessary for oxidation to take place or add a fuel to the combustion chamber.

The oxidation may be either thermal at a high temperature of 850 to 1000 °C or catalytic at a lower temperature of 250 to 400 °C.

Thermal oxidation (incineration)

For safety reasons, the concentration must be below 25% of the lower explosion limit if an oxidation process is used for control of air pollutants, and usually the concentration is well below this limit.

It is necessary to heat the polluted gas before the combustion and this is usually done by heat exchange between the incoming cool gas with the cleaned hot gas, however it may be necessary to supply additional fuel to the process. There is a gas and oil fired burner to supply additional heat to the system and to be used during start-up. A conventional plant for oxidation of pollutants consists of a heat exchanger and a combustion chamber with a burner. In Figure 4.8 it is seen that the polluted gas enters the shell side of a tube and shell heat exchanger. The vertical part of the plant contains many steel tubes, and the incoming gas flows downwards on the outside of the tubes and the cleaned and hot gas flows upwards on the inside of the tubes. In this way the hot gas is cooled and the incoming gas is heated in this counter current heat exchanger. The heated gas flows into the combustion chamber, usually a steel containment with fire resistant bricks on the inside. At the end of the combustion chamber there is a burner to supply additional energy and heat the polluted gas to the combustion temperature if necessary.

The cleaning efficiency depends on a number of parameters, and they are not independent: combustion temperature, residence time, flow conditions and concentration, and type of pollutants. The conversion of pollutants increases with temperature and residence time, but depends also on the design of the combustion chamber, because the flow field depends on the design. It is important to have a good mixing of the polluted gas and the flue gas from the burner and to avoid by-pass. The important parameters are often termed the three T's: Temperature, Time and Turbulence. It is important to have high values of these parameters to get complete combustion.

The detailed design of the combustion chamber and the heat exchanger is not an easy task, and textbooks on chemical kinetics of combustion (Turns,

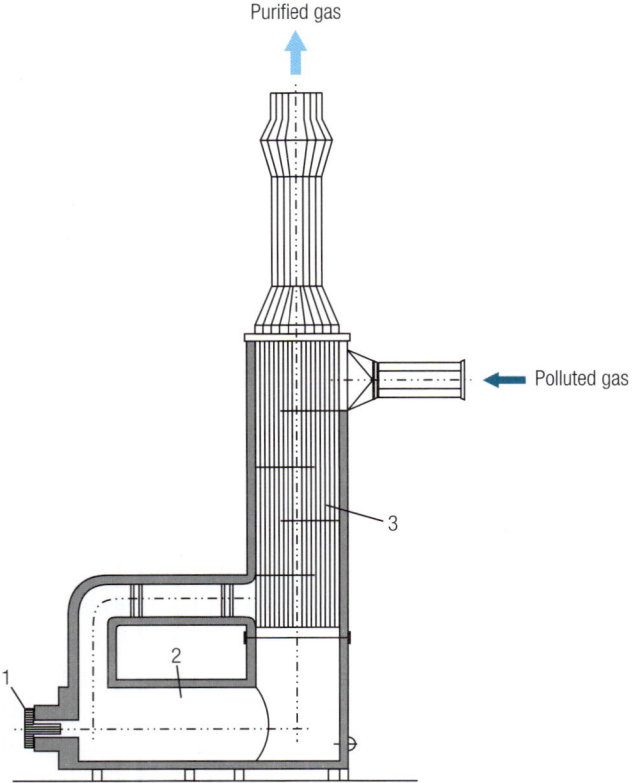

Purified gas

Polluted gas

Figure 4.8 Combustion process with burner (1), combustion chamber (2) and heat exchanger (3). (After VDI 1987).

1996), chemical reaction engineering (Fogler and Gürman, 2005; Levenspiel, 1999) and unit operations of chemical engineering (McCabe et al., 2005) should be consulted.

If the concentration of pollutants is above 10-15 g m^{-3} of solvents the adiabatic temperature rise of combustion may be 200-350 °C and the hot cleaned gas may be used for heating of water or steam production in a so-called waste heat recovery boiler.

If the concentration of pollutants is below about 5 g m^{-3} the fuel consumption is high, because it is not possible to preheat the gas to a high temperature in a counter current heat exchanger. The heat exchanger should be very large and expensive and combustion would initiate within the heat exchanger possibly causing damaging due to high temperatures. In this case other ways of heat exchange may be considered e.g. using regenerative heat exchangers with very high effectiveness and thus the use of additional fuel can be minimised. Figure 4.9a shows the principles of combustion with regenerative heat exchange.

The box may have a content of sand or ceramic packing. In one mode the gas enters in the top and is heated by contact with the material in the box, and additional heat may be added in the centre, when the gas is halfway through. On its way out the gas is cooled by contact with the material in the box and

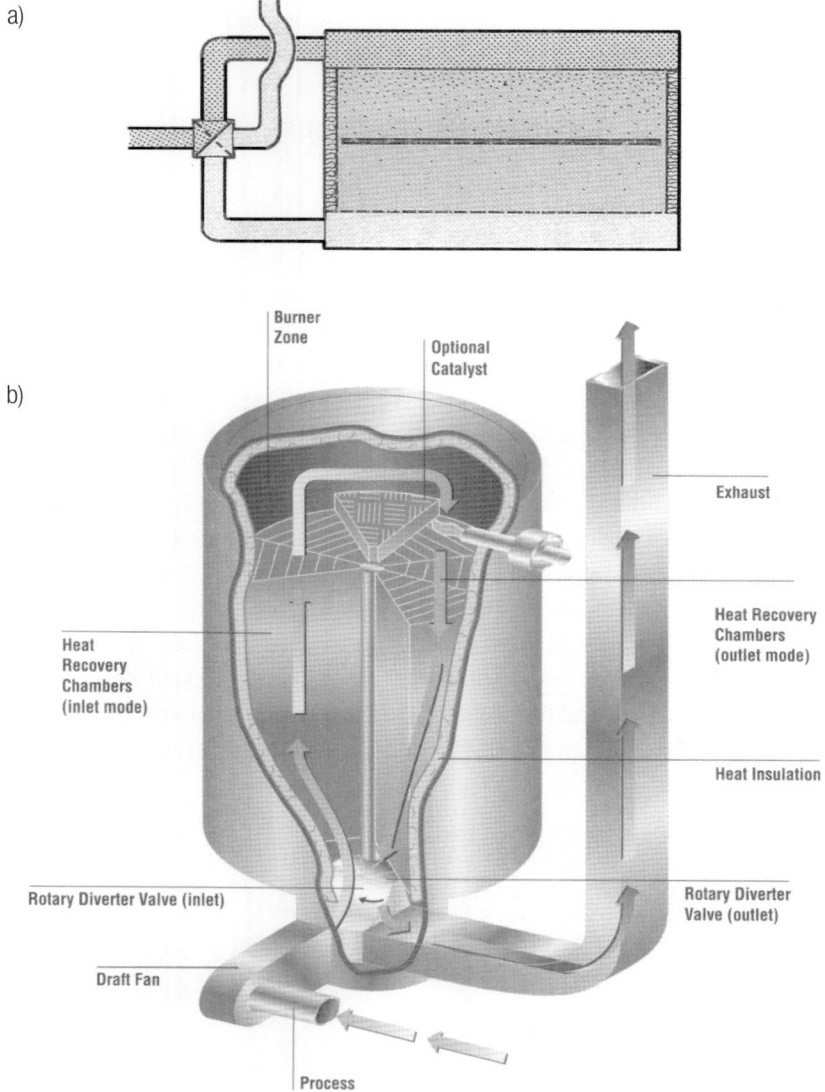

Figure 4.9 a) Principles of combustion with regenerative heat exchange (Courtesy of LESNI). b) An industrial combustion chamber with regenerative heat exchange (Courtesy of Reeco Stroem A/S).

leaves the unit in the bottom. During this mode the material in the top is cooled by the inlet gas in the upper part, and heated in the lower part by the outlet gas. After a few minutes the direction of flow is reversed and now the gas is heated in the lower part and cooled in the upper part. The efficiency of this heat exchange is very effective, and even at relatively low pollutant concentrations of 2 g m^{-3} of solvents there is no demand for additional fuel.

Figure 4.9b shows an example of an industrial unit with regenerative heat exchange. In this case the gas enters in the bottom and passes a slice of the ceramic material on its way to the combustion chamber in the top, and then turns around and leaves the unit via another slice. A rotating valve in the bottom reverses the flow in the ceramic packing after a preset time.

Catalytic combustion

By the use of a catalyst the rate of reaction can be increased, and the oxidation can be carried out at a lower temperature and in a more compact unit with a lower residence time. The catalytic material can be noble metals like Pt or Pd or it can be oxides of the transition metals like CuO, Cr_2O_3, or NiO. The catalysts are typically shaped spheres, cylinders or rings, but they may also be monoliths, i.e. box-like structures with channels to lower the pressure drop and reduce deposition of dust in the catalyst (Figure 4.10).

The principles of the catalytic unit are shown in Figure 4.11a. The gas is preheated in the heat exchanger and additional heat may be added by the burner before the gas flows into the catalyst. Again the clean gas is used to preheat the inlet gas before it leaves the unit.

The major difference to a thermal combustion unit is the lower temperature, 250-350 °C compared to 900-1000 °C. This means that construction materials are cheaper and the additional fuel consumption is lower compared to thermal combustion, however the cost of a catalyst must be added.

Monolithic catalysts   Pellet catalysts

Figure 4.10. Catalyst pellets and monolith catalysts (Courtesy of Haldor Topsøe).

Typical values for space velocity (m^3 gas per second per m^3 catalyst) are about 4-5 s^{-1}, or a residence time of about 0.2 s. For thermal combustion a typical residence time is about 0.5-1 s. An important issue is the lifetime of the catalyst, and in general catalyst lifetimes of 25,000 hours can be obtained.

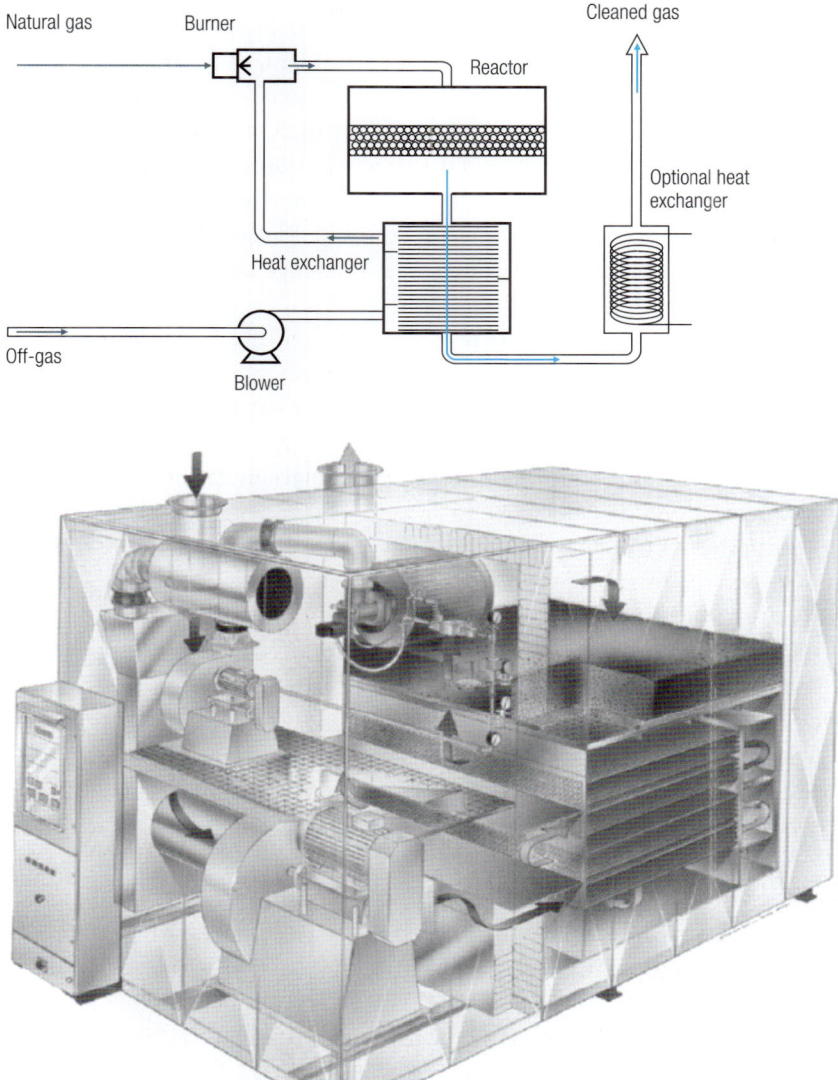

Figure 4.11 a) Principles of catalytic combustion with heat exchanger for raw gas heating. b) An industrial catalytic combustion unit (Courtesy of Haldor Topsøe).

Figure 4.11b shows the catalytic unit with blower, heat exchanger and catalyst layer.

It should be mentioned that catalytic combustion with regenerative heat exchange is also an option. In this case catalyst packing also serves as the heat exchanger, and the principles are the same as described in connection with Figure 4.9.

Summing up – oxidation

Oxidation is an attractive process because very high removal efficiencies can be obtained, typically above 99%. If the concentration of air pollutants is high enough to give an adequate adiabatic temperature rise, no additional fuel is needed. If the concentration is low, fuel must be burned to heat the gas and compensate heat losses. Regenerative heat exchange may be used to reduce the fuel costs.

There are several advantages of oxidation processes:
- It is possible to obtain very good cleaning for organic and combustible inorganic compounds
- The process is not sensitive to moderate changes in gas flow or pollutant concentration
- The performance is not dependent on the type of pollutant if the temperature and the residence time are high enough
- Heat recovery may be possible
- Fuel consumption may be low at high pollution concentration or regenerative heat exchange

But also disadvantages:
- Relatively large investment
- High operating costs if additional fuel is needed and heat recovery is not an option
- Regeneration of valuable compounds is out of the question
- Secondary pollutants may be formed if the pollutants have a content of S or Cl as SO_2 and HCl is formed and may have to be removed before emission
- There may be formation of NO_x if the temperature in the burner zone is high or the pollutant has a content of nitrogen

Thermal oxidation or incineration has been used in the chemical industry for many years, and now catalytic oxidation has a wide range of applications also.

Absorption

The principle of absorption is that components in the gas phase may go into solution in a liquid phase. In air pollution control the most common liquid is water or water with chemicals in solution or suspension. The method is widely used, because it is relatively cheap and very robust. In practice the control of gaseous pollutants by absorption or wet scrubbing involves bringing the gas into contact with the scrubbing liquid and subsequently separating the cleaned gas from the contaminated liquid.

The method is especially well suited for very soluble gases like NH_3, HCl and acetone, but it is also useful for slightly soluble gases, which can react with chemicals in the scrubbing liquid. A good solubility or a fast reaction in the liquid is important to reduce the amount of scrubbing liquid and thereby reducing the amount of wastewater, and to get a fast absorption and thereby reducing the size of the absorption tower. Table 4.4 shows a list of air pollutants and chemicals.

Pollutant	Absorption liquid or reactive compound
HCl, HF, NH_3	Water
SO_2, NO_2	NaOH, NH_3, $CaCO_3$, $Ca(OH)_2$
H_2S	NaOH, NaOCl
NH_3	H_2SO_4, HNO_3
Organic acids	NaOH
Phenols, cresols	NaOH
Phosgene	NaOH, vand
Cl_2	NaOH, $NaSC_3$, $Na_2S_2O_3$
Mercaptans	NaOCl, H_2O_2 O_3, $KMnO_4$, Cl_2
Amines	NaOH
Polar solvents	Microorganisms
Non-polar solvents	Oil
Odorous compounds	Water, NaOCl, H_2O_2, O_3, $KMnO_4$, Cl_2, microorganisms

Table 4.4. Absorption liquids and reactive compounds used for air pollution control

This method of absorption combined with chemical reactions is very useful to solve air pollution problems. Common absorption liquids are acids and bases and chemicals with a strong oxidation potential.

The oxidizing chemicals are often used in acidic or basic solution to ensure that the oxidation products are not stripped from the absorption liquid. An important example of absorption with chemical reaction is the removal of SO_2 from flue gas by scrubbing with a solution or slurry of $CaCO_3$ in water,

(Takeshita and Soud, 1993). Oxidising compounds like ozone and NaOCl are used for removal of odorous pollutants in the food and fish meal industry.

There are many types of absorbers or scrubbers and the spray tower was described in the section about particle removal. It is important to have a large contact area between gas and liquid, and in the spray tower many nozzles at several levels are used to spray and distribute the liquid over the total cross section of the tower. It is often a counter current process and the gas enters the tower in the bottom. Absorption towers remove gaseous components and particulates as well. However an absorption tower for gas absorption usually will have a relatively low pressure drop to reduce the energy consumption, and so the removal efficiency for particles will not be high, as a large pressure drop is needed to remove fine particles.

The scrubber may be without a packing as shown in Figure 4.4 or with a packing as shown in Figure 4.12. The purpose of the packing is to give a large interfacial area between the gas and the liquid and to ensure an even distribution of liquid over the cross section, to improve the cleaning efficiency. However, the installation cost and the pressure drop is higher than for a spray scrubber without packing, and there is a risk of contamination with dust at high dust loadings.

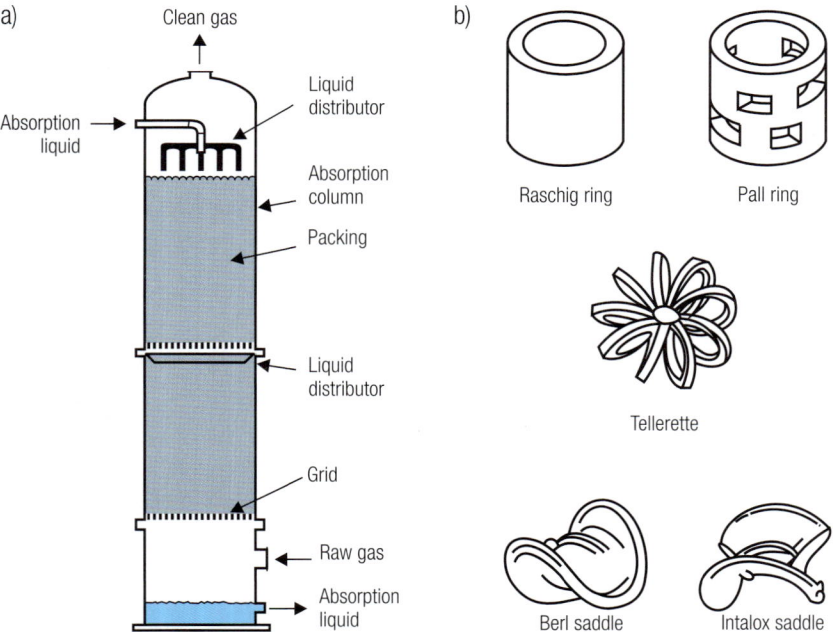

Figure 4.12 a) Counterflow absorption tower with packing. b) Various packing types. (Adapted from Wark et al., 1998).

Spray towers are widely used in air pollution control, and important reasons are:
- Spray towers are not clogged by particles, which may happen for absorption towers with packing
- The pressure drop is low, usually below 1000 Pa
- The gas velocity may be above 3 m s^{-1} and the column diameter may be smaller than for packed towers.

Design

The traditional chemical engineering literature about absorption treats the case of physical absorption, with relatively high concentrations in the gas and liquid phase and equilibrium at the interface (McCabe et al., 2005). In air pollution control the concentrations are usually low, and so the attainable solubility is low. In physical adsorption a large amount of water is needed to remove the pollutants from the gas and addition of chemicals to the absorption liquid is often necessary. In this case it is absorption with a chemical reaction with different principles of design, and textbooks on chemical reaction engineering must be consulted (e.g. Levenspiel, 1999).

The advantages of absorption with a chemical reaction are:
- The absorption is fast, and the dimensions of the absorption tower become small compared to physical absorption
- The capacity of the absorption liquid is not limited by the solubility of the gas in water because the pollutants react in the liquid phase, and the water consumption is reduced

The advantages of absorption in general are:
- The method is relatively simple and cheap compared to oxidation and adsorption
- The pollutant may be regenerated and recycled to the process, or a useful by-product may be formed
- The method is not sensitive to the humidity in the gas

Disadvantages of absorption processes are:
- The cleaned gas is saturated with water
- Wastewater must be treated
- There is consumption of chemicals
- There is a consumption of water depending on the temperature and humidity of the polluted gas and the amount of wastewater

Absorption is used in air pollution control in many industrial processes and in combustion processes as well. The size of the absorption tower may vary from a few meters in height in small scale industry to 50 m height for a flue gas desulphurization column on a power plant.

Adsorption

The principle of an adsorption process is to utilize the ability of solid surfaces to remove one or more gases or liquids selectively from a mixture. In air pollution control the gaseous pollutant, the adsorbate, is adsorbed on the solid phase, the adsorbent. The method is widely used for drying gases, but is also important for solving air pollution problems. Adsorption may be physical or chemical (chemisorption). In physical adsorption the molecules stick to the surface because of van der Walls forces. The adsorption process is exothermic and the enthalpy change is 1.5-2 times the enthalpy change of condensation, i.e. about 2-20 kJ mol^{-1}. An important feature of the physical adsorption process is the reversibility. By increasing the temperature or lowering the concentration the adsorption process is reversed, and the pollutant is desorbed from the surface without any chemical conversion. In this way the pollutant may be obtained as the pure compound and may be recycled to the production process. In chemisorption a chemical reaction occurs and regeneration is more complicated.

There are many types of adsorbents available: activated carbon, zeolites, polymers and silica gel, but in air pollution control activated carbon is the most common adsorbent. Activated carbon is a porous material with an extremely high specific surface of 500-1000 m^2 g^{-1} in small pores. Figure 4.13a shows a sketch of activated carbon. The macro-pores facilitate transportation of the pollutants to the micro-pores where adsorption takes place. The finest micro-pores have a size of a few molecule diameters. The adsorption is not restricted to mono-layers on the surface, and the fine pores may become filled with pollutants.

In air pollution control the activated carbon is used either as tablets or as a granulated material with a size of 2-5 mm. Activated carbon is widely used to remove non-polar solvents from polluted gases. There are two different kind of plants, depending on the concentration of the pollutant:

• At low concentrations the adsorbent is not regenerated on site, but disposed of or regenerated at the supplier
• At high concentrations the adsorbent is regenerated regularly on site and the activated carbon may be used for several years

a)

b)

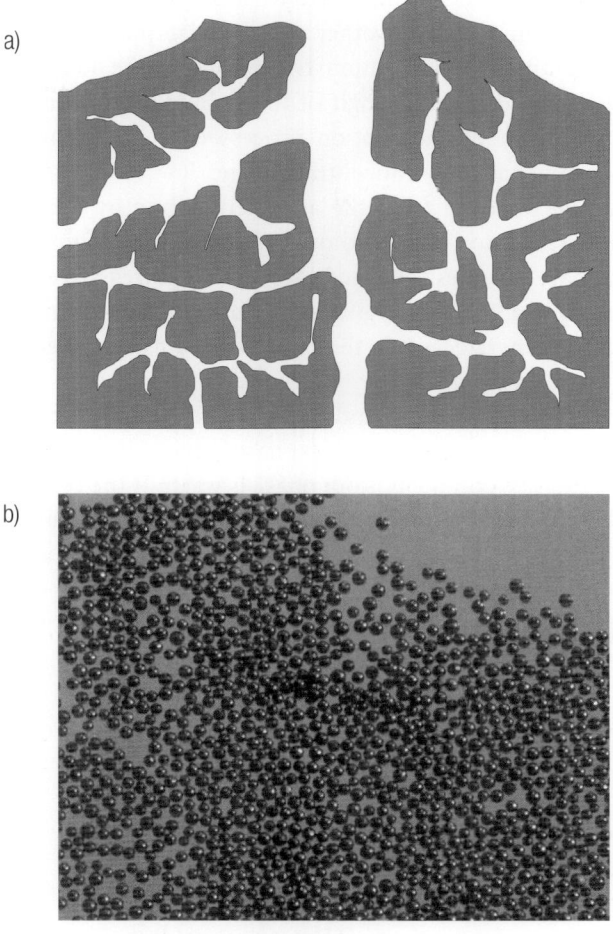

Figure 4.13 a) Sketch of the cross section of an activated carbon. b) Activated carbon spheres.

The gas passes through a layer of activated carbon particles, and the height may be in the range of 0.2-1 m depending on the concentration of pollutants and the lay out. At low pollutant concentrations of 1-10 ppmv a shallow layer with a low pressure drop is chosen. In this case adsorption can go on for several months before the activated carbon is saturated and must be changed. The carbon may be placed in a cylindrical vessel as shown in Figure 4.14, but at low concentrations vertical panels with a depth of about 0.2 m may be used.

In an industrial plant with a high concentration of pollutants, the activated carbon is placed in a vessel as seen in Figure 4.14a. In this case the carbons are regenerated with steam and the pollutant may be recycled. In a continuous process several parallel adsorbers are needed as shown in Figure 4.14b. In this case one adsorber is regenerated while the other is in the adsorption mode.

The adsorption can go on for several hours, but eventually the activated carbon becomes saturated and breakthrough of the pollutants will occur. Now, an important feature of the adsorption process is its reversibility. Adsorption is most effective at high concentration and low temperature, and the pollutants can be removed from the activated carbon by heating with steam. In practice adsorption is carried out at room temperature and when the activated carbon becomes saturated, desorption or regeneration is done with steam at 100 to 150 °C. In Figure 4.14b two parallel adsorbers are shown, one is being regenerated while the other is used as adsorber, as describe above.

The mixture of steam and pollutant is condensed in a heat exchanger and is lead to a gravimetric separation vessel. The top phase contains the organic compounds with a low density, and the bottom phase is water. If the organic compound is an insoluble solvent we have a complete separation, and the sol-

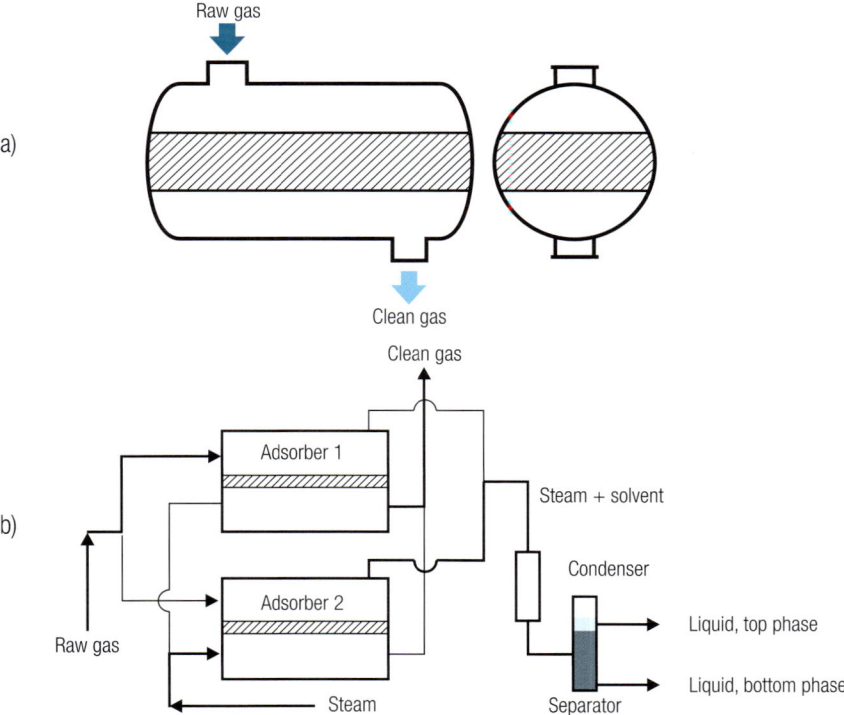

Figure 4.14 Adsorption process. a) Vessel with fixed bed of adsorbent. b) Parallel adsorbers with regeneration equipment. (Adapted partly from Wark et al., 1998).

vent can be recycled to the production process. If the adsorbed compound is water soluble the liquids must be processed e.g. by distillation to get the pure compounds for recycling. A typical time between adsorption and regeneration is a few hours, however in some cases adsorption takes place during the day and the adsorber is regenerated automatically during night and ready for the next day production period next morning. In this case only one adsorption unit is needed.

Design of an activated carbon adsorption unit depends heavily on the adsorption capacity of the carbon. For easily adsorbed compounds like benzene and toluene the carbon may take up 30% of their own weight, but the capacity depends on the type of compound, the concentration and the temperature. The relationship between the concentration in the gas phase and the solid phase at a given temperature is named the adsorption isotherm. In general, the concentration in the solid phase increases with the molecular weight and an increasing concentration of the pollutant in the gas and

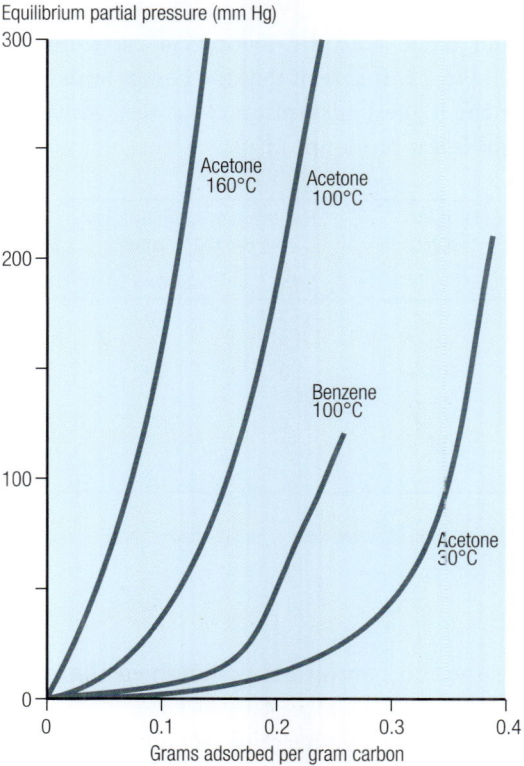

Figure 4.15 Adsorption isotherms for acetone and benzene on a given activated carbon. (Adapted from Wark et al., 1998).

with decreasing temperature. The adsorption is most effective for non-polar (hydrophobic) organic compounds like hydrocarbons, and less effective for polar (hydrophilic) compounds like organic acids and lower alcohols. Figure 4.15 shows adsorption isotherms for acetone and benzene, and it is seen that benzene adsorbs better than acetone, and also the important influence of temperature is seen, acetone adsorbs much stronger at 30 °C compared to 160 °C.

The curves in Figure 4.15 can be described by a Freundlich isotherm:

$$y_a = \alpha \cdot x_a^{\beta} \qquad\qquad\qquad (4.10)$$

y_A og x_A are the molar fractions of the compound in the gas and on the carbon respectively, and α og β are constants characterizing the adsorbent and the adsorbate. The Freundlich isotherm can be used for calculating a rough estimate of the amount of carbon needed. However, more carbon is usually needed, because of the breakthrough curve leaving part of the outlet section unsaturated, as described by Wark et al. (1998) and Nevers (1995).

Suppliers of activated carbon often state the theoretical capacities of activated carbons as shown in Table 4.5. It is seen that the compounds with the highest boiling points have the highest adsorption capacities, while the dependence on molecular weight is less pronounced.

Component	Boiling point °C	Molecular weight	Theoretical adsorption capacity kg per 100 kg carbon		
			10 ppmv	100 ppmv	1000 ppmv
Benzene	80.1	78.1	13.3	18.6	28.0
Methyl chloride	40.2	84.9	1.3	2.6	9.3
Freon115	37.7	154.5	1.3	4.0	7.2
Styrene	145.2	104.1	26.0	34.7	42.0
Toluene	110.6	92.1	21.0	26.7	34.0
Trichloro ethylene	87.2	131.4	18.6	33.0	48.0
Vinyl chloride	13.9	62.5	0.7	1.3	2.7

Table 4.5 Theoretical adsorption capacity for a given activated carbon for selected organic compounds at concentrations in the gas given in ppmv = $cm^3\ m^{-3}$.

Adsorption is especially well suited to control air pollution in the cases when:

• The pollutant is not combustible or forms a harmful secondary pollutant upon combustion
• The pollutant is valuable and recycling to the production process is possible

- Adsorption without regeneration can be used at low pollutant concentrations

There are many advantages of the adsorption process:
- High removal efficiency for non polar compounds
- Recycling of the pollutants to the production process, at high pollutant concentration the pay back time of the plant may be only a few years
- No chemicals needed

There are also disadvantages.
- High investment and operating costs in units with regeneration.
- The activated carbon must be supplemented or changed; the lifetime is a few years
- Waste water from condensation of steam must be treated if the pollutants are (partly) water soluble

In general, adsorption by activated carbon is a very useful process for non polar compounds, because they are removed very effectively and are easy to separate from the condensed steam for direct recycling in the process.

Biological methods

The principle of biological air pollution control is the same as for biological wastewater treatment, i.e. the air pollutants are removed by biological decomposition by microorganisms. The primary products are CO_2 and H_2O but there is also a production of biomass:

$$\text{Air pollutants} + O_2 \xrightarrow{\textit{microorganisms}} \text{Biomass} + CO_2 + H_2O \qquad (4.11)$$

The method is especially well suited to solve problems with odorous compounds at relatively low concentration. There are two different principles of biological air pollution control: biological filters and biological scrubbers.

In biological filters the microorganisms are fixed to a porous organic material. The gas flows through a layer e.g. peat moss mixed with twigs and small branches to give stability and porosity. The odorous substances are adsorbed to the organic material and used as feedstock by the microorganisms. The microorganisms adapt to the pollutants after some time, but in some cases special microorganisms have been grown in the laboratory to ensure the filter performance. The principle of a biological filter is shown in figure 4.16.

The contaminated air stream initially enters a humidification vessel and is drawn upward through a plastic packing material. As the air flows upward through the packing water flows downward over the packing. This counter current operation saturates the contaminated air with water vapour, once saturated; the contaminated air enters the upper chamber of the biological oxidation vessel. The contaminated air passes downwards through the biologically active media. As this occurs, the contaminants in the gas stream are transferred to a water film surrounding the biologically active media. Microorganisms present in this water film oxidise the contaminants to CO_2, H_2O and salts. The cleaned air leaves the filter in the bottom, enters a process blower, and is finally discharged to the atmosphere through a stack.

When a new filter is installed, the microorganisms adapt to the environment, and the activity of the microorganisms increases with time. After some weeks the filter approaches steady state. In principle the microorganisms

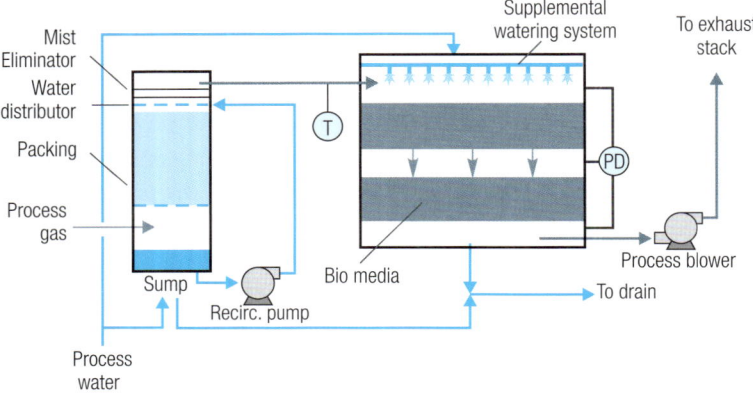

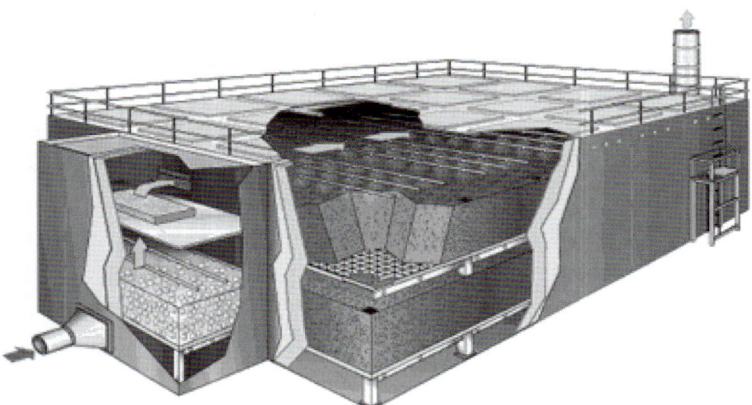

Figure 4.16 Biological filter. The humidification vessel is to the left and the filter section to the right (McGrath, 1998).

feed upon the pollutants, but at the same time the biological filter material degrades slowly, and must be changed regularly, however, the lifetime may be several years.

There are several advantages of biological filters:
• The biological filter has a high efficiency
• The biological filter has a low pressure drop
• There are no expenses to chemicals
• The operation is simple

There are also some disadvantages:
• The biological filter takes up a lot of space
• The filter material may need a change after a few years
• The clean gas is saturated with water vapour and has ambient temperature, which means that the thermal rise of the plume is low and reheating may be necessary before emission

Biological scrubbers are of the same type as the scrubbers described earlier, but in this case microorganisms are suspended in the scrubbing liquid or fixed to the packing in the scrubber. The analogy to biological waste water treatment is obvious, but in many cases N and P has to be added to the scrubbing liquid as nutrients for the microorganisms. The absorption process is absorption with chemical reaction, but in this case the conversion of contaminants is due to biochemical reactions. The advantages and disadvantages of this process are similar to those stated about absorption.

Summing up – gases and odorous compounds

The choice of gas cleaning method for gases and odours depends on many parameters as mentioned in the introduction. Table 4.6 gives a comparison of the different methods of air pollution control described above, and the table may be used as a first guideline to choose the methods for further consideration. More information about design and operation of gas cleaning equipment for gases and odours may be found in Theodore and Buonicore (1994), Nevers (1995), Wark et al. (1998), Heinsohn et al. (1999) and VDI (2005).

Properties of the gas at inlet	Oxidation		Absorption (Scrubbers)	Adsorption	Biological methods	
	Thermal	Catalytic			Scrubbers	Filters
Components removed with high efficiency	VOC and other combustibles (e.g. hydrocarbons, solvents, odorous substances)	VOC and other combustibles (e.g. hydrocarbons, solvents, odorous substances)	Water soluble compounds (e. g. SO_2, HCl, NH_3, organic acids, mercaptans, amines	VOC, hydrocarbons, nonpolar solvents and odorous substances	Water soluble and biodegradable compounds, odorous substances	Biodegradable compounds, odorous substances
Dust content	A low dust load is tolerable	A low dust load is tolerable	No harm	Dust must be removed before adsorption	No harm	Dust must be removed before the filter
Water content	Increases the cost of operation	Increases the cost of operation	Beneficial, lowers the water consumption	A maximum of 70% relative humidity is tolerable	Beneficial, lowers the water consumption	Beneficial, lowers the water consumption
High temperature	Beneficial for heat recovery	Beneficial for heat recovery	May lower the efficiency	Cooling is necessary	Cooling is necessary	Cooling is necessary
Advantages	Potential for heat recovery	Potential for heat recovery	Simple process	Possibilities for reuse of adsorbed compounds	Simple process	Simple process
Disadvantages	High cost High fuel costs in some cases	High cost Risk of catalyst poisoning in special cases	Waste water and/or sludge	High cost in plants with regeneration	Waste water and/or sludge	Takes up a large space

Table 4.6 Comparison of methods to control gaseous air pollutants

4.3 Flue gas cleaning

One way of lowering the impact from heat and power production on the air pollution in cities was to centralise the production in large power stations with very tall stacks. But already in the 1970's it was found that a local air pollution problem was solved at the expense of a regional air pollution problem, the acid rain. As described in Section 3.2, it is important to look at the whole life cycle to make a Life Cycle Assessment of the whole process, and take all environmental impacts into account. The most important local and regional air pollutants from heat and power production are SO_2, NO_x and dust, and as seen in the emission inventories in Chapter 12, heat and power production are among the major sources of emission for these components.

There is a long tradition for controlling dust emissions from power stations with electrostatic precipitators, to avoid local problems, and at the same time regional air pollution problems due to dust have been reduced. Since the beginning of the 1980's flue gas desulphurization and NO_x abatement has been introduced on modern European power plants to improve the local air quality and to reduce trans-boundary air pollutant transport. Figure 3.2 showed a sketch of a modern power plant with flue gas cleaning. The combined heat and power production was explained in Section 3.1 about clean technology, and now the principles of the flue gas cleaning processes to remove SO_2 and NO_x will be described.

Flue gas desulphurization (FGD)

There are many commercial FGD processes, but wet scrubbers have a share of about 88% of the total FGD capacity worldwide. Other processes are spray dry scrubbers (semi-dry systems), sorbent injection processes, regenerable processes and combined SO_2/NO_x removal processes as desribed by Soud (2000) and Soud and Fukusawa (1996).

A short description of the wet scrubber FGD process will be given. The wet scrubber or absorption tower for removal of gaseous pollutants has been shown earlier, but there are some special features of FGD scrubbers. Sulphur dioxide has a low solubility in water and a reaction with a basic reagent is necessary to obtain a high degree of desulphurization without the need of an absorption tower of enormous size and an excessive water consumption and wastewater production.

The most common choice of reagent is limestone with a high content of $CaCO_3$, often above 95 wt%, because it is cheap and because it is possible to produce saleable gypsum as a by-product. The overall chemical reaction can be written as:

$$SO_2(g) + \tfrac{1}{2}O_2(g) + CaCO_3(s) + 2H_2O(l) \rightarrow$$

$$CaSO_4 \cdot 2H_2O(s)(gypsum) + CO_2(g) \tag{4.12}$$

In practice there are many reactions which influence the performance, and the optimal process conditions are difficult to assess. The objectives of the process are at the same time: to obtain a high degree of desulphurization, to have a low consumption of $CaCO_3$ and to produce saleable gypsum. Figure 4.17 shows a process flow sheet for a limestone based FGD plant on a coal fired boiler.

The inlet gas temperature is typically 100-120 °C, and 99% or more of the

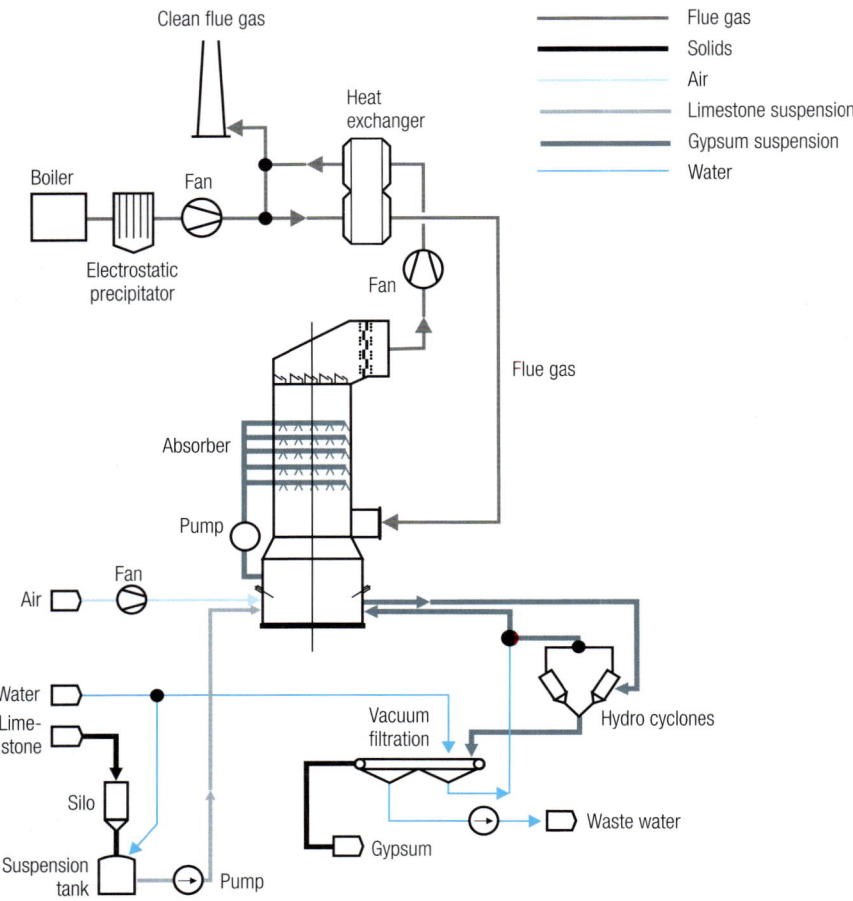

Figure 4.17 Flow sheet of a limestone based FGD plant on a coal fired boiler (Johnsson, 1998).

fly ash (dust) has been removed in the electrostatic precipitator upstream the FGD plant. The flue gas is cooled to 70-80 °C in a heat exchanger counter currently with the cleaned flue gas and enters in the bottom (counter current) or in the top (concurrent) of the absorption tower. The tower may be with or without packing. In the process in Figure 4.17 there is a counter current absorption tower without packing, a spray tower. The temperature of the gas drops to about 50 °C due to evaporation of water from the scrubbing liquid, and simultaneously a number of important steps in the absorption process take place (Kiil et al., 1998; Johnsson and Kiil, 2000):

- SO_2 from the gas phase is absorbed by the scrubbing liquid
- HSO_3^- is oxidized to SO_4^{2-}
- $CaCO_3$ is dissolved
- Crystals of $CaSO_4 \cdot 2H_2O$ (gypsum) are formed

Limestone is added to the bottom of the absorption tower and is partly dissolved in the scrubbing liquid, which is pumped to the top of the scrubber and distributed over the cross section using spray nozzles. The liquid is recycled over the tower and part of the limestone dissolves during the contact between gas and liquid in the tower. Air is blown to the bottom of the absorber to ensure complete oxidation of sulphite to sulphate. A side stream is taken out for removal of gypsum using hydro cyclones and vacuum filters. All the sulphur removed from the flue gas leaves the process as gypsum of high quality. The gypsum is washed with water, and this water is partly recycled to the process and partly taken out for waste water treatment where heavy metals, mainly from the small amount of fly ash, are removed. The cleaned flue gas is heated to 70 to 80 °C by heat exchange with the incoming flue gas to ensure a proper plume rise and good dispersion of the remaining air pollutants in the atmosphere. The efficiency of flue gas desulphurisation with scrubbers is typically about 95% removal of SO_2, but in newer plants 99% removal efficiency has been obtained.

As a side effect there is a removal of fine particles remaining after the electrostatic precipitator and an almost complete removal of HCl and HF. Flue gas desulphurization and simultaneous production of gypsum is a gas cleaning process, but it is in a way also a clean technology, because the gypsum is sold and used as a raw material in the production of cement and wallboards. In this way it is an example on the use of by-products.

NO_x removal from flue gas

The formation of NO_x in combustion processes is complicated and depends on a number of parameters as illustrated in Equation 3.1. Because of this dependence, the formation of NO_x can be lowered by proper design and operation of the power plant and other combustion systems. The important parameters are temperature, O_2-concentration and residence time in the hot zone. By staging the air to the burner, the temperature is lowered and reducing conditions are obtained in the primary combustion zone. Figure 4.18 shows a low-NO_x-burner.

Secondary and tertiary air or over-fire air is then used to ensure complete burn-out of the fuel (Figure 4.19). This method is often called the use of low-NO_x-burners, even if more than burner design and operation is involved. It is

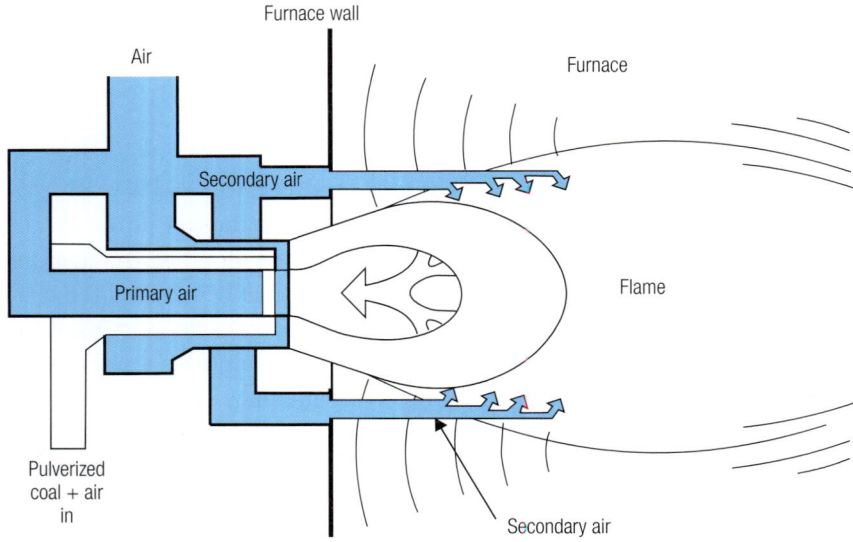

Figure 4.18 Low-NO$_x$-burner. A mixture of coal dust and air enters from the left and the flame is seen to the right. (Adapted from Strauss, 1992).

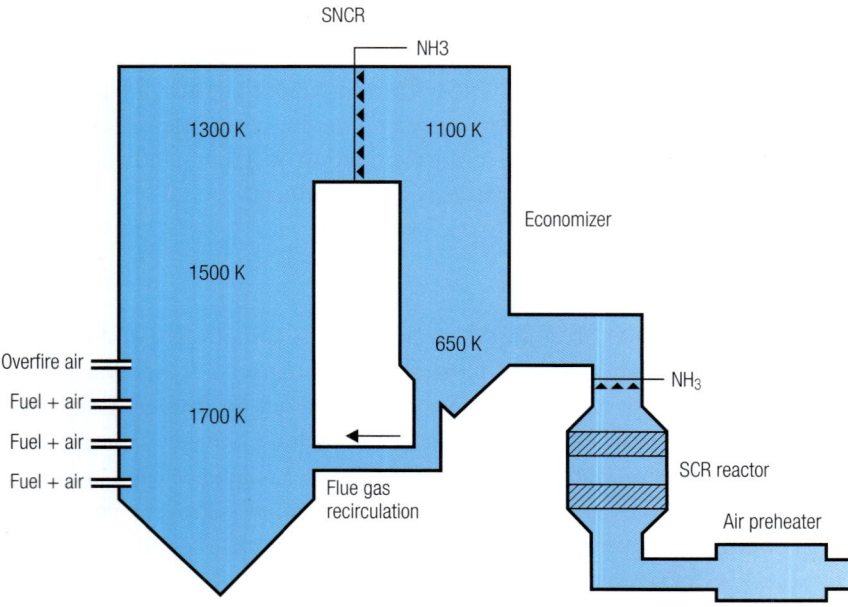

Figure 4.19. Schematic diagram of a coal fired boiler showing the position of over-fire air, SNCR and SCR.

really a clean technology and not a gas cleaning method and it is widely used to lower the formation of NO_x on power stations.

Unfortunately, it is not possible to eliminate NO_x from the flue gas with low-NO_x-burners, and therefore flue gas cleaning is still necessary. More than 95% of the NO_x in flue gas is NO and only a minor part is NO_2. As a consequence NO_x is not removed from the flue gas by conventional flue gas desulphurisation with scrubbers, because NO has a very low solubility in water and is not an acid. The commercial processes for NO_x removal use the possible reduction of NO and NO_2 to N_2. The most common reactant for this reduction is NH_3, and the reaction can be carried out at a relatively low temperature (300-400 °C) using a catalyst, the SCR-process (Selective Catalytic Reduction) or at a higher temperature (850-1000 °C) without a catalyst present, the SNCR-process (Selective Non Catalytic Reduction). A short description of the two processes will be given below (Soud and Fuku-sawa, 1996).

The principle of the SCR-process is that NO and NO_2 reacts with NH_3 over a solid catalyst and N_2 and H_2O is formed. The main reactions are:

$$6\,NO + 4\,NH_3 \rightarrow 5\,N_2 + 6H_2O \qquad (4.13)$$

$$6\,NO_2 + 8\,NH_3 \rightarrow 7\,N_2 + 12\,H_2O \qquad (4.14)$$

There is also an undesirable side-reaction:

$$3\,O_2 + 4\,NH_3 \rightarrow 2\,N_2 + 6\,H_2O \qquad (4.15)$$

The overall reaction is often written:

$$4\,NO + 4\,NH_3 + O_2 \rightarrow 4\,N_2 + 6\,H_2O \qquad (4.16)$$

in good agreement with an observed almost stoichiometric consumption of NH_3. The reaction takes place on the surface of a solid catalyst. The catalyst is porous to give a large specific surface and most of the reaction takes place on the interior surfaces. The most common catalyst consists of a carrier of TiO_2 with V_2O_5 and WO_3 as the catalytic material. Because the optimum temperature is 350-370 °C, the catalyst is placed in the flue gas stream before the air preheater where usually the flue gas has this temperature. At this position the flue gas has a high content of fly ash, because it is before the electrostatic precipitator, see Figure 3.2, and therefore the catalyst is in blocks with small channels, to ensure that the fly ash is carried through and not deposited in the catalyst (see Figures 4.20 and 4.10).

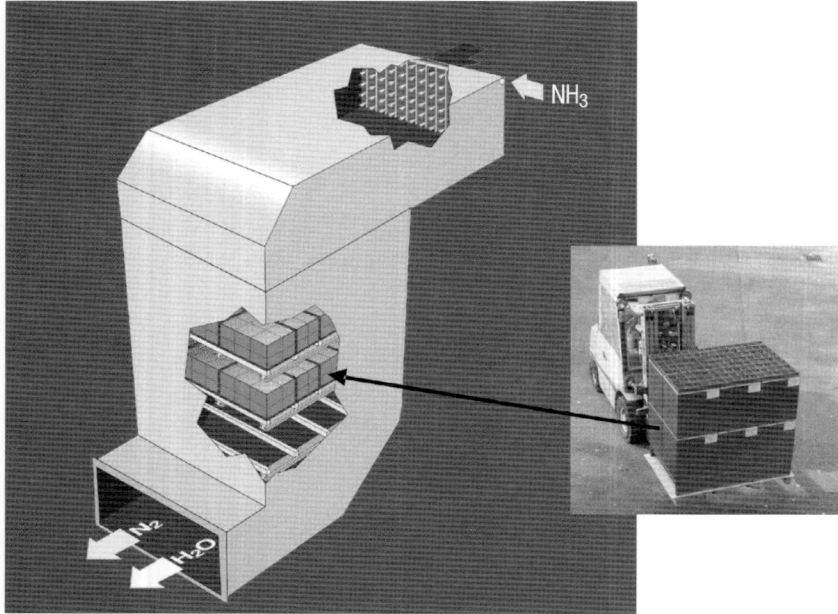

Figure 4.20 SCR catalyst units and modules and the SCR reactor. (Courtesy of Haldor Topsø).

In some cases, especially when retrofitting an existing power plant, the SCR-process is placed after the FGD plant, a tail end SCR, but then the flue gas has to be heated again to at least 300 °C. The catalyst lifetime is an important parameter for the economy of the SCR-process. Industrial experience from a number of European power stations burning coal indicate that the average life time is 4 to 6 years, but it is difficult to give a precise figure, because the catalyst is replaced gradually over many years of operation.

The principle of the SNCR- process is the same as described above in reaction 4.13 and 4.14, but without a catalyst the reaction temperature is higher, 900-1000 °C, and another reaction is possible:

$$4\,NH_3 + 5\,O_2 \rightarrow 4\,NO + 6\,H_2O \tag{4.17}$$

The undesirable and detrimental oxidation of NH_3 to NO takes place at temperatures above 950 °C and dominates at 1000 °C. At temperatures below 900 °C the reaction rate in Reaction 4.13 becomes low and un-reacted NH_3 may be present in the flue gas. The conclusion is, that it is extremely important to inject the NH_3 at a proper place in the boiler, where optimum temperatures of about 950 °C prevail (Kasuya et al., 1995; Østberg et al., 1997). The temperature changes with fuel type, boiler load and fouling of heat transfer surfaces, and therefore the NH_3 consumption is higher and the efficiency is lower compared to the SCR-process. This method is not used on power plants with coal combustion.

In recent years there has been a growing interest in biomass combustion because of possible climate effects from the emission of CO_2 from combustion of fossil fuels. Biomass fuels like straw and wood have a content of potassium and unfortunately this element is a catalyst poison for the vanadium based SCR catalyst. The deactivation is fast, but it may be possible to reactivate the catalyst by washing with water or acid (Zheng et al., 2004, 2005). In waste incinerators the flue gas also has a content of potassium and other catalyst poisons, and for that reason in waste combustion it is not possible to place the SCR process before the particle and gas cleaning processes. In this case the cleaned and cooled gas must be reheated before the SCR reactor, a so-called tail-end placement of the SCR process. This configuration has higher operating cost, and many waste incinerators use the SNCR process to avoid problems with catalyst poisons and reheating of the flue gas.

Flue gas cleaning summary

Flue gas cleaning is costly, and Table 4.7 gives a summary of the costs involved. Low-NO_x-burners and fly ash removal by electrostatic precipitators are in the low cost end while SCR and especially desulphurization are costly processes. On a modern power station flue gas cleaning may add 19 to 38% to the cost of electricity production (Takeshita, 1995; Soud, 2000).

Air pollution control costs for coal-fired power stations	Capital costs		Running costs	
	ECU per kW_e	As a proportion of the capital cost of the plant %	0.001ECU per kWh	As a proportion of the cost of electricity %
Low-NO_x-burners + over-fire air	20-40	2-3	1-2	1-4
SCR (Selective Catalytic Reduction of NO_x)	70-110	5-8	4-5	6-12
Particulate control	40-50	3-4	2-3	4-8
Flue gas desulphurization	130-220	10-17	5-6	8-14
Total	260-420	20-32	12-16	19-38

Table 4.7 Air pollution control costs for coal-fired power stations. (Adapted from Takeshita, 1995).

It is common practice to calculate the necessary stack height based on emissions without flue gas cleaning, because situations might occur, where the flue gas cleaning systems are not functioning, but the power plant still has to produce heat and electricity because of public demand. This means that during normal operation of a modern power station with efficient flue gas cleaning techniques the impact on the local air pollution level will be very small.

4.4 Literature

Brauer, H. and Varma, Y.D.G. (1981): *Air Pollution Control Equipment.* Springer-Verlag, Boston-Heidelberg-New York.

Fogler, H.S. and Gürman, N.H. (2005): *Elements of chemical reaction engineering*, University of Michigan.

Heinsohn, R.J. and Kabel, R.L.(1999): *Sources and control of air pollution*, Prentice Hall, USA.

Johnsson J. E. and Kiil, S. (2000): *Sulfur Chemistry in Combustion II – Flue Gas Desulphurization*, Ed. C. Vovelle, NATO Science Series C: Mathematical and Physical Sciences – Vol. 547, 2000, Kluwer Academic Publishers, p. 283-301.

Johnsson, J.E. (1998): *Stationary sources*, pp 35-62 in Urban Air Pollution – European Aspects, eds. Fenger, J., Hertel, O. and Palmgren, F., Kluwer Academic Publishers, Dordrecht.

Kasuya, F. et al. (1995): *The Thermal DeNO$_x$ Process: Influence of Partial Pressures and Temperature*, Chemical Engineering Science, *50, 9*, 1455-1466.

Kiil, S. et al. (1998): *Experimental and Theoretical Investigations of Wet Flue Gas Desulphurisation Pilot Plant*, Ind. Eng. Chem. Res., *37, 7*.

Levenspiel, O. (1999): *Chemical reaction engineering*, 3. ed., John Wiley & Sons, USA.

McCabe, W.L. et al. (2005): *Unit operations of chemical engineering*, 7. ed., McGraw-Hill Int. Ed., Singapore.

McGrath, M.S. (1998): *Solving VOC and odor problems*, Plant Services Magazine Article, May 1998, http://www.enviro-chem/airpol/bioton/article.htm.

Nevers, N. de (1995): *Air pollution Control Engineering*. McGraw-Hill, USA.

Soud, H.N. (1995): *Developments in particulate control systems for coal combustion*, IEA Coal Research, IEACR/78.

Soud, H. N. (2000): *Developments in Flue Gas Desulfurization*, IEA Coal Research,. CCC/29.

Soud, H.N. and Fukusawa, K (1996).: *Developments in NO$_x$ Abatement and Control*, IEA Coal Research, IEACR/89.

Strauss, K. (1992): *Kraftwerkstechnik*, Springer-Verlag, Berlin, (in German).

Takeshita, M. and Soud, H.N. (1993): *FGD performance and experience on coal-fired plants*, IEA Coal Research, IEACR/58.

Takeshita, M. (1995): *Air pollution control costs for coal-fired power stations*, IEA Coal Research, IEAPER/17.

Theodore, L. and Buonicore, A. (eds.) (1994): *Air Pollution Control Equipment*, Springer-Verlag, Berlin, Heidelberg.

Turnes, S.R. (1996): *An Introduction to Combustion.* Mc Graw-Hill Inc. Singapore.

VDI (1987): Waste Gas Cleaning by Thermal Combustion.

VDI/DIN (2005): *Handbuch Reinhaltung der Luft Band 6 – Abgasreinigung – Staubtechnik (VDI/DIN manual Air Pollution Prevention Volume 6: Waste Gas Cleaning – Dust Technology)*, Kommision Reinhaltung der Luft / KRdL) im VDI und DIN – Normenausschus.

Wark, K. et al. (1998): *Air Pollution – Its Origin and Control*, Addison, Wesley, Longmann, USA.

Zheng, Y. et al. (2004): *Laboratory investigations of selective catalytic reduction catalysts: Deactivation by potassium compounds and catalyst regeneration*, Ind. Eng. Chem. Res., 43, 941-947.

Zheng, Y. et al. (2005): *Deactivation of V_2O_5-WO_3-TiO_2 SCR catalyst at a biomass-fired combined heat and power plant*, Applied Catalysis B, Environmental, 60, 253-264.

Østberg, M. et al.(1997): *Influence of Mixing on the SNCR Process*, Chemical Engineering Science, *52,11,* 1715-1731.

5 Mobile sources

Jesper Schramm

Mobile sources contribute significantly to air pollution at a local, national and global scale. The origin of the air pollution is almost exclusively from combustion engines, varying with respect to type and size, from very small engines with an output of only a few kW to large marine engines in the range of 50 MW.

5.1 Road vehicles

The most commonly applied engines for road vehicles are the traditional internal combustion (IC) engines, and among these the gasoline (Otto) engine and the diesel engine are the most common ones. The emissions and the composition of emissions from combustion engines are much dependent on the type of engine in question and thereby the fuel that is used. This chapter therefore gives an overview of the different basic types of engines that are commonly seen. "Basic" means that the description of accessory equipment, like fuel supply system and exhaust after treatment is omitted.

Gasoline engines

The engine consists of a piston, located in a cylinder, and connected to the crankshaft via the connecting rod. The piston moves up and down inside the cylinder, and this movement makes the crank rotate. The rotation in the end makes the wheels of the vehicle rotate.

Normally, a 4-stroke engine is applied. The name refers to the fact that the process, which takes place inside the engine, can be divided into four steps in two crank rotations. The first process is the intake stroke, or intake process, in which a fresh charge of fuel and air is sucked into the cylinder volume. The piston at the same time moves from the top position to the bottom position, with an open intake valve. In the second process the fresh charge is compressed. At the same time the piston moves from the bottom position to the top position with both intake and exhaust valve closed. The third process is the combustion and expansion, in which the fresh charge is ignited with the spark plug, causing an almost instantaneous combustion of the fuel and an increase in temperature and pressure. The piston moves from the top to the bottom position with both valves closed during the expansion. The fourth process is the exhaust stroke, in which the combustion products are forced out through the open exhaust valve, while the piston moves from the bottom to the top.

The whole process is cyclic, meaning that after the exhaust stroke, a new process begins all over again, starting with the intake of a fresh charge of fuel and air. The open/close periods for the intake and exhaust valves, called "valve timing", is very important. The described timing only outlines the principle. In a real engine there will be some "overlap" and other deviations from the described pattern.

The combustion process takes place over a very short period, with the piston in almost the same position all the time. The process is initiated by the spark plug, and a well-defined flame spreads out from this position and penetrates the combustion chamber. Ahead of the flame front is unburned charge and behind the flame are combustion products. The mixture of fuel and air is homogeneous because the fuel is evaporated before the combustion takes place and thereby allows a very efficient mixing process. This is called premixed combustion. The process requires a fuel that is resistant against self-ignition during the high temperatures that occurs during the compression before the spark plug ignites the mixture. Otherwise the phenomenon called "knocking" will occur. Gasoline engines therefore have some limitations on the compression ratio, which is the ratio between the cylinder volume above the piston, when the piston is in the bottom position, and the volume when the piston is in the top position. The ability to resist "knocking" is determined by the fuel and characterized by the so-called octane number. A picture of the important elements of the gasoline engine is shown in Figure 5.1.

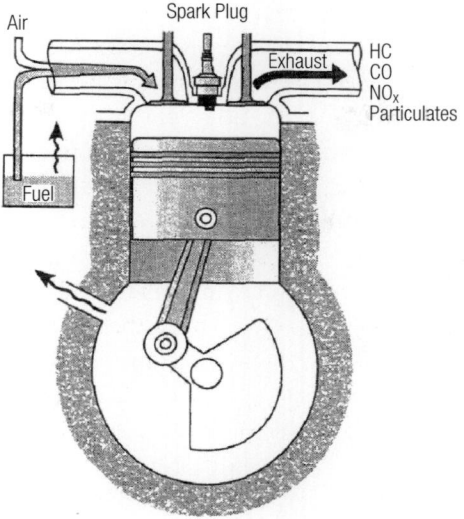

Figure 5.1 Schematic picture of the gasoline engine showing the basic elements of the gasoline engine: piston, connecting rod, crankshaft, spark plug, intake and exhaust valves.

Diesel engines

Four-stroke diesel engines are different from gasoline engines mainly in the way ignition takes place. In a diesel engine, the fuel is ignited by the heat, which is generated during the compression. Since the compression process in this way is responsible for the ignition, a diesel engine needs a higher compression ratio in order to work properly.

The fuel consists of larger hydrocarbon molecules which don't evaporate as easily as gasoline. Therefore, it is atomised by injection at high pressures through a nozzle into the air, either directly into the combustion chamber, which is the most common today, or into a smaller pre chamber. Due to the poorer mixing circumstances, the diesel engine needs a high surplus of air. During the atomisation a fuel spray is formed as a mixture of small fuel droplets and vapour. The vapour is combusted with the air that surrounds the spray. In order to keep the combustion process going, the fuel vapour and air must meet at the flame zone. This process is controlled by diffusion of fuel vapour and air and is thus named diffusion combustion. A picture of the important elements of the diesel engine is shown in Figure 5.2.

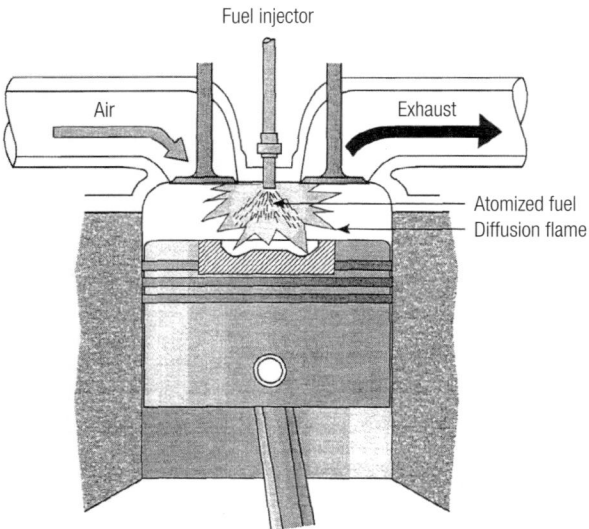

Figure 5.2 Schematic picture of the diesel engine showing piston, fuel injection nozzle, and intake and exhaust valves.

Due to the high compression ratio and the large amount of surplus air, which requires a larger cylinder volume, diesel engines are larger, heavier and more expensive than the similar gasoline engines. It is however, an advantage with the high compression ratio with respect to the fuel economy, which improves with higher compression ratio. Diesel engines are therefore traditionally used in vehicles where the engine weight is of less importance, but where the fuel costs are an important part of the total costs. This is typically the case with trucks, buses, taxis etc, which are generally equipped with diesel engines. Fast going sporty vehicles, on the other hand, are equipped with gasoline engines due to the higher ratio between power and engine weight.

In recent years, however, more and more passenger cars are equipped with diesel engines due to intensive engine development, especially with the diesel fuel injection system. The injection pressures of the newest engines are much higher than just a few years ago. Today's modern high-pressure injection common rail diesel engines have a very efficient fuel air mixing process and are associated with higher power per cylinder volume than earlier. This makes diesel engines more suited for small vehicle purposes, and the advantage with better fuel economy is still valid.

Two stroke engines

The processes in a two-stroke engine are thermodynamically the same as the corresponding four stroke processes, but the time frame is only one crank rotation per cycle. The word "two-stroke" thus refers to the fact that the piston undergoes two strokes in one cycle. Figure 5.3 illustrates the processes in cross section views of a two-stroke Otto engine during one cycle.

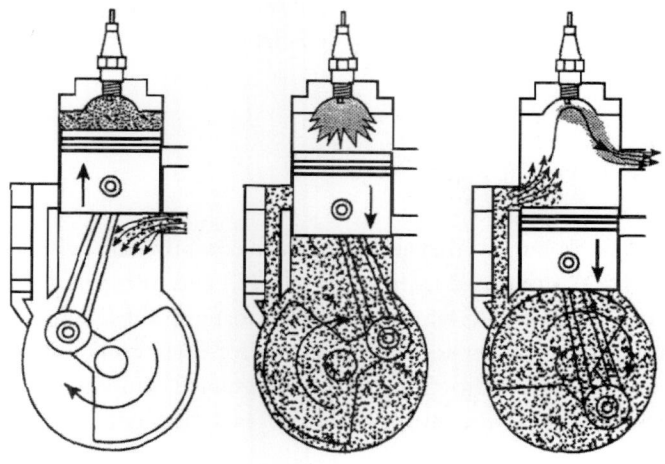

Figure 5.3 Cross section view of the two-stroke Otto engine cycle showing two different compartments, one above the piston and one below the piston. The compartments are connected via the scavenging channel.

When the piston moves upwards fresh fuel/air charge enters the chamber below the piston through the intake port. At the same time a fresh charge that previously has been sucked into the engine is compressed above the piston. The compressed charge is ignited by the spark plug, and the piston moves downwards at the same time as the fresh charge below the piston is pushed into the chamber above the piston through the tube that connects the two chambers. The combustion products are thus forced out through the exhaust port above the piston by the entraining fresh charge, and the cyclic process starts over again.

The obvious advantage of the two-stroke engine is the high power/weight ratio. This occurs because fresh charge is introduced for every crank rotation. Compared to the four-stroke engine, where fresh charge is introduced only every second rotation, this results in principally twice as much engine power. Two stroke engines are therefore well suited for smaller vehicles like scooters.

Two stroke engines also have a potential for better fuel economy, because the relatively small engine size means less mechanical losses. However, scooters have a very poor fuel economy, on weight basis, compared to passenger cars with four stroke engines, because of the difficulties with separation of combustion products and fresh charge when the combustion products are pushed out by the fresh charge above the piston. This results in much fuel loss directly out of the exhaust port. In large two-stroke marine diesel engines on the other hand the principle works well, without fuel loss due to this effect. These engines today have a maximum energy efficiency above 50%.

5.2 Formation of pollutants

The important gaseous pollutants that are formed in a combustion engine are carbon monoxide (CO), unburned hydrocarbons (HC) and nitrogen oxides (NO_x). Furthermore particulate emissions are important, but the formation processes are less well understood. Regulations for these emissions are introduced in many countries. The emissions of unregulated compounds like PAHs (Polycyclic Aromatic Hydrocarbons) and SO_x (sulphur oxides) is only discussed briefly.

Carbon monoxide

Carbon monoxide (CO) is mainly formed in gasoline engines, because CO formation takes place under conditions, where oxygen supply is insufficient. Since diesel engines operate at a large surplus of air, the CO emissions from these engines are very limited compared to gasoline engine emissions. In order to characterize the formation picture the emission is often plotted as a function of the excess air ratio, λ. This parameter has a value of 1 at stoichiometric combustion. There is a surplus of air, when λ is bigger than 1, and a lack of air, when λ is smaller than 1. In Figure 5.4 the trends of gaseous emissions during combustion are shown. When λ is above 1 all the fuel carbon is ideally combusted into carbon dioxide (CO_2). When there is a lack of oxygen CO is formed instead of CO_2. CO formation is thus a result of the conditions for the chemical reaction alone.

As later explained the formed CO can be oxidised to CO_2 in a catalytic converter, located in the exhaust stream. This requires extra oxygen addition in the exhaust when the CO concentration is very large.

Unburned hydrocarbons

The emission of unburned hydrocarbons (HC) follows the same tendency as

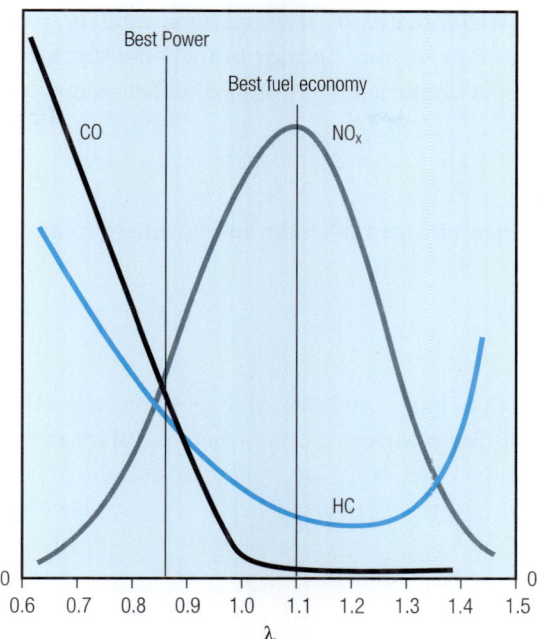

Figure 5.4 The gaseous emissions of CO, HC and NO_x vs. excess air ratio for the premixed combustion process. At an excess air ratio of 1, the amount of air corresponds exactly to amount needed for a complete combustion. As the excess air ratio increases there is a surplus of air, below 1 there is a lack of air.

CO in Figure 5.4, since HC is formed when there is not enough oxygen to support a complete combustion of the fuel into carbon dioxide and water. The lower the supply of oxygen, the more unburned fuel is left in the exhaust.

As we can see in Figure 5.4, the HC curve increases when the excess air ratio becomes too big. This is because the temperature at these conditions can be so low that the fuel is not ignited at all or the combustion is not completed during some of the engine cycles. This also results in larger amounts of unburned fuel in the exhaust.

Even in a well functioning engine with surplus of air there will still be some unburned fuel left in the exhaust. This is mainly due to small unavoidable crevices, particularly between the piston and the cylinder liner, and around the spark plug. In these crevices the fuel can hide during the combustion, because the flame will be extinguished before entering the crevice (due to cold surfaces). When the fuel mixture leaves the crevice again, the temperature in the combustion chamber is too low for an efficient oxidation of the fuel. The formation process in this case is a physically controlled process rather than a chemically controlled.

Like CO, HC can be almost eliminated by introducing a catalytic converter in the exhaust of the gasoline engine. Because of the large surplus of air, there is almost no emission of unburned fuel from a diesel engine.

Nitrogen Oxides

Nitrogen oxides are formed when nitrogen from the air is oxidised to nitrogen oxide. The overall reaction is:

$$N_2 + O_2 \rightarrow 2NO_2 \qquad (5.1)$$

This is not an elementary reaction, but a summary of the real reactions that take place on a molecular level. The primary reactions that take place are:

$$O + N_2 \leftrightarrow NO + N \qquad (5.2)$$

$$N + O_2 \leftrightarrow NO + O \qquad (5.3)$$

These reactions are the primary reactions of interest according to Zeldovich´s free radical chain mechanism. The formed NO can further be oxidised to NO_2, either in the engine or later in the exhaust pipe or in the atmosphere. The reactions 5.2 and 5.3 require high activation energy, therefore they do not occur at room temperature, but only at rather high temperatures. The experiences show that at temperatures below 1500 °K no further reaction is possible. This means that NO formation takes place near the peak temperatures during combustion, i.e. when the piston is very close to the top of the cylinder (± a few crank angle degrees). The reactions then freeze at high temperature equilibrium concentrations when cooled rapidly.

The pattern seen in Figure 5.4 reflects the fact that NO_x formation requires oxygen, therefore the NO_x concentrations increase with λ at values below 1 and a little above 1. Then the excess of air causes the temperature to be the limiting factor, and the concentrations decrease as a consequence of lower peak temperatures. NO_x can be reduced to nitrogen in a catalytic converter in the exhaust stream.

Particulate matter

Particulate matter consists of a solid fraction called soot and an organic fraction called SOF (Solid Organic Fraction). Furthermore the particulate matter contains water and other volatile inorganic matter like sulphur oxides. This chapter will focus on the formation processes responsible for the presence of these compounds. In contrast to most of the mechanisms behind the formation of gaseous emissions, the particulate formation mechanisms are not completely understood, but this chapter explain the theory that most researchers believe in today.

Experimentally it has been verified that the soot formation in a premixed combustion process occurring when the C/O ratio exceeds 0.5 (Haynes & Wagner, 1981). The combustion of a hydrocarbon C_nH_m, including soot formation, is then:

$$C_mH_n + y(O_2 + 3,76N_2) \rightarrow 2yCO_2 + \frac{n}{2}H_2O + 3,76yN_2 + (m-2y)C_s \quad (5.4)$$

where C_s is the soot. One might believe that soot formation will occur at an excess air ratio just below stoichiometric combustion, where the oxygen concentration becomes insufficient for a complete combustion into carbon dioxide and water. However the excess carbon is partly converted into carbon monoxide, and partly simply exhausted as unburned fuel.

It is invalid to consider the diffusion combustion in the same way as the premixed combustion, since at some locations in the diffusion combustion the excess air ratio is above 1, and at other locations the excess air ratio is below 1.

Looking at the particulate matter from a diesel engine in a microscope, it looks like shown in Figure 5.6, as a cluster of spherical solid material with adsorbed and condensed material on the surface.

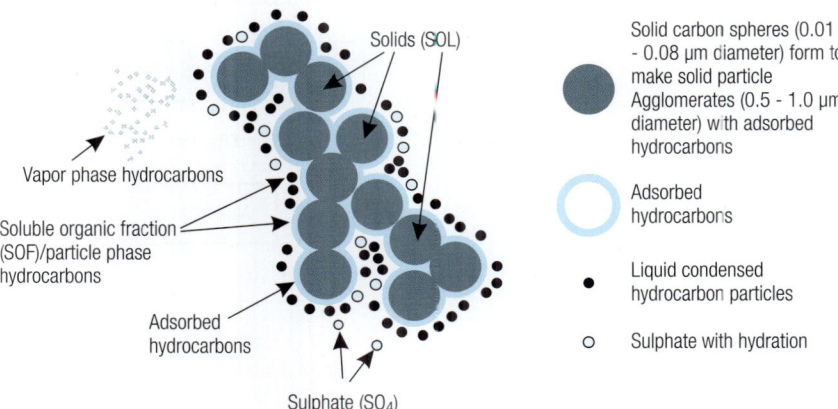

Figure 5.6. Enlarged picture of diesel exhaust particulate matter (Johnson, 1994).

Particulate formation follows successive steps: pyrolysis and oxidative pyrolysis, nuclei formation, particulate growth and agglomeration, aggregation, adsorption, condensation and oxidation.

The initial step is pyrolysis in which precursors for nuclei formation are formed. The presence of radicals like O and OH are enhancing nuclei formation. It is known that especially poly-acetylens and poly-aromatic hydrocarbons are present when nuclei formation occurs. These formed nuclei will agglomerate and grow, resulting in spherical soot particles. When these spherical particles are stable, they aggregate to chain-like structures and eventually condensed hydrocarbons and inorganic material will be adhered to the surface. The surface will also be covered with adsorbed, primarily organic, material. As long as the temperature is sufficiently high and oxygen is present, oxidation of the formed particulate matter will take place along with the formation processes. When the particulate matter is diluted and the temperature decreases, hydrocarbons primarily from the lubricant, but especially in diesel engines also fuel hydrocarbons and hydrocarbons formed during combustion will condense on the surface of the particulate matter. The processes can schematically be illustrated as shown in Figure 5.7.

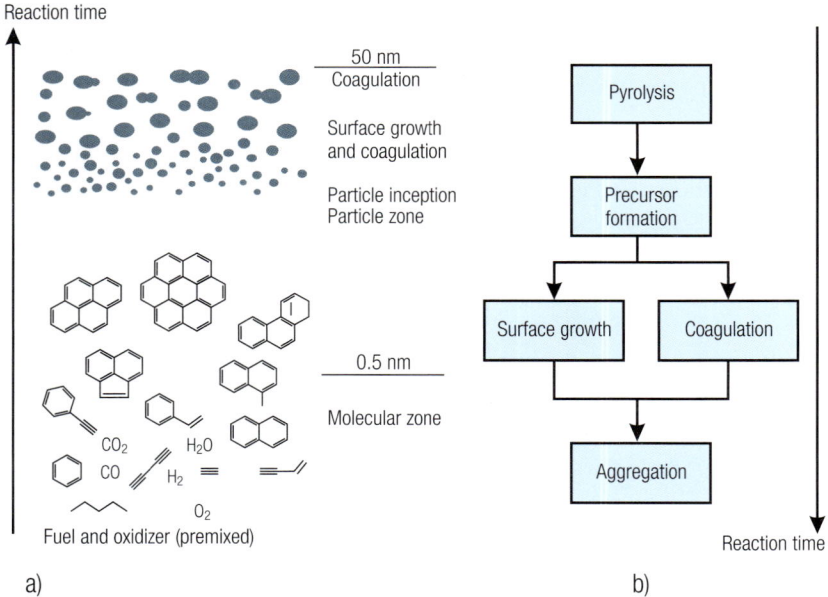

Figure 5.7 Schematic illustration of particulate formation in diesel combustion.
(a) Soot formation, (b) Steps in soot formation.

Unregulated emissions

For some pollutants no regulatory action on the emissions has been established. Among these are the PAHs (Polycyclic Aromatic Hydrocarbons) of which many are carcinogenic. The PAH in the exhaust arise from the fuel, which contains a multitude of aromatic compounds. Only a small fraction of the exhaust PAH is generated during combustion. The authorities therefore try to limit the PAH emissions by regulating the allowed PAH content of the fuel.

Some of the PAHs are more dangerous than others. One of these is benzo(a)pyrene, which is a 5-ring compound, shown in Figure 5.8 together with pyrene, a 4-ring compound, which is found in considerable amounts in normal diesel exhaust (Schramm, 1994). Pyrene is mostly associated with the particulate matter, adsorbed or condensed on the surface. Generally, the larger the molecule is, the higher the fraction found on the particulate matter. PAH's with up to 3 rings are mostly associated with the gas phase. PAHs with more than 4 rings are all in the particulate phase.

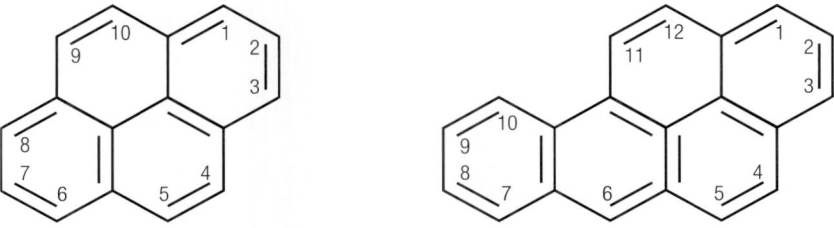

Figur 5.8. Structures of Pyrene (left) and Benzo(a)pyrene (right).

Other unregulated emissions include sulphur oxides, which is a result of the oxidation of the sulphur in the fuel. This is therefore regulated by limiting the fuel sulphur content. From 2005 the maximum allowed sulphur content in gasoline and diesel in EU is 50 ppm. At the same time the EU member countries should take actions to make sure that 10 ppm sulphur gasoline and diesel is widely available at fuel stations. In some member countries, like Denmark, local tax policy has resulted in the fact that only 10 ppm sulphur fuels have been sold at fuel stations since 2005.

Several selected hydrocarbons like benzene and heavy metals like lead are also treated with special concern, and the emissions are mainly regulated by limiting the fuel content of these compounds.

5.3 Emission reduction for road vehicles

Problems with traffic emissions became clear in cities like Tokyo and Los Angeles already in the 1960s. It was, however, with the introduction of catalytic converters in the late 1970s in Japan and USA that the revolution in

emission reduction started. Europe was a bit slower, and the catalytic converters were introduced here as late as the end of the 1980's.

Catalytic converters

The catalytic converter (CAT) is located in the exhaust pipe of the vehicle.
Here the pollutants are converted into harmless components like nitrogen
(N_2), carbon dioxide (CO_2) and water (H_2O). The CAT has a honeycomb
structure, which is shown in Figure 5.9. A typical volume of the CAT is a
few litres, but the long parallel channels, which the exhaust passes through,
have a very large surface on which the catalytic material is spread out. The
catalytic materials are precious metals like Platinum (Pt), Rhodium (Rh) and
Palladium (Pd)

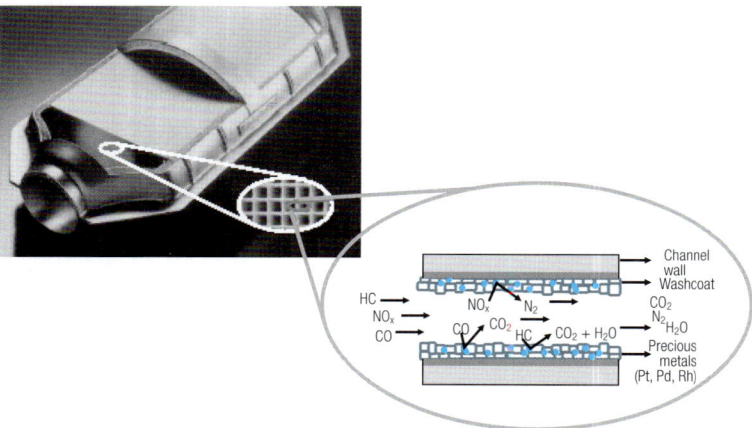

Figure 5.9 Catalytic converter and an enlargement of the inner surface (AECC, 2006).

The first type of CATs were oxidation CATs, that required a surplus of oxygen
in order to oxidize CO and HC according to reactions:

$$CO + \frac{1}{2}O_2 \rightarrow CO_2 \tag{5.5}$$

$$C_m H_n + (m + \frac{n}{4})O_2 \rightarrow \frac{n}{2}H_2O + mCO_2 \tag{5.6}$$

Today's CATs for gasoline engines are so-called three way catalysts (TWC),
referring to all three components that is converted: CO, HC and NO_x. NO_x is
reduced mainly according to the reaction:

$$2NO + 2CO \rightarrow N_2 + 2CO_2 \tag{5.7}$$

This reaction requires a reducing chemical atmosphere, meaning no oxygen. The only situation having a surplus of oxygen and a reducing atmosphere is close to the stoichiometric combustion. Therefore the engine has to operate very close to an excess air ratio of 1. This is clearly seen in Figure 5.10, which shows a typical conversion efficiency of a TWC versus the excess air ratio. In order to keep the excess air ratio close to 1, a sensor that measures the excess air ratio is placed in the exhaust. This sensor, which is called a "Lambda Sensor", gives a signal to the engine control unit which adjusts the air and fuel flow to the desired excess air ratio value. As seen in Figure 5.10 the conversion efficiency is close to 90%. This is valid for a new TWC. One must expect a somewhat lower efficiency during the whole lifetime of a CAT.

Due to the high excess air ratio, a TWC cannot work with a diesel engine.

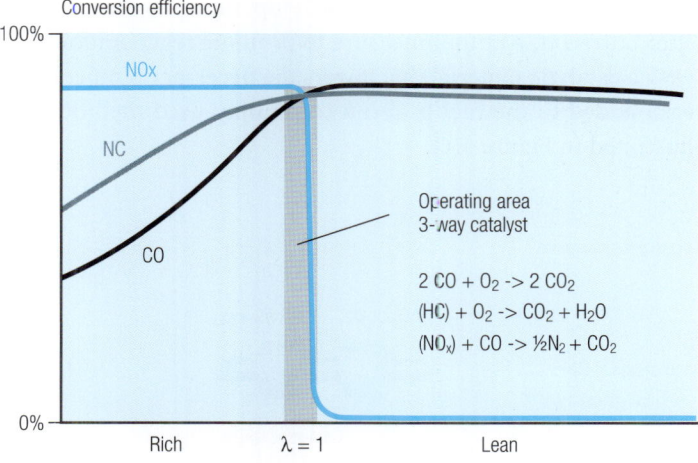

Figure 5.10. The conversion efficiency of a TWC for CO, HC and NOx versus excess air ratio (λ) (AECC, 2006).

Exhaust gas recirculation.

As mentioned earlier, NO_x is formed at the peak temperature during the combustion. A way to reduce the peak temperature is to re-circulate 10-20% of the exhaust to the intake air. In this way a part of the energy, which is released during the combustion, is used to heat up the inert gases CO_2 and

H_2O. This results is lower maximum temperatures. With this technique about 50% reduction in NO_x emissions can be achieved.

Particulate filters

A particulate filter can be mounted in the exhaust pipe to collect the particulate matter, while the gaseous components are passing through the filter. However, the size of the filter limits the amount of exhaust that can pass. The first filters were therefore quite big, so that they could collect particulates during the whole day. During night time the filter was burned clean by electrical heating.

Today's filter technology includes the regeneration of the filter during operation of the vehicles. This can be done in several ways. The burning of particulate matter may occur spontaneously, if the temperature is sufficiently high. It is, however, often necessary to promote this process by addition of heat, most commonly by a late pilot injection of fuel in the engine. A part of the fuel will then burn in the filter or in an oxidation catalyst located upstream the filter. This will increase the catalyst temperature to a level where particulate matter burns off. Another measure to promote regeneration is to coat the filter with a catalytic material that lowers the necessary temperature. Catalytic material added to the fuel is also a commonly used method. The methods are illustrated in Figure 5.11.

- Engine control
    - To increase exhaust gas temperature

- Continuously regenerating CR-DPF

- Catalysed C-DPF

- DPF + fuel borne catalyst (FBC)

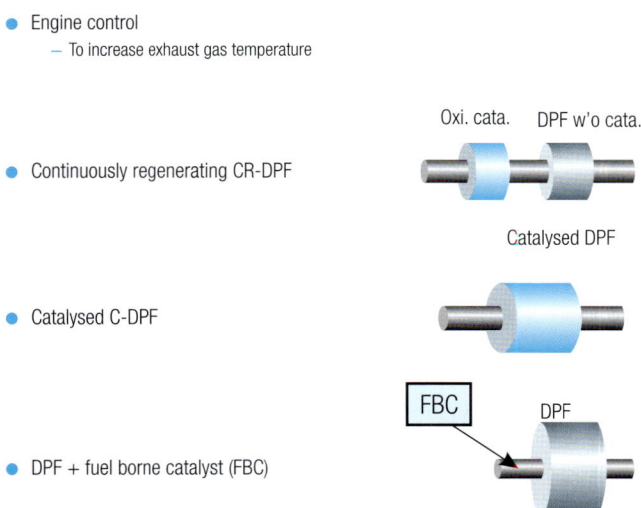

Figure 5.11 Three different systems for collection and burn off of particulate matter during engine operation (AECC, 2006). The upper one doesn't involve catalytic material, the particulate burn off is created by increasing the temperature in the oxidation catalyst. In the lower system catalyst is supplied with the fuel, i.e. fuel borne catalyst (FBC).

Recent technology

Diesel engines often operate at very lean conditions. Many new gasoline cars are at least partly operating with an excess of air, due to better fuel economy. This gives a problem with NO_x conversion in the CAT. As seen in Figure 5.10 the TWC efficiency is very low with respect to NO_x in this situation. To overcome these problems Lean NO_x Traps (LNT's) and Selective Catalytic Reduction (SCR) have been introduced.

In *SCR technology* a chemically reducing agent is introduced in combination with a catalytic converter system. The chemical that is used is urea, $(NH_2)_2CO$. This is introduced between an oxidising catalyst and the SCR catalyst. The urea is converted into ammonia by hydrolysis, and the ammonia is then reducing NO and NO_2 into pure nitrogen. Remaining ammonia is subsequently eliminated in an oxidising catalyst. The system is shown in Figure 5.12.

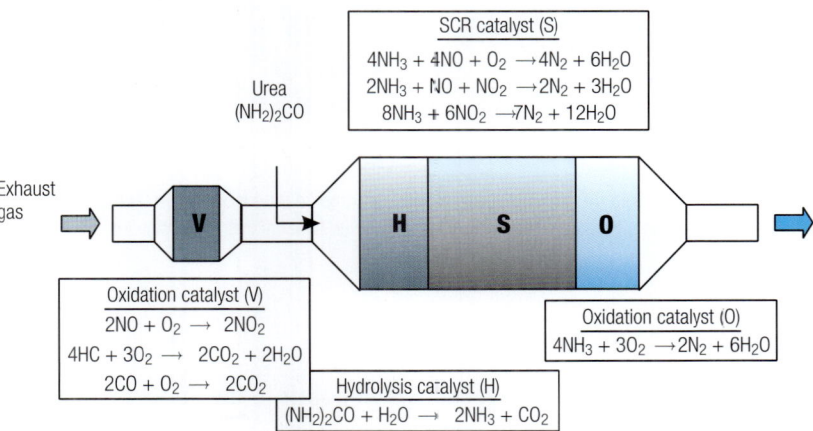

Figure 5.12 The SCR Catalyst System (AECC, 2006). The system consists of several catalytic sections in which selected reactions occur. The catalytic sections are oxidation, hydrolysis, selective catalytic reduction (SCR) and oxidation again.

The LNT (Lean NO_x Trap) oxidises NO catalytically into nitrate, which is held by an adsorbing material during lean operation. When the engine later operates with an excess air ratio below 1, the nitrate is released and the catalyst then reduces the released nitrate into nitrogen, which leaves with the exhaust. The process is shown in Figure 5.13. Recent LNTs convert a small amount of some of the NO_x into nitrate which is used for catalytic reduction of the remaining NO_x. This is particularly interesting in diesel engine context.

LNC (Lean de-NO$_x$ Catalyst), also known as "hydrocarbon SCR" systems use advanced structural properties in the catalytic coating to create a rich 'microclimate' where hydrocarbons from the exhaust can reduce the nitrogen oxides to nitrogen, while the overall exhaust remains lean. The hydrocarbon may come with the exhaust gas ("native") or it may be added to the exhaust gas through injection of a small amount of additional fuel. This has the advantage that no additional reductant source (urea) needs be carried but these systems do not presently offer the same performance as ammonia-SCR systems (AECC, 2007).

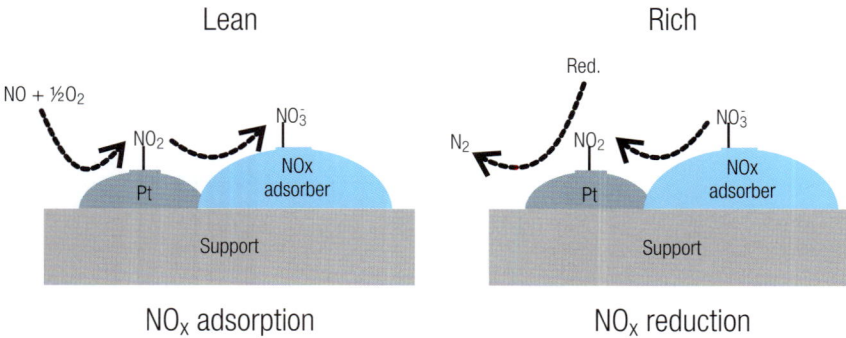

Figure 5.13 Lean NO$_x$ Trap (AECC, 2006). During lean operation nitrate is adsorbed by the adsorber, during rich operation the nitrate is released again, and followed by a catalytic reduction into nitrogen.

Basic engine development

The development of newer diesel engines all points in the direction of higher fuel injection pressure. In diesel engines new "Common Rail" systems operate with injection pressures up to 2000 bars. These systems have a common fuel supply line and computer controlled injection timing. Pre chamber engines were often preferred in passenger cars due to better NO$_x$ performance. However, with the recent development in NO$_x$ control technology; this is no longer a crucial problem for direct injection engines.

With the ability to produce a higher fuel injection pressure, the atomisation of the fuel inside the combustion chamber becomes much more efficient, and the mixing of fuel and air is enhanced at the same time. Furthermore, the direct injection principle results in a better fuel economy than the pre chamber engine.

In gasoline engines direct injection engines are becomimg more frequent, combined with increasingly lean combustion. This improves the fuel economy and reduces NO$_x$ emissions, but some deterioration is seen with respect to particulate emissions. Direct injection seems to promote the formation of very small particles, and may thus be more harmful to human health than traditional indirect injection.

Emission regulations in Europe

Regulation of emissions from road vehicles started in USA and Japan in the 1960s, and in the 1970s in Europe. In the European Union the regulations were overcome just by introducing minor engine modifications until 1989, where the first catalytic converters were seen in larger passenger cars. In 1992 where "EURO 1" was introduced, all passenger cars with gasoline engines had TWCs. In recent years the European situation is a result of EURO 2-5 stages, which entered into force in 1996, 2000, 2005 and 2008 respectively. This development from the beginning for diesel and gasoline engines is seen in Table 5.1. The emissions are expressed as grams per kilometre, referring to an artificial driving pattern, which is seen in Figure 5.14. During driving the exhaust is collected continuously in a bag, which is subsequently analysed for its content of pollutants. The emission factors are thus calculated as an average of the whole driving cycle.

Tier	Date	CO	HC	HC+NO$_x$	NO$_x$	PM
Diesel						
Euro 1 **	1992.07	2.72 (3.16)		0.97 (1.13)		0.14 (0.18)
Euro 2, IDI	1996.01	1.0		0.7		0.08
Euro 2, DI	1996.01a	1.0		0.9		0.10
Euro 3	2000.01	0.64		0.56	0.50	0.05
Euro 4	2005.01	0.50		0.30	0.25	0.025
Euro 5 ***	mid-2008	0.50		0.25	0.20	0.005
Petrol						
Euro 1 **	1992.07	2.72 (3.16)		0.97 (1.13)		
Euro 2	1996.01	2.2		0.5		
Euro 3	2000.01	2.30	0.20		0.15	
Euro 4	2005.01	1.0	0.10		0.08	
Euro 5 ***	mid-2008	1.0	0.075		0.06	0.005[b]

* Before Euro 5, passenger vehicles > 2,500 kg were type approved as Category N1 vehicles
** Values in brackets are conformity of production (COP) limits
*** Draft proposal, July 2005
a - until 1999.09.30 (after that date DI engines must meet the IDI limits)
b - applicable only to vehicles using lean burn DI engines

Table 5.1 Development in emission standards for passenger cars with diesel and gasoline engines in Europe. Units are in g km^{-1} (Dieselnet, 2007)

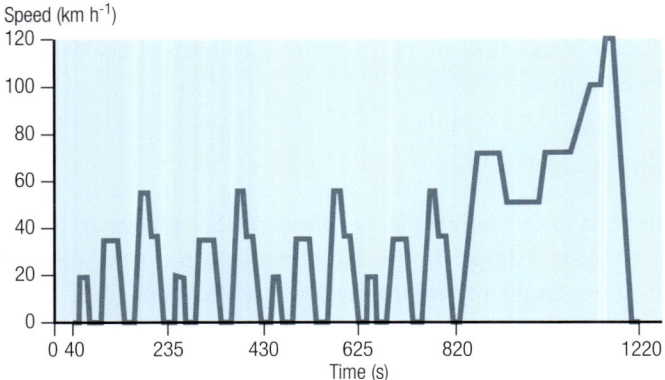

Figure 5.14. Driving-pattern for the European emission test procedure. The driving speed is shown as a function of time during the test.

The emissions from heavy duty vehicles are, for practical reasons, measured on the engine alone. The emissions measured during a specified engine load/speed programme are used for regulatory purposes. The emissions are thus specified in g (kWh)$^{-1}$. The development in heavy duty vehicle emission standards in Europe (>3.5 tons and engine power >85kW) are seen in Table 5.2.

Tier	Date & Category	Test Cycle	CO	HC	NO$_x$	PM	Smoke
Euro I	1992, <85 kW	ECE R-49	4.5	1.1	8.0	0.612	
	1992, >85kW		4.5	1.1	8.0	0.36	
Euro II	1996.10		4.0	1.1	7.0	0.25	
	1998.10		4.0	1.1	7.0	0.15	
Euro III	1999.10, EEVs only	ESC & ELR	1.5	0.25	2.0	0.02	0.15
	2000.10	ESC & ELR	2.1	0.66	5.0	0.10 0.13*	0.8
Euro IV	2005.10		1.5	0.46	3.5	0.02	0.5
Euro V	2008.10		1.5	0.46	2.0	0.02	0.5

*for engines less than 0.75 dm^3 swept volume per cylinder and a rated power speed of more than 3000 min^{-1}

Table 5.2 The development in heavy duty diesel engine emission standards in Europe. Units are in g (kWh)$^{-1}$ (Dieselnet, 2007)

Real life emission factors

As mentioned earlier vehicles are not normally driven as given in the driving patterns for the emission regulation procedures. This means that the real emission factors are different from the emission factors specified in the regulations, shown in the Table 5.1. Real driving is much more transient than the pattern shown in Figure 5.14. This generally results in higher emission factors in real life.

The driving pattern in congested cities can be very slow, compared to the EU driving pattern and expressway driving is of course much faster. Emissions under these conditions are very different from the regulated emissions.

Typical variations in the emissions as a function of the average speed of the driving pattern are seen in Figure 5.15. At very low average speed the emissions increase almost exponentially due to much idling, accelerations and decelerations. At high speed with high engine load, the emissions increase again.

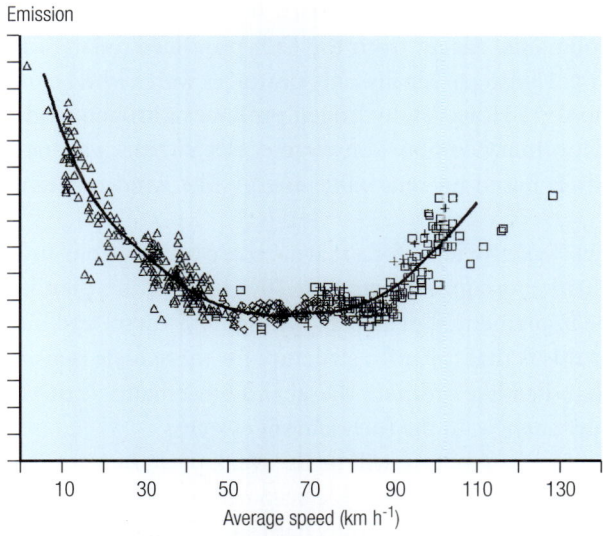

Emission

Average speed (km h⁻¹)

Figure 5.15 Typical variations in emission with average speed. Emission trends are shown vs. average speed for different realistic driving patterns (Jensen, 1992)

Another important factor is the cold start contribution to the emissions. If the vehicle is used for a very short trip, the engine will be cold in a larger part of the driving pattern. This has a great influence since the emissions are higher with cold engine. In the beginning of the era of the catalytic converters this was particularly important, since the CAT needs some time to heat up to the working temperature. In recent years this effect has been smaller, since much effort has been put into reducing the emissions in the cold phase of the engine operation, for example by introducing NO_x adsorbers which can adsorb NO_x during the cold start, and later convert the material when released through the CAT.

Alternative fuels and carbon dioxide

The global resources of crude oil are becoming smaller, and expensive alternative fuels become more compatible with price. At the same time politicians favour a sustainable energy policy. The future will experience introduction of more alternative fuels for transportation. The most relevant alternatives to gasoline and diesel oil seems to be natural gas, biodiesel, methanol, ethanol, DME (DiMethylEther), LPG (Liquefied Petroleum Gas), hydrogen and electricity (which, of course, can be produced from many sources, fossil, nuclear and various alternative energy sources).

Many of these fuels traditionally demonstrate advantages with respect to emissions. However, recent gasoline and diesel engine technology is so clean that this is no longer a strong argument.

If we consider CO_2 emissions, all biomass based fuels can be considered as approaching zero pollutants, except from the CO_2 produced during fuel production and handling. Hydrogen ideally only produces water when combusted and therefore no CO_2. However, hydrogen produces nitrogen oxides when combusted in an ordinary combustion engine, electric cars produces no CO_2 if the electricity is based on renewable energy like wind power or hydropower.

The European Biofuel Directive aims for a market share of 5,75% bio-fuels for transportation in 2010 and commitments for 10% in 2020 are planned (EU, 2003). This will be implemented primarily through bio-diesel and bio-ethanol. The reason for this is that the infra structure for these fuels is available. Biodiesel can be handled like ordinary diesel and bio-ethanol is mixed into gasoline without any changes in the fuel delivery systems.

It is very likely that the infrastructure will be the weak point in introducing new fuels, and therefore the most probable scenario will be a development towards new liquid fuels produced from natural gas (GtL fuels) and coal (CtL), followed by liquid fuels produced from biomass (BtL fuels). These fuels have a low content of aromatic hydrocarbons. Therefore lower PAH emissions and particulate emissions from these fuels are expected.

It is also evident, that hydrogen for fuel-cells will be introduced. The problem with hydrogen is again the lack of infra structure, and a decision on how to get the hydrogen is needed. Fuel-cell driven cars do not produce nitrogen oxides, and therefore the only "pollutant" formed is water.

5.4 Non-Road Vehicles

Off road vehicles

The equipment covered by this group include devices as agricultural and forestry tractors, industrial drilling rigs, compressors, construction wheel loaders, bulldozers, non-road trucks, highway excavators, forklift trucks, road maintenance equipment, snow ploughs, ground support equipment in airports, aerial lifts and mobile cranes. The listed vehicles are mainly equipped with diesel engines, but there are examples of off road vehicles equipped with other type of engines. Like heavy-duty vehicles, the emission regulations are based on engine tests. The emission regulations are listed in Table 5.3.

Stage III A Standards for Nonroad Engines

Cat.	Net Power	Date *	CO	NO_x + HC	PM
	kW			g $(kWh)^{-1}$	
H	$130 \le P \le 560$	2006.01	3.5	4.0	0.2
I	$75 \le P < 130$	2007.01	5.0	4.0	0.3
J	$37 \le P < 75$	2008.01	5.0	4.7	0.4
K	$19 \le P < 37$	2007.01	5.5	7.5	0.6

* dates for constant speed engines are: 2011.01 for categories H, I and K; 2012.01 for category J.

Stage III B Standards for Nonroad Engines

Cat.	Net Power	Date	CO	HC	NO_x	PM
	kW				g $(kWh)^{-1}$	
L	$130 \le P \le 560$	2011.01	3.5	0.19	2.0	0.025
M	$75 \le P < 130$	2012.01	5.0	0.19	3.3	0.025
N	$56 \le P < 75$	2012.01	5.0	0.19	3.3	0.025
P	$37 \le P < 56$	2013.01	5.0	4.7*		0.025

* NOx + HC

Stage IV Standards for Nonroad vehicles

Cat.	Net Power	Date	CO	HC	NO_x	PM
	kW				g $(kWh)^{-1}$	
Q	$130 \le P \le 560$	2014.01	3.5	0.19	0.4	0.025
R	$56 \le P < 130$	2014.10	5.0	0.19	0.4	0.025

Table 5.3 The development of non-road diesel engine emission standards in Europe. (Dieselnet, 2007)

Aircrafts

Two types of engines are generally used for aircrafts: gasoline engines, similar to those used for road transportation and more importantly gas turbines.

In a gas turbine the air is compressed before entering the combustion chamber where the fuel is added. The most common fuel is liquid kerosene. The heated combustion products expand through a turbine, which drives the compressor. The major energy part of the expanding exhaust stream is used for propulsion in a turbojet engine. In a turbofan or turboprop, the turbine uses most of the energy in the hot exhaust to drive a propeller for propulsion.

Emissions from aircrafts are generally treated in two different modes with large differences in emission factors. One set of emission factors apply when the aircraft operates during landing and take-off (LTO) and another during cruise. Table 5.4 shows average emission factors for the different flying modes as given by (Rypdal, 2002) for an average aircraft and for old aircrafts.

Domestic	Fuel	SO_2	CO	CO_2	NO_x	NMVOCs	CH_4	N_2O
LTO (kg/LTO) – Average fleet	850	0.8	8.1	2680	10.2	2.6	0.3	0.1
LTO (kg/LTO) – Old fleet	1000	1.0	17	3150	9.0	3.7	0.4	0.1
Cruise (kg/ton)		1.0	7	3150	11	0.7	0	0.1
International	Fuel	SO_2	CO	CO_2	NO_x	NMVOCs	CH_4	N_2O
LTO (kg/LTO) – Average fleet	2500	2.5	50	7900	41	15	1.5	0.2
LTO (kg/LTO) – Old fleet	2400	2.4	101	7560	23.6	66	7	0.2
Cruise (kg/ton)		1.0	5	3150	17	2.7	0	0.1

Table 5.4 Emission factors for different flying modes (Rypdal, 2002).

Since the start of the jet age aviation fuel consumption is reduced by about 70% (IATA, 2005). This is a consequence of the natural competition between airline construction companies, because the fuel price is important to the economy. This in turn has reduced emissions as well.

The European Commission has proposed that aviation emissions in the future should be included in the EU Greenhouse Gas Emissions Trading Scheme, since this is the most cost-efficient way to reduce emissions (IHS, 2007).

Ships

Engines for ships include a variety of diesel engines from high- to low-speed diesel engines as well as steam and gas turbines. Of special concern for this type of transportation are the emissions of sulphur dioxide, SO_2, nitrogen oxides, NO_x and particulate matter, PM.

The emission of SO_2 is mainly determined by the fuel content of sulphur. In shipping matters it is usual to distinguish between two main categories of fuel: heavy bunker fuel oil (HFO) and marine distillates. The latter is divided into two groups: marine diesel oil (MDO) and marine gas oil (MGO).

Large vessels usually operate on cheap HFO on open sea, but switches to a lighter fuel for their auxiliary engines when they are in a harbour. Marine distillates are then used, as they are for the engines of small vessels (EEB, 2004).

HFO contains several percent of sulphur. The MARPOL (International Convention for the Prevention of pollution from ships) 73/78 Convention, Annex VI from 1997 establishes a sulphur cap of 4.5% for HFO. This convention was the result of the work in the Marine Environment Protection Committee of the UN International Maritime Organization (IMO). The convention also establishes two "sulphur emission control areas", the Baltic Sea and the North Sea, where sulphur in fuel used by ships must be below 1.5%. The 1.5% limit applies to all ships in the Baltic Sea from May 2006 and in the North Sea from 2007.

The MARPOL Annex VI also prescribes emission standards for NO_x for diesel engines with an output of more than 130 kW. But even with the enforcement of the MARPOL Annex VI, the development in the total emissions of SO_2 and NO_x from ships will exceed the total emissions from all land based mobile sources in 2020 in Europe (Figure 5.16) (EEB, 2004).

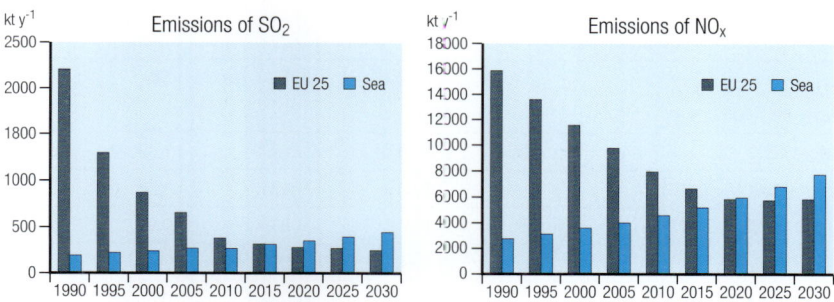

Figure 5.16 Development in emissions of SO_2 (left) and NO_x (right) from land based mobile sources and international shipping around Europe (EEB, 2004).

Particulate emissions from ships are considerable. PM formation is enhanced both by the high polycyclic aromatic hydrocarbon content of the fuel in use and by the high sulphur content.

Polycyclic aromatic hydrocarbons (PAH) emissions from ships are not studied very much, but they are obviously very high due to the high concentrations in the fuel.

A comparison of emissions from different types of cargo transportation, expressed in grams per ton-kilometre is shown in Table 5.5. The comparison relates the emissions from two different means of transportation, heavy truck with trailer and different types of cargo vessels. It is noted that the emissions of SO_2 and PM are much higher with cargo vessels.

An EU strategy for future cleaner ships was adopted by the European Commission in 2002 and has been formulated in directive 2002/80/EC of the European Parliament. It contains a broad series of objectives, proposed actions and recommendations for bringing about emission reductions over a period of 5-10 years. Some essential examples of actions are given in (EEB, 2004):

- International actions within the International Maritime Organization towards reduction of ship emissions. Recommendations of member countries to ratify MARPOL Annex VI.
- Regulations of fuel content of sulphur and development of standards for emissions of NO_x, PM and CO for new non-road engines marketed in EU.
- Economic incentives in a differentiated transport charge policy, which take into account the value of environmental performance.
- Urging the international bunker industry to make significant quantities of HFO with sulphur content of less than 1.5%, and urging port authorities to consider introducing voluntary speed reductions and provide incentives for ships to use land based electricity while in port.

	CO_2	PM	SO_2	NO_x	VOCs
Heavy truck with trailer:					
Before 1990	50	0.058	0.0093	1.00	0.120
Euro 0 (1990)	50	0.019	0.0093	0.85	0.040
Euro 1 (1993)	50	0.010	0.0093	0.52	0.035
Euro 2 (1996)	50	0.007	0.0093	0.44	0.025
Euro 3 (2000)	50	0.005	0.0093	0.31	0.025
Cargo vessel:					
large (>8000 dwt)	15	0.02	0.26	0.43	0.017
medium (2000-8000 dwt)	21	0.02	0.36	0.54	0.015
small (<2000 dwt)	30	0.02	0.51	0.72	0.016
RoRo (2-30 dwt)	24	0.03	0.42	0.66	0.029

*Emissions are average in each case. Trucks: maximum overall weight 40 tons, loading 70 per cent, operating on diesel with a sulphur content of 300 ppm
Cargo vessel: bunker oil with an average sulphur content of 2.6 per cent, no cleaning of NO_x.

Table 5.5 Comparison of emissions from trucks on long hauls with different EU standards for emissions and cargo vessels of various sizes (EEB, 2004). Numbers in grams per ton-kilometre.

In 2007 legislation in EU limits emission factors in Directive 2004/26/EC, which applies to non-road diesel engines. This directive includes standards for inland waterway vessels. As seen in Table 5.6, the emission regulations are phased in from 2007-2009.

Cat.	Displacement (D) dm³ per cylinder	Date	CO	NO$_x$ + HC g (kWh)$^{-1}$	PM
V1:1	D ≤ 0.9, P > 37 kw	2007.01	5.0	7.5	0.40
V1:2	0.9 < D ≤ 1.2		5.0	7.2	0.30
V1:3	1.2 < D ≤ 2.5		5.0	7.2	0.20
V1:4	2.5 < D ≤ 5	2009.01	5.0	7.2	0.20
V2:1	5 < D ≤ 15		5.0	7.8	0.27
V2:2	15 < D ≤ 20, P ≤ 3300 kW		5.0	8.7	0.50
V2:3	15 < D ≤ 20, P > 3300 kW		5.0	9.8	0.50
V2:4	20 < D ≤ 25		5.0	9.8	0.50
V2:5	25 < D ≤ 30		5.0	11.0	0.50

Table 5.6 Emissions standards for inland waterway vessels. (Dieselnet, 2007)

Trains

Trains are generally either powered by electric engines or diesel engines. Diesel fuelled trains have engines similar to trucks. EU emission factors for these categories of vehicles have been given in (Colls, 2002) and shown in Table 5.7. Also given are predicted emission factors for 2020.

Pollutant	Electrical power generation g (kWh)$^{-1}$		Diesel locomotive emissions g (kWh)$^{-1}$	
	1998	2020	1998	2020
SO$_2$	2.7	0.8	1.0	0.03
NO$_x$	1.2	0.35	12	3.5
HC	1.1	0.55	1.0	0.50
CO	0.08	0.04	4.0	0.50
PM	0.14	0.07	0.25	0.08

Table 5.7 Estimated emission factors for trains (Colls, 2002)

No regulations have been applied to this category of vehicles before 2006. However, in 2006 legislation in EU gives emission factors in Directive 2004/26/EC, which applies to non-road diesel engines. As seen in Table 5.8 the emission regulations are phased in from 2006-2012.

Cat.	Net Power	Date	CO	HC	HC + NO$_x$	NO$_x$	PM
	kW				g (kWh)$^{-1}$		
RC A	130 < P	2006.01	3.5		4.0		0.2
RL A	130 ≤ P ≤560	2007.01	3.5		4.0		0.2
RH A	P > 560	2009.01	3.5	0.5*		6.0*	0.2

*HC = 0.4 g (kWh)$^{-1}$ and NO$_x$ = 7.4 g (kWh)$^{-1}$ for engines of P > 2000 kW and D > 5 liters/cylinder

Cat.	Net Power	Date	CO	HC	HC + NO$_x$	NO$_x$	PM
	kW				g (kWh)$^{-1}$		
RC B	130 < P	2012.01	3.5	0.19		2.0	0.025
R B	130 < P	2012.01	3.5		4.0		0.025

Table 5.8 Emissions standards for Rail Traction Engines in EU. (Dieselnet, 2007).

5.5 Concluding Remarks

The introduction of more stringent emission regulations for road vehicles, especially since the late 1980s, has led to a dramatic technological development in diesel and gasoline engine technology as well as in fuel technology. Diesel engines have become more frequent due to their better fuel economy. At the same time the diesel engines and exhaust after treatment equipment have undergone a dramatic technological improvement. Especially the latter has led to demand for higher fuel quality, since after-treatment often requires a "cleaner" fuel – low sulphur content being the most important factor.

Diesel engines needed to be improved in order to be comparable to new gasoline engines, which have low particulate emissions and high engine output. Therefore the introduction of high performance fuel systems, based on very high injection pressures. This leads to a better atomisation of the fuel and thereby to a better fuel combustion. Further improvements have been seen in electronic control, both in combination with the engine control and exhaust after treatment control. Single components, like turbo chargers, have been improved as well.

5.6 Literature

AECC (2006): *The Association for Emissions Control by Catalyst*, www.aecc. be

Colls, J. (2002): *Air Pollution* Clay's Library of Health and the Environment, Spon Press, London.

Dieselnet (2007): www.dieselnet.com.

EEB (The European Environmental Bureau) **(2004):** *Air Pollution from Ships*, Briefing Document.

European Union (2003): *Directive 2003/30/EC. The Biofuel Directive.*

Haynes, B.S. and Wagner, H.G. (1981): *Soot Formation* Progress in Energy and Combustion Science 7, 229-273, Pergamon Press Ltd.

IATA (International Air Transport Association) **(2005):** *Policy Options for Addressing Aircraft Emissions,* www.iata.org/whatwedo/environment/ emissions_policy.htm.

IHS (2007): *EC Proposes Bringing Air Transport into EU Emissions Trading Scheme*, http://aero.ihs.com/news.

Jensen, S.S. (1992): *Driving Patterns and Air pollution – In Provincial Areas.* Danish Road Administration, 1992 (In Danish).

Johnson, J.H. et. al. (1994): *A Review of Diesel Particulate Control Technology and Emission Effects.* Society of Automotive Engineers International, paper no. 940233.

Rypdal, K. (2002): *Aircraft Emissions".* Proceedings of IPCC (Intergovernmental Panel on Climate Change) Expert Meeting on Good Practice Guidance and Uncertainty Management in National Intergovernmental Management in National Greenhouse Gas Inventories, pp 93-102.

Schramm, J. et al. (1994): *The Emission of PAH from a DI Diesel Engine Operating on Fuels and Lubricants with Known PAH Content.* Society of Automotive Engineers International Special Publication (1028): Diesel Combustion Processes, pp. 1-16, paper no. 940342.

6 Agricultural sources

Carsten Ambelas Skjøth

Agricultural activities emit a series of pollutants from machinery, vehicles, heating, manure and soil management etc. In countries such as Denmark, the large agricultural sector is one of the large green house gas emitters, due to emission of CO_2 and the more powerful climate gases methane (CH_4) and nitrous oxide (N_2O) – see Chapter 20. Emissions originate directly from machinery and heating, but also from manure management as well directly from the soils. Agriculture also emits pollutants such as pesticides where a large number of substances are used in modern agriculture. However, the most characteristic agricultural pollutant is ammonia.

Ammonia emissions have a large environmental impact in Europe with respect to PM formation in the atmosphere and nitrogen deposition to natural surfaces. On local scale ammonia deposits very fast to most surfaces and may have an effect on ecosystems close to the sources. Within the local area around large animal farms, the atmospheric load to ecosystems can be around 100 kg ha^{-1} y^{-1} (Sutton *et al.*, 2003), depending on the distance to the source as well as the source strength. However, the alkaline NH_3 is also efficient for incorporation into acidic aerosols forming secondary atmospheric components containing NH_4^+. Components like ammonium-hydrogen-sulphate, NH_4HSO_4, ammonium-sulphate, $(NH_4)_2SO_4$ and ammonium-nitrate, NH_4NO_3 are quickly converted into aerosols. The mass size distribution of these nitrogen (N) containing aerosols has its peak in the accumulation mode where the dry deposition velocity is very low (Hov and Hjøllo, 1994). Therefore, the main removal path from the atmosphere of the N containing aerosols is through wet deposition (see Chapter 10). Thus, NH_3 and its reaction products can either be deposited near the emission source or in some cases transported more than 1000 km (Hov et al., 1994) before wet deposition takes place.

6.1 Sources of ammonia

The largest sources of ammonia is to air primarily related to agricultural activities (Bouwman *et al.* 1997), which in the western countries constitute from 85% to 100% (Anderson *et al.*, 2003, Sutton *et al.*, 2000). For the Danish area approximately 98% is related to agricultural production, where the remaining 2% is emitted from petrol cars with catalysts.

Emission of ammonia to the atmosphere is a physical process taking place from wet surfaces (Elzing and Monteny, 1997). The process is highly temperature dependent and varies therefore during the day and over the seasons (e.g. Gyldenkærne *et al.*, 2005, Ambelas Skjøth *et al.*, 2004). Furthermore, as the emission of ammonia is related to different agricultural activities and sources throughout the year the overall variation is highly variable and depends on location and time of the year due to meteorology and different production methods within the different regions. Here is discussed various agricultural sources of ammonia from agriculture.

• Point sources such as animal stables and manure stores
• Manure and mineral fertilisers
• Grazing animals
• Other sources: plants

Animal stables, barns and manure stores

Groot Koerkamp *et al.* (1998) found a high variation in ammonia emission between animal species and between stable types. However, emission variation is primarily related to the temperature variations within the stables. Pigs and poultry have an elevated lower critical temperature compared to cattle, and consequently stables in Northern Europe are ventilated and heated during winter. Annual emission from manure stores primarily reflects storage method and the temporal variation reflects ambient temperatures. As a consequence all buildings in warm areas and warm periods have a temporal variation reflecting outside temperatures. However, stables for pig and poultry will during cold periods therefore continue to emit ammonia at a high level contrary to the emission rate from cattle barns and storage facilities, where the emission will be considerably reduced. Based on a survey by Seedorf *et al.* (1998a, 1998b) a simple connection between indoor and outdoor temperatures can be established (Gyldenkærne *et al.*, 2005) and the temporal variation of ammonia emission thus be calculated. This is clearly illustrated in Figure 6.1, which shows the calculated temporal variation of ammonia emission in the Tange area (Denmark).

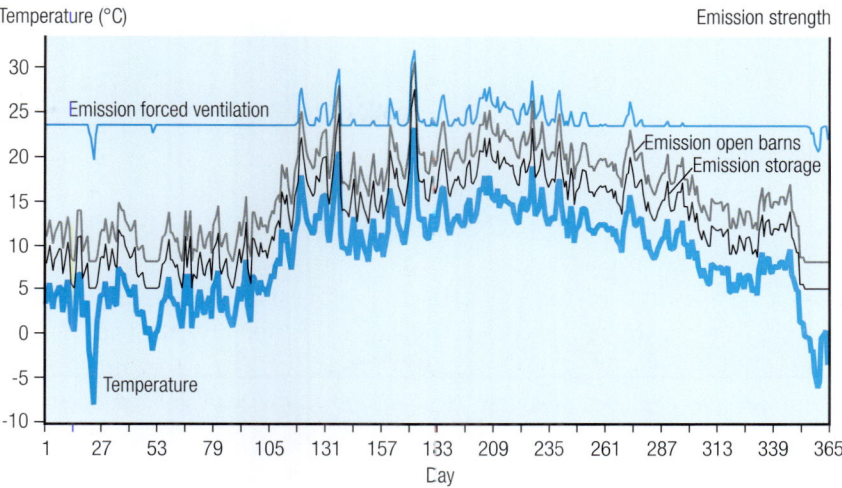

Figure 6.1 Calculation of emission strength from the three typical agricultural sources: heated animal stables with forced ventilation, open barns and manure storage. Calculations are based on the model by Gyldenkærne et al. (2005) and measured daily mean temperatures

Manure and mineral fertilisers

Ammonia emission from application of manure or fertiliser takes place at distinct times of the year with relatively short duration compared to emission from the point sources. The emission varies greatly between application methods ranging from broad spreading methods with large emission to soil injection methods with very low emission. Furthermore national regulations govern application time. In some countries almost no regulations are present, whereas in other countries, such as Denmark, application during wintertime is forbidden. Finally, farmers usually apply fertiliser to the crops in the beginning of the growth season. A result is that farmers in the southern part of Europe initiate application earlier, than farmers in Northern Europe. In the autumn, storages are emptied making space within the storage to the accumulated manure during winter time. Depending on national regulations and local production methods the application can be divided into several periods. Figure 6.2 shows the temporal variation and amount of emission from four typical application times within the Danish area Tange for the year 2000.

Similar variations of ammonia emissions are also found from application of mineral fertiliser (Gyldenkærne et al. 2005).

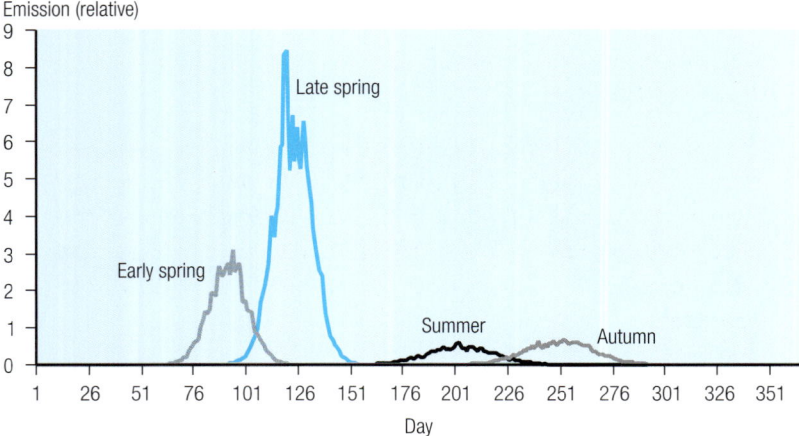

Figure 6.2 Emission strength and temporal variation of ammonia emission from application of manure during springs, summer and autumn in the Tange area (Denmark).

Grazing animals

Emission from grazing animals depends mainly on two parameters: Outside temperature and amount of time in the field. In the southern part of Europe, animals are in the field most of the year, whereas in Northern Europe the animals are within stables approximately half of the year. The grazing time varies between species, where sheeps are found in the field most of the time compared to dairy cattle, which in many countries are inside stables approximately half of the year (see e.g. the RAINS database, http://www.iiasa. ac.at/rains/index.html). Therefore, during winter the emission patterns from grazing animals approximates the patterns from open barns: low emission emitted as point sources, compared to summer emission patterns, where the emission approximates the pattern from manure storages but emitted as an area source due to the large grazing field areas.

Other sources

Fertilised legumes and plants are believed to emit ammonia to the atmosphere (e.g Larsson *et al.*, 1998). The emission depends on the enrichment of the apoplast with NH_4^+, and the compensation point (Farquehar *et al.*, 1980), which again depends on the plant status including growth, stress etc. Therefore, this type of emission is still highly unknown, with respect to amount, temporal and spatial variation. For the Danish area rough estimates are 15% of the total emission of NH_3 to air (Ambelas Skjøth *et al.*, 2006).

6.2 Spatial and temporal variation of emission

Variation over Europe

On the European scale ammonia emission data from agriculture is collected by the organisations EMEP (http://www.emep.int) and CORINAIR. Data reported by individual countries on either 50 km by 50 km or national level. Table 6.1 shows the annual emission of ammonia to the atmosphere from a selected number of European countries for the years 1985, 1990, 1995 and 2000, respectively.

	1985	1990	1995	2000
Denmark	113	109	92	83
Germany	706	630	523	514
Netherlands	204	187	153	126
France	642	628	624	649
Sweden	44	42	50	46
Norway	19	19	21	21
Finland	35	31	29	27

Table 6.1 Annual emission of ammonia (ktons N) from selected European countries during the period 1985 to 2000.

From Table 6.1, it is seen that the large countries Germany and France are large emitters of ammonia compared to other countries. However, it is also seen that the relatively small countries Denmark and Netherlands have larger national ammonia emissions compared to the large Scandinavian countries, Sweden, Norway and Finland. Thus there is a significant regional difference between high and low emitting areas. This can clearly be seen in Figure 6.3 (Gyldenkærne *et al.* 2005).

Figure 6.3 shows that the total annual ammonia emission has a non-uniform distribution throughout Europe. Areas with very large emission rates are found in parts of Germany, Belgium, the Netherlands and the Po-Valley in Italy. Moderate levels are found in most countries and very low levels are found in most of the Scandinavian countries.

As already discussed in previous chapters, the temporal distribution of ammonia emission varies throughout Europe due to climatic differences as well as agricultural practices and national regulations. This is exemplified in Figure 6.4 showing the variation in the Danish Tange area of the accumulated ammonia emission from all agricultural sources for the year 1989 (Ambelas Skjøth *et al.* 2008). In 1989 almost no regulations were present.

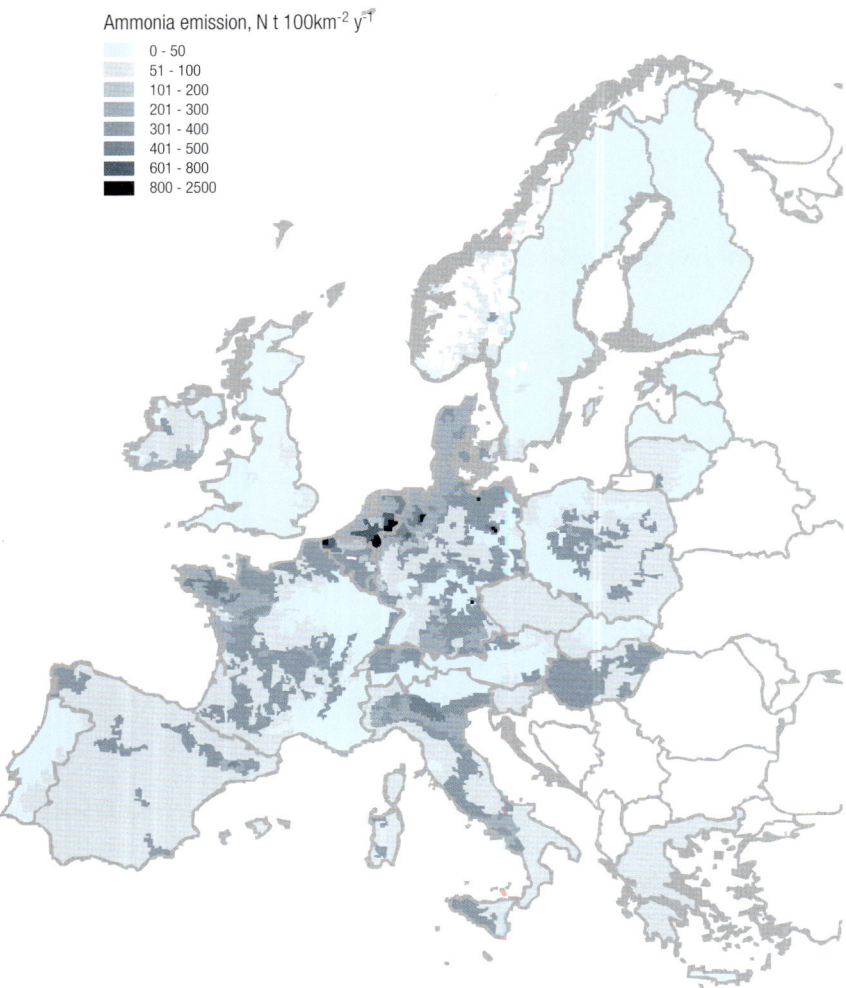

Figure 6.3 Annual ammonia emissions in Europe.(Gyldenkærne *et al.* 2005).

During winter the emission is relatively low due to low activity and low temperatures. The manure, which has accumulated during winter, was applied to crop fields in the spring and grass fields in summer. Finally, the manure storages were emptied during autumn, represented by the large peak shown in Figure 6.4. This temporal variation can be considered a typical variation for areas in Northern Europe with moderate to large agricultural activity and limited national regulations.

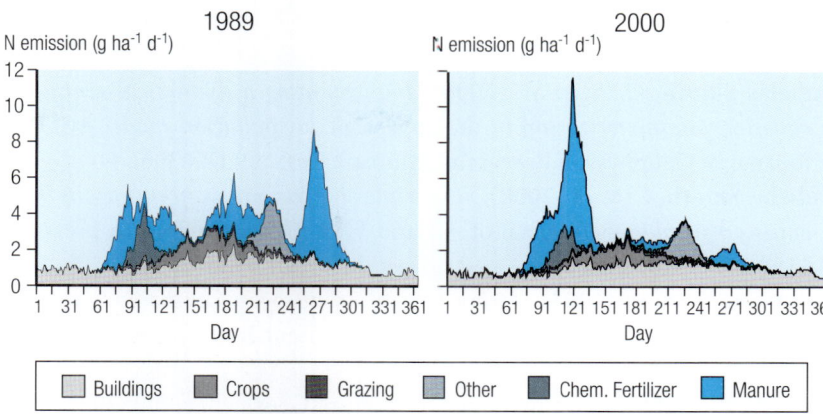

Figure 6.4 Temporal variation in daily ammonia emission from different ammonia sources in the Danish area Tange for the years 1989 and 2000. (Ambelas Skjøth et al., 2008).

During 1990s a vast number of regulations were initiated to reduce air and soil pollution from Danish agriculture (Grant et al. 2004, Ambelas Skjøth et al. 2008). Among these reduction methods was improved efficiency within the entire production chain with respect to emitted ammonia. This has forced the farmers to increase the percentage of applied animal manure during spring and decrease percentage of the summer and autumn applications. This is also seen in Figure 6.4. The relatively large spring emissions have increased compared to the 1989 level and the summer and autumn level has decreased. Overall the amount of lost ammonia to the atmosphere has been reduced significantly for the Danish area (see Table 6.1) during that period, even though the farmers have managed to increase production during the same time (Ambelas Skjøth *et al.* 2008).

In Europe regulations have been introduced which have controlled emissions from agriculture, although large variations exists throughout the countries (Table 6.1). The regulations are carried out through agreements such as the Gothenburg protocol. Contrary in the United States where agriculture is hardly regulated. The US Environmental Protection Agency (EPA) has not applied federal regulations to agriculture. This may be one of the reasons, that ammonia emissions in intensive husbandry regions in the United States are particular problematic (Aneja et al., 2008)

6.3 Literature

Ambelas Skjøth, C. A. et al. (2004): *Implementing a dynamical ammonia emission parameterization in the large-scale air pollution model ACDEP:* Journal of Geophysical Research- Atmospheres, 109 D06306,.

Ambelas Skjøth, C. et al. (2008): *Footprints on ammonia concentrations from emission regulations.*Journal of Air and Waste Management, In press 58, 1158-1165

Anderson, N. et al. (2003): *Airborne reduced nitrogen: ammonia emissions from agriculture and other sources:* Environment International, *29*, 277-286.

Aneja, V. P., (2008): Farming pollution. Nature Geoscience, July 2008, 409-411.

Bouwman, A. F. et al. (1997): *A global high-resolution emission inventory for ammonia:* Global Biogeochemical Cycles, *11*, 561-587.

Elzing, A., and Monteny, G. J. (1997): *NH3 emission in a scale model of a dairy-cow house,* Am. Soc. Agric. Eng., *40*, 713–720.

Farquhar, G. D. et al. (1980): *On the Gaseous Exchange of Ammonia between Leaves and the Environment – Determination of the Ammonia Compensation Point:* Plant Physiology 66, 710-714.

Grant, R. and Blicher-Mathiesen, X. (2004): *Danish policy measures to reduce diffuse nitrogen emissions from agriculture to the aquatic environment,* Water Science and Technology *49*, 91-100,.

Groot Koerkamp, P.W.G., et al. (1998): *Concentration and emissions of ammonia in livestock buildings in northern Europe,* J. Agric. Eng. Res., *70*, 79–95.

Gyldenkærne, S. et al. (2005): *A dynamical ammonia emission parameterization for use in air pollution models:* Journal of Geophysical Research-Atmospheres, *110*, D07108.

Gyldenkærne S. et al. (2005): *A high resolution ammonia emission inventory for regional scale air pollution.* Proc. 1st ACCENT Symposium, Urbino, Italy, 12th-16th September 2005.

Hov, O. and Hjollo, B.A., (1994): *Transport Distance of Ammonia and Ammonium in Northern Europe .2. Its Relation to Emissions of SO$_2$ and NO$_x$:* Journal of Geophysical Research-Atmospheres, *99*, 18749-18755.

Hov, O. et al. (1994): *Transport Distance of Ammonia and Ammonium in Northern Europe .1. Model Description:* Journal of Geophysical Research-Atmospheres, *99*, 18735-18748.

Larsson, L. et al. (1998): *Ammonia and nitrous oxide emissions from grass and alfalfa mulches,* Nutr Cycles Agroecosyst., *51*, 41–46.

Seedorf, J. et al. (1998a): *A Survey of Ventilation Rates in Livestock Build-*

ings in Northern Europe: Journal of Agricultural Engineering Research, *70*, 39-47.

Seedorf, J. et al. (1998b): *Temperature and Moisture Conditions in Livestock Buildings in Northern Europe*: Journal of Agricultural Engineering Research, *70*, 49-57.

Sutton, M.A. et al. (2000): *Ammonia emissions from non-agricultural sources in the UK*: Atmospheric Environment, *34*, 855-869.

Sutton, M.A. et al. (2003): *Establishing the link between ammonia emission control and measurements of reduced nitrogen concentrations and deposition*: Environmental Monitoring and Assessment, *82*, 149-185.

7 Natural Sources

Gitte Brandt Hedegaard

A large variety of natural emitting sources exist which can influence the global radiation balance and thus the climate systems. Through respiration and photosynthesis plants e.g. emit many different Volatile Organic Compounds (VOCs). The type and amount of the chemical species emitted depend strongly on plant type, growth conditions (e.g. nutrition, local climate, etc) and season. Another main contributor to the natural emitted species is the degradation and bacterial processes going on in soil. By these processes species like nitrous oxide (N_2O), nitrogen oxides (NO_x) and methane (CH_4) are emitted to the atmosphere. Methane is particularly interesting because of its capacity as a greenhouse gas giving rise to many different feedback mechanisms in the climate system. Also the ocean is acting both as a chemical sink and source for the atmosphere. Carbon dioxide (CO_2) e.g. is stored in large amounts in the ocean. Cool waters are a large sink of CO_2 in contrast to the equatorial warm waters, which release CO_2 to the atmosphere.

With respect to emissions in general most concerns regard the human-made emissions from e.g. traffic, power plants and industry. However, the natural sources contribute significantly to the total emissions of chemical species to the atmosphere, and their output participates in the chemistry of the atmosphere. In the following some of the most important naturally emitted species will be discussed and a brief overview of the problems in modelling natural emissions is given.

7.1 Volatile Organic Compounds – VOCs

Volatile Organic Compounds (VOCs) are defined as all gaseous organic compounds. Methane is included as a VOC. However, here the text will concentrate on the non-methane VOCs (NMVOC), which are highly reactive species in the atmosphere. These NMVOCs are split into two main groups – the anthropogenically emitted NMVOCs and the biogenically emitted NMVOCs. The most prominent biological NMVOCs are isoprenoids (isoprene and monoterpenes) but also alkanes, alkenes, carbonyls, alcohols, esters, ethers and acids are included (Kesselmeier and Staudt, 1999). Volatile Organic Compounds are relatively short-lived in the atmosphere and they participate in a large number of chemical reactions especially in the lower part of the troposphere. In table 7.1 estimates are shown of the lifetime against reaction with OH, O_3 and NO_3 together with the average atmospheric concentrations. Isoprene is by far the most important biological NMVOC. Global biogenic emissions are estimated to be 220-750 Tg y^{-1} that is several times greater than the total anthropogenic emissions of VOCs (Millet et al., 2008).

BVOC	Lifetime against OH	Lifetime Against O_3	Lifetime against NO_3	Atmospheric Concentration
Isoprenoids				ppt to several ppb
Isoprene	1.4 h	1.3 days	1.6 h	
α-Pinene	2.6 h	4.6 h	11 min	
β-Pinene	1.8 h	1.1 day	27 min	
Limonene	49 min	2.0 h	5 min	
α-Terpinene	23 min	1 min	0.5 min	
Sesquiterpenes				Not detectable due to high reactivity
β-Caryophyllene	42 min	2 min	3 min	
Oxygenates				
Acetone	61 days	>4.5 years	>8 years	2-30 ppb
Methanol	12 days	>4.5 years	2.0 years	2-30 ppb
2-Methyl-3-buten-2-ol	2.4 h	1.7 days	7.7 days	1-3 ppb

Table 7.1 Lifetimes and average concentrations of biological NMVOC, data from (Atkinson and Arey, 2003) and from (Kesselmeier and Staudt, 1999)

The amount and type of VOC emitted from the different plants vary a lot. As an example almost half of the global isoprene emissions originate from tropical broadleaf trees because these trees are often exposed to conditions, which are highly conducive for isoprene emissions (Guenther et al., 2006). In contrast monoterpenes are mainly emitted from coniferous trees. The main isoprene emissions originate from "woody" plants like trees and scrubs and only minor amounts stems from other plants like grass and crops. Besides plants, animals (including human), microbes and aquatic organisms contribute with a minor amount (less than 10%) of the total isoprene emission (Guenther et al., 2006). The weather conditions especially temperature, availability of light, the age and location of the leaf in the canopy, the availability of nutrients and finally the chemical composition of the atmosphere are other determining factors for the size of the isoprene emission (Guenther et al., 2006). As an example, isoprene emissions increase significantly with increasing temperatures. Since isoprene enhances the production of ozone in the troposphere, the emissions and the dependency on temperature makes isoprene a very important specie with respect to global warming. Additionally, recent studies indicate that e.g. monoterpenes contribute to the formation of secondary organic aerosols (Atkinson and Arey, 2003), which influences the radiation balance of the atmosphere probably as a cooling effect.

7.2 Nitrogen containing compounds

Nitrogen cycles appear in the atmosphere with different nitrogen containing compounds and in the different compartments. The key nitrogen containing compounds include nitrogen oxides (NO_x), ammonia (NH_3) and the ammonium ion (NH_4^+), also called reduced nitrogen (NH_x), and finally the greenhouse gas nitrous oxide (N_2O).

NO_x

The term NO_x includes nitric oxide (NO) and nitrogen dioxide (NO_2) which are both very important compounds with respect to atmospheric chemistry. The concentration of NO_x in the atmosphere strongly influences the chemistry in the atmosphere especially regarding the ozone production. NO is naturally emitted into the atmosphere from processes in the soil. The highest emissions are over agricultural and grasslands, forest poses moderate NO emissions, whereas wetlands and tundra emit rather low NO amounts. NO emissions from soil depend on many parameters like e.g. soil texture, temperature, pH and nitrate levels. In general, the emission increases with increasing temperature and increasing nitrate content of the soil (Guenther et al., 2000). The nitrate content is obviously strongly dependent on the use of nitrogen fertiliser. With respect to the issue of natural and anthropogenic part of the emissions, the individual contributions can be difficult to distinguish. Since the largest NO contributions from soil originate from farming, one could argue that the source is anthropogenic. On the contrary, it is natural processes in the soil that leads to the emissions. Also the use of nitrogen-based fertilisers raises the discussion of anthropogenic versus natural effect and emissions. The most important point here is always to define, what is considered natural and what is concidered anthropogenic in any particular study or experiment. A common definition is to consider the emissions originating from soils under natural vegetation as natural emissions.

Another source of NO_x in the atmosphere is lightning and cosmic radiation (Seinfeld and Pandis, 2006). Molecular nitrogen is – by lightning or cosmic rays – dissociated and reacts with oxygen to form NO (Guenther et al., 2000, Seinfeld and Pandis, 2006). In some cases lightning leads to wildfires. Biomass burning, which wildfires are a subset of, is another source of NO_x in the atmosphere. Emissions from wildfires are discussed in section 7.4 in more detail. Finally, conversion by oxidation of NH_3 in the atmosphere is partly a natural source of NO_x.

Reduced nitrogen – NH_x

The reduced nitrogen (NH_x) consists of the two species ammonia (NH_3) and the ammonium ion (NH_4^+). Ammonia is naturally emitted from soils, animal waste and the ocean. In the atmosphere the ammonium ion is a result of absorption of ammonia into water droplets. The residence time of NH_3 in the atmosphere is rather short – approximately 10 days and therefore the spatial variation in concentration is rather large and depends strongly on the proximity of a source-rich region (Seinfeld and Pandis, 2006).

Application of NH_3 in fertiliser and manure leads to an emission flux from the soil to the atmosphere. These fertilisers can be inorganic chemicals or application of animal manure. The emission of ammonia from animal waste originates from decay of urea and uric acid in animal manure (Simpson et al., 1999). Since manure from domestic animals often is collected and kept liquid for longer periods, the dominating source becomes anthropogenic. However, also areas where wild animals live closely together give rise to a significant natural source of ammonia. As an example large bird breeding colonies at small islands could be mentioned (Simpson et al., 1999).

The ocean comprises a significant source (and sink) for ammonia in the atmosphere. Ammonium is biologically produced by zooplankton in the ocean. The size and direction of the ammonia flux depend on both the ammonia content of the seawater and the atmosphere (Simpson et al., 1999). According to the newly released AR4 IPCC report (Denman et al., 2007) three quarters of the natural global source of atmospheric ammonia originates from the oceans.

Nitrous oxide – N_2O

Nitrous oxide (N_2O) is like the other nitrogen containing compounds mainly emitted from soil and waters (ocean and fresh water reservoirs like estuaries, rivers etc.). Tropical soils are by far the largest contributors to the total natural emitted amount of N_2O and the majority of these sources is situated in the Northern Hemisphere (Seinfeld and Pandis, 2006, Denman et al., 2007).

Nitrous oxide is uniformly distributed in the troposphere and have a chemical lifetime of approximately 120 years. Thus N_2O is a relatively stable compound in the troposphere, and its largest sink is transport into the stratosphere, where photo-dissociation and oxidation occur (Seinfeld and Pandis, 2006). N_2O is a greenhouse gas and it is the fourth largest positive radiative forcing element in the global warming (Denman et al., 2007).

7.3 Methane – CH_4

Natural wetlands is the largest single source of CH_4 emissions in the atmosphere (Denman et al., 2007). Other major anthropogenic and natural sources for CH_4 are the ocean, ruminants, termites, rice cultivation, biomass burning, landfills and fossil fuel production. As mentioned it can be difficult to distinguish between natural and anthropogenic sources. As an example, CH_4 emissions from landfills originate from bacterial degradation of the organic material. One could argue that this is a natural process and thereby a natural source of methane. However, the landfills exist due to anthropogenic activities. The same issue is maintained with respect to rice cultivation, which is a result of human activity, but the emissions have natural sources. In addition phenomena like biomass burning consist of both an anthropogenic and a natural component (wildfires) and here the individual contributions are very difficult to determine. In Figure 7.1 the natural sources are defined to include: wetlands, termites, the ocean, hydrates, geological sources and wildfires. The total natural CH_4 emission in this study is 145 Tg y^{-1} which corresponds to about 30% of the total annual CH_4 emissions.

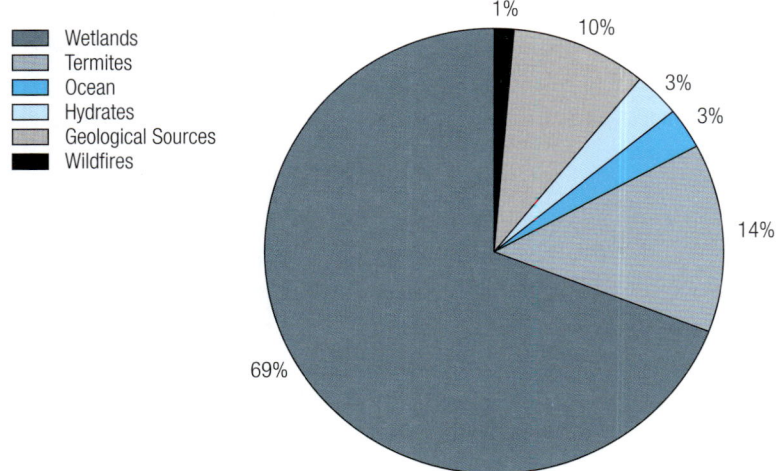

Figure 7.1 Percentage distribution of the average annual natural methane emissions. The diagram is based on data from the 4AR IPCC report 2007 (Denman et al., 2007).

In addition to the natural sources already mentioned, growing tropical trees have recently been identified as a significant contributing source to the natural CH_4 emissions. The global contribution is in different studies ranging from 10-262 Tg y^{-1} out of a total CH_4 emission of 503 Tg y^{-1} (Denman et. al., 2007).

Generally, the emission of CH_4 depends on both temperature, precipitation, and the amount of water stored in the soil. Moreover, CH_4 is a strong greenhouse gas and possesses both direct and in-direct effects on the climate system. Therefore also methane concentration and emissions play important roles with respect to global warming.

7.4 Other naturally emitted species

This section concentrates mostly on aerosol species emitted naturally from wildfires, volcanoes, dust storms etc.

Sulphur containing compounds

The principal sulphur containing compounds in the atmosphere are H_2S, CH_3SCH_3, CS_2, OCS and SO_2 (Seinfeld and Pandis, 2006). Every year large amounts of these species are emitted to the atmosphere. The total annual emitted sulphur (S) is in the range 98-120 Tg y^{-1} and out of this the anthropogenic sources account for approximately 75% and the natural sources for the remaining 25% (Seinfeld and Pandis, 2006).

As a natural emitting sulphur source volcanoes account for 43% of the total natural S flux (Andres, 2000). Another very large single source is the ocean where the naturally formed dimethylsulphide (DMS) occurs. The DMS cycle is described in Section 7.5.

The sulphur burden from volcanic eruptions is often rather large, since the volcanoes usually are situated in cleaner and more remote areas, where the saturation point for radiation balance for sulphur is not yet reached. General circulation models predict that the direct and indirect radiative forcing due to volcanic emitted sulphur is comparable in size to the radiative forcing due to anthropogenic emitted sulphur (Andres, 2000). Further volcanoes are often powerful enough to emit species up to the higher troposphere and even into the stratosphere, where the degradation processes are much slower. Therefore it can take up to two years for the atmosphere to return to its natural state after a large volcanic eruption. A point source like a volcano can result in hemispheric or even global perturbations of the chemical composition and the radiative properties of the atmosphere.

Mineral dust and sea salt

The size of the emission of mineral dust and sea salt is highly dependent on meteorological parameters like wind speed, stability of the atmosphere, friction, air temperature and humidity. In the case of sea salt also the sea surface condition and sea surface temperature are determining factors for the amount of sea salt being emitted. Mineral dust emissions depend on the humidity of the surface and the chemical and physical composition of surfaces.

Desert areas are very large contributors to the global aerosol emission budget. They are very dry and usually un-vegetated and the particle size distribution in the area is rather wide and includes small particles, which can be transported far away from the source. As an example, it is not unusual to observe Saharan dust particles in the summertime in Northern Europe. Examples of significantly reduced visibility across the Atlantic Ocean in Miami have also occurred several times.

Sea salt is a key aerosol in the marine boundary layer. Sea salt is mainly produced over the open ocean by bursting bubbles made by wave activity. A more local source of sea salt occurs in the surf-zone in coastal areas. The emission from the surf-zone can have dominating effects up to 25 km from the coast (Athanasopoulou et al., 2008). Sea salt forms particles acting as cloud condensation nuclei resulting in cloud and precipitation formation. Secondly, sea salt functions as a sink for some reactive gases and small particles and it is responsible for a large fraction of the non-sea salt sulphate formation (Denman et. al., 2007).

Other species

Earlier it was discussed how lightning can produce NO_x species in the atmosphere. Another secondary effect of lightning is the possible initiation of wildfires. Wildfires are a subset of biomass burning and the major emission products are CO_2 and water vapour (Simpson et al., 1999). A large variety of gas-phase species and aerosols like carbon monoxide (CO), molecular hydrogen (H_2), nitrogen- and sulphur-compounds are usually also emitted from biomass burning. The single largest cause of fires in Europe is arson and accidents and is therefore defined as an anthropogenic source. In the populated areas of Russia the same applies. 70% of all fires occur within 5 km of a road and 60% within 10 km of populated areas, while lightning induced fires in the European part of Russia are estimated to account for only 3% of the total fires. In comparison to the more remote areas like the Asian part of Russia, lightning accounts for over 50% of induced fires (Simpson et al., 1999).

In temperate regions tropospheric ozone production tends to be NO_x-limited during the warmer months. However, it is also during these periods the highest frequency of lightning induced fires occurs and therefore the largest NO_x emissions from wildfires take place. For this reason lightning could be a significant precursor to tropospheric ozone production.

Another example of the consequences of the species emitted from biomass burning is the emission of molecular hydrogen. Molecular hydrogen leads to two primary effects. Increased concentration of H_2 results in a lowering of the global oxidation capacity of the atmosphere, since H_2 is a sink for OH in the atmosphere. Presently, molecular hydrogen comprises 5-10% of the global average OH sink (Denman et al., 2007). The increased formation of water vapour can also lead to increased cirrus cloud formation and therefore to an increase in the presence of polar stratospheric clouds (PSC). An increase in PSC can initiate cooling of the stratosphere and more efficient ozone depletion (Denman et al., 2007). See also Chapter 19.

7.5 Natural cycles

The atmosphere is an integral part of the climate system with various natural cycles. They originate from natural mechanisms keeping a given compound in balance by self-regulating processes. The hydrological cycle that is described in Chapter 1 is a good example of this kind of mechanism. Other examples are the carbon cycle, nitrogen cycle or the sulphur self-regulating cycle based on the compound Di-Methyl-Sulphide DMS, the so-called DMS cycle. All these cycles are important sub-mechanisms of the climate system and are therefore important with respect to climate change. Details about the carbon cycle can be found in Chapter 20. In this section an example of the natural cycles are given by summarising the processes in the DMS cycle.

The DMS cycle

The main natural source of tropospheric sulphur on a global scale is the biological production of the volatile compound dimethylsulphide (DMS) contributing 23% of the total sulphur emissions while 7% originate from volcanic eruptions (Simó, 2001).

In the ocean DMS is produced by marine phytoplankton. By several sea-air exchange processes this DMS is emitted into the atmosphere and converted to gas-phase. In the atmosphere oxidation takes place and sulphate particles are created. These particles alter the radiation balance by reducing the amount of in-coming solar radiation at the earth surface. Sulphate particles pose both a direct and an in-direct effect on the radiation balance. The particles scatter the in-coming solar energy directly and are indirectly acting as cloud condensation nuclei, which affect the cloud albedo. The resulting change in the cloud albedo and the scattering effects implies a temperature perturbation, which potentially affects the marine biosphere and thereby the concentration of DMS (Kloster et al., 2006). Again, a natural cycle is finalised. The DMS cycle is sketched in Figure 7.2.

7.6 Modelling the natural emissions

In general, naturally emitted species account for a substantial part of the totally emitted species in the atmosphere. Many of these species are also strongly influencing the atmospheric chemistry by altering the composition and physical properties of the atmosphere. In addition, many of the emission processes are highly sensitive to changes in the various meteorologi-

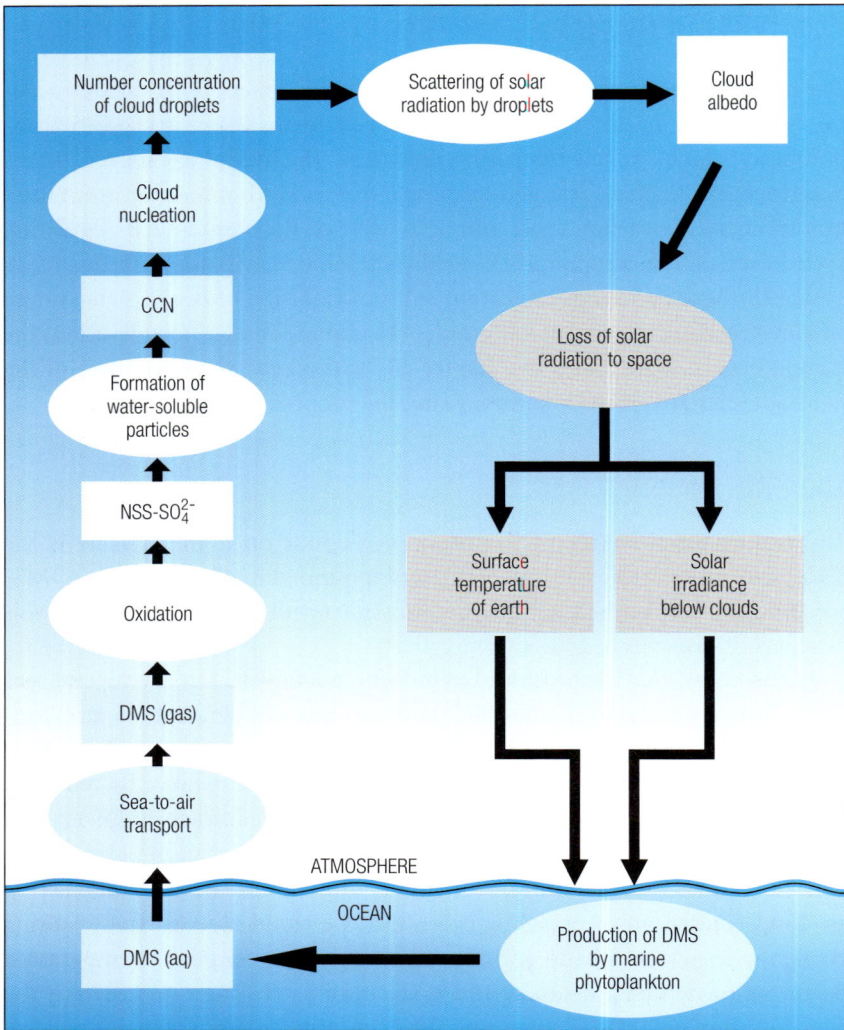

Figure 7.2 The Di-methyl-sulphide (DMS) cycle.
Based on http://me-www.jrc.it/dms/dms.html

cal parameters, such as temperature, humidity, wind velocity and direction
etc. Moreover, a strong dependency on the surface of the Earth is known for
the vegetation and soil dependent emissions. These are all good reasons for
including natural emissions in attempts to model the chemical composition
and processes of the atmosphere. However, many difficulties are connected
to the inclusion of natural emissions in a model.

First of all many of the naturally emitted species are still poorly understood
and the processes that are understood are often very complicated and therefore

difficult to parameterise with a satisfactory result. Based on observations and field experiments inventories for natural emissions are created. For many species a decadal, annual or monthly average concentration for a given grid cell with a specific land-use category are given with some reasonable variability.

Some natural emitted species like for example many VOCs are so complicated in their chemistry, that they are not even included in the models. If they were, the uncertainties related to the results would be so significant that the model results could *not* be interpreted in any scientifically sound way. Alternatively, the effects from the given compound are simply excluded from the model calculation, which also introduces errors of unknown size. However, when interpreting model calculations modellers have these uncertainties in mind and try to include the current knowledge in a more intuitive way. Currently, with the fast increasing computer capacity and greater understanding of the various species and processes, the inclusion of natural species and processes are improving every year.

Within the last five years a new cross-cutting discipline has been developed between the field of atmospheric chemistry modelling and climate modelling. A steadily increasing number of research groups are working on coupling hemispherical or global chemical transport models to climate models in order to include the mutual effects of global warming, climate change and chemical composition of the atmosphere (see e.g. Langner et al., 2005, Murazaki and Hess, 2006 or Hedegaard et al., 2008).

The emission and the chemistry of most chemical compounds are strongly dependent of meteorology and climate change has proven to perturb many, if not all, meteorological parameters. This gives a good reason to believe that a great amount of feedback mechanisms between atmospheric chemistry, and other climate parameters will be altered in the future. These feedback mechanisms are of great importance when trying to predict the future climate and air pollution levels and therefore important to have represented in the models.

7.7 Literature

Andres, R.J. (2000): *A time-averaged inventory of subaerial volcanic sulphur emissions. Prepared for the GEIA*, IGACtivities Newsletter, 22, December 2000.

Athanasopoulou, E. et al. (2008): *The role of sea-salt emissions and heterogeneous chemistry in the air quality of polluted coastal areas,* Atmospheric Chemistry and Physics Discussion, 8, 3807-3841.

Atkinson, R. and Arey, J. (2003): *Gas-phase tropospheric chemistry of biogenic volatile organic compounds: a review*, Atmospheric Environment, *37*, 197-219.

Denman, K.L. et al. (2007): *Couplings between Changes in the Climate System and Biogeochemistry, p. 499-588* in: Climate Change 2007: The Physical Science Basis. Contribution of Working Group I to the Fourth assessment Report of the IPCC, .

Guenther, A, et al. (2000): *Natural emissions of non-methane volatile organic compounds, carbon monooxide, and oxides of nitrogen from North America*, Atmospheric Environment, *34*, 2205-2230.

Guenther, A. et al. (2006): *Estimates of global terrestrial isoprene emissins using MEGAN (Model of Emissions of Gases and Aerosols from Nature)*, Atmospheric Chemistry and Physics, 6, 3181-3210.

Hedegaard, G.B. et al. (2008): *Impacts of climate change on air pollution levels in the Northern Hemisphere with special focus on Europe and the Arctic*, Atmospheric Chemistry and Physics, 8, 3337-3367.

Kesselmeier, J. and Staudt, M. (1999): *Biogenic Volatile Organic Compounds (VOC): An overview on Emission, Physiology and Ecology*, Journal of Atmospheric Chemistry, *33*, 23-88.

Kloster, S. et al. (2006): *DMS cycle in the marine ocean-atmosphere system – a global model study*, Biogeosciences, *3*, 29-51.

Langner, J. et al. (2005): *Impact of climate change on surface ozone and deposition of sulphur and nitrogen in Europe*, Atmospheric Environment, *39*, 1129-1141.

Millet, D.B. et al. (2008): *Spatial distribution of isoprene emissions from North America derived from formaldehyde column measurements by OMI satellite sensor*, Journal of Geophysical Research, *113*.

Murazaki, K. and Hess, P. (2006): *How does climate change contribute to surface ozone change over the United States*, Journal of Geophysical Research, *111*, D05301.

Seinfeld, J.H. and Pandis, S.N. (2006): *Atmospheric Chemistry and Physics, From Air Pollution to Climate Change*, second edition. Wiley and Sons Inc., Hoboken, New Jersey, Canada.

Simó, R. (2001): *Production of atmospheric sulfur by oceanic plankton: biogeochemical, ecological and evolutionary links*, Trends Ecology and Evolution, *16*, 287-294.

Simpson, D. et al. (1999): *Inventorying emissions from nature in Europe.* Journal of Geophysical Research, D7, 8113-8152.

III

Processes in the atmosphere

The size of emissions is of course important for the degree of damage that air pollution can cause. But equally important is what happens in the atmosphere between the emission and the impact.

In principle the air pollutants follow the air movements, but since the pollution normally starts from a point the dispersion processes (Chapter 8) can be more complicated and depend upon a series of parameters as release height and emission temperature. This must be taken into account, especially in evaluating local, short range pollution.

During the dispersion the emitted so-called *primary* pollutants can react with other components of the atmosphere, often in photochemical processes under the influence of sunlight (Chapter 9). Thus both the general pollution and the meteorological conditions influence the lifetime and ultimate fate of the pollutants including the transformation to other, so-called secondary pollutants. Ultimately the pollutants, primary or secondary, are removed from the atmosphere by deposition processes (Chapter 10).

The whole system of processes is complicated. Therefore often a simple relation between local emissions and pollution levels can not be established.

8 Atmospheric physics

Ole Hertel, Carsten Ambelas Skjøth and Jes Fenger

In Chapter 2 the composition of, and the processes in the clean atmosphere were described. This chapter explains how air pollutants take part in these processes. The chapter is fairly short because further and more detailed information on physical processes can be found in Chapter 10 on deposition and Chapter 14 on modelling of air pollution.

8.1 Dispersion and transport

When air pollutants are released to the atmosphere they follow the air flow and take part in chemical and physical transformations. A plume from a chimney will thus be advected with the wind. The rate of dispersion depends on the turbulence (Section 8.2) that again may be expressed in terms of the atmospheric stability (Section 8.3).

The aerodynamic conditions may play a crucial role for the dispersion of a plume. The zone nearby a chimney and just after a building is often a critical area. A strong down draft may in some cases bring the plume all the way down to the ground and lead to high pollution levels near the source. On the other hand a high temperature of the plume leads to so-called buoyancy lifting, which may mean that the plume rises considerably just after leaving the chimney thus reducing the local impact. Dispersion in plumes is treated in detail in Chapter 14, Section 14.2.

The presence of building obstacles and deep street canyons pose special conditions to the urban areas. Pollutants can be trapped inside the street canyons which generally have the highest pollutant levels. The topography around the urban area is naturally also important. The high pollutant levels e.g. in cities like Los Angeles are caused by the fact that the city is placed inside a valley with generally low wind speed and little mixing with the air outside the valley. This may further promote the formation of photochemical air pollution.

Many pollutants have a long lifetime in the atmosphere and may be transported over long distances. This applies especially for particulate pollution that may stay in the air for up to 10 days in case the air mass does not meet a precipitation event. This leads to transboundary air pollution and means e.g. that emissions in central Europe can affect ecosystems in northern Scandinavia. The transport distance is in this case governed by the dry deposition velocity (Chapter 10), whereas for many of the gaseous pollutants it is governed by the rate of transformation (Chapter 9).

8.2 Turbulence

The dispersion of pollutants in the atmosphere is driven by *small scale eddies* in the mixing (or Ekman) layer, that is the lowest part of the atmosphere, where the friction against the surface is felt. Typically the mixing layer has a height of about 1 to 1.5 km on a summer day (see later Figure 8.4).

These small scale eddies are termed turbulence, and they are defined as three dimensional motions with local velocities and pressures that cannot be predicted in time and space. However, the statistical properties like time and space averages of the turbulent parameters show regular behaviour, and the motion is stochastic and can thus be characterised.

In the description of turbulence, it is common to apply an averaging approach defined by Reynolds in 1895. The turbulent parameter – in this case the wind velocity – is broken down into a mean and a fluctuation component:

$$u = \bar{u} + u' \qquad\qquad\qquad (8.1)$$

where u is the wind speed, and $\bar{u}$ is the mean and u' the fluctuation component of the wind.

This simple decomposition is powerful for a number of applications. An example is in the calculation of the flux of a given pollutant, which may be expressed as:

$$F_c = u \cdot c \tag{8.2}$$

where F_c is the flux of the pollutant, u is the wind speed and c is the concentration of the pollutant.

If the wind speed and the concentration are both broken down into a mean and a fluctuation component – again applying the approach of Reynolds – it may be shown that the flux can be expressed as:

$$F_c = \overline{u \cdot c} + \overline{u' \cdot c'} \tag{8.3}$$

This means that the flux is equal to the sum of the product of the mean values and the product of the fluctuations components. We will return to the description of concentration fluxes in Chapter 10 on deposition processes.

Mechanical and convective turbulence

A commonly used parameter for the stability of the atmosphere is the so-called Monin-Obukhov length that formally expresses the height at which the production of mechanical turbulence equals the production of convec-

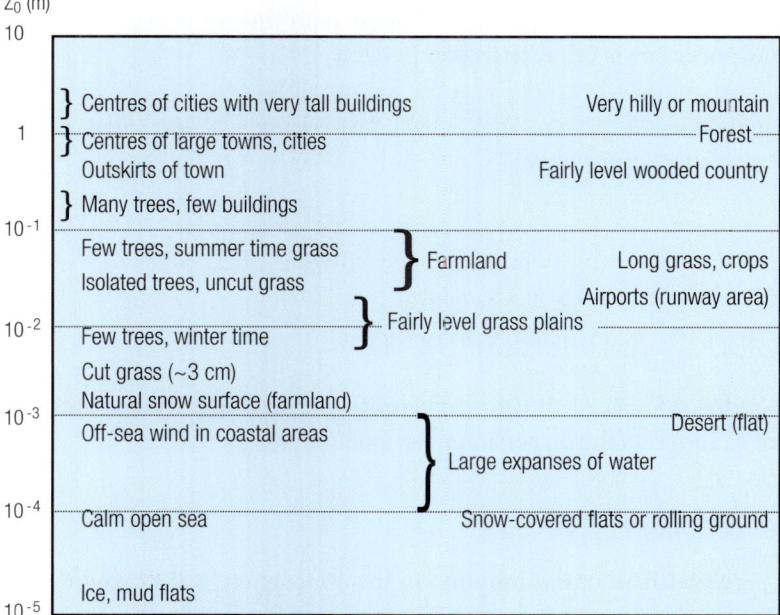

Surface roughness values (Z_0) for typical terrain types

Figure 8.1 Roughness elements for different types of surfaces. The ordinate is a typical size of a whorl.

tive or thermal turbulence. These two types of turbulence are defined in the following.

The mechanical turbulence is formed by the breaking of the wind towards the surface or obstacles associated with the surface. For a smooth surface e.g. grass field or water (the roughness of various types of surface is shown in Figure 8.1), the production of mechanical turbulence is low, whereas it is considerably larger for a rough surface like a forest or an urban area. Mechanical turbulence is characterised by relatively small whorls (characteristic length scale of cm to few meters), especially near the surface.

Convective or thermal turbulence is a result of the solar radiation. The sun heats the surface that again heats the air just above. The heated air is lifted due to buoyancy effects, and is replaced by colder air. The convective turbulence is characterised by large and relatively slowly moving "convective cells" consisting of updrafts and downdrafts that may have a characteristic time scale of 10 to 20 minutes.

Turbulence must continuously be driven otherwise it will decay, returning the fluid motion to a laminar state. The initially created turbulent whorls (or eddies) will after a while be broken down (dissipate) into smaller whorls, and this process continues until the smallest whorls end up as heat in the atmosphere. This chain brake down of the turbulence was expressed poetically by one of the most important researchers in the physical processes of the atmosphere Lewis Fry Richardson in 1922:

Big whorls have little whorls, which feed on their velocity; and little whorls have lesser whorls and so on to viscosity.

8.3 Stability and vertical motions

The so-called *stability* of the atmosphere is of crucial importance and will enhance or suppress the dispersion of air pollutants.

Neutral conditions

An air parcel with an upward motion in the atmosphere will expand due to the decreasing surrounding pressure. The pressure is caused by the weight of the overlying atmosphere, and decreases exponentially by a factor e for each so-called scale height H. At a height of z the atmosphere has a pressure P_z related to the pressure P_0 at sea level:

$$P_z = P_0 e^{\left(-\frac{z}{H}\right)} \tag{8.4}$$

Given that no heat is exchanged with the surroundings, the temperature of the air parcel will decrease with the expansion. This is called the adiabatic temperature decrease. In an atmosphere which is unsaturated with water vapour, the dry adiabatic temperature decrease (also called the dry adiabatic lapse rate) will have a rate of 9.8 °C per km, or about 1°C per 100 m. In air saturated with water vapour, "evaporation heat" is released from the condensation of water vapour. This release of heat reduces the cooling with height in the atmosphere, and in the atmosphere saturated with water vapour; the adiabatic temperature decrease is therefore in average only about 6.5 °C per km. This rate of decrease is termed the saturated adiabatic lapse rate. When the actual temperature profile (the environmental lapse rate) follows the dry adiabatic lapse rate, the stability conditions in the atmosphere are *neutral* (Figure 8.2).

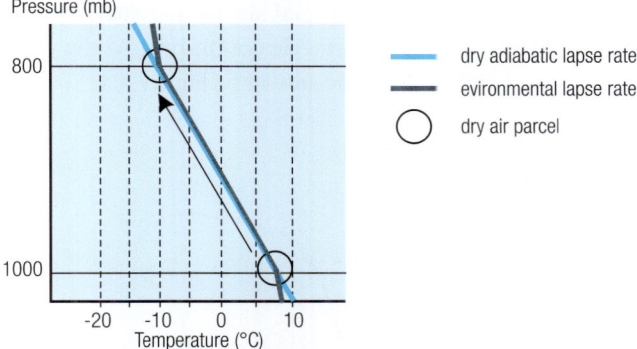

Figure 8.2 Vertical temperature profiles in the atmosphere during neutral conditions.

Unstable conditions

An air parcel with an upward motion in an atmosphere, where the temperature decreases in the vertical direction (the environmental lapse rate) by more than the adiabatic lapse rate is in a different situation. Hot air is lighter than cold air. The lifting of the air parcel that may result from its higher temperature compared to the surroundings is called buoyancy. In this case the air parcel will continue or may even accelerate the upward motion, since during the vertical movement it will go on meeting colder surrounding air. If the air parcel had a downward motion, this would likewise continue or even be accelerated. The conditions in the atmosphere are in this case called *unstable* and they are promoting a fast vertical mixing of pollutants (Figure 8.3).

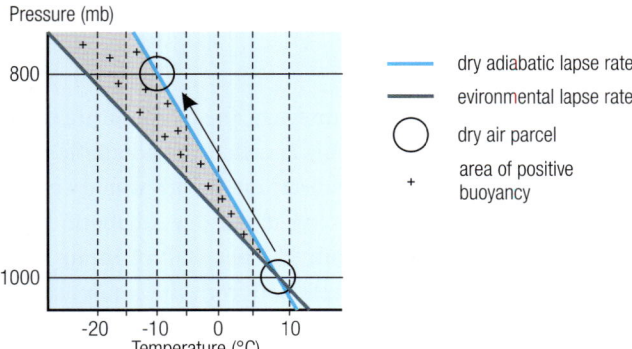

Figure 8.3 Vertical temperature profiles in the atmosphere during unstable conditions.

Unstable conditions in the atmosphere are common at daytime during summer season, when the Earth surface is heated by the solar radiation. The result of this heating is a warm surface that again heats the air mass just above the surface. The buoyant lifting of this heated air mass creates convective turbulence leading to the unstable atmospheric conditions. At the same time it leads to a high mixing height as the one shown in Figure 8.4.

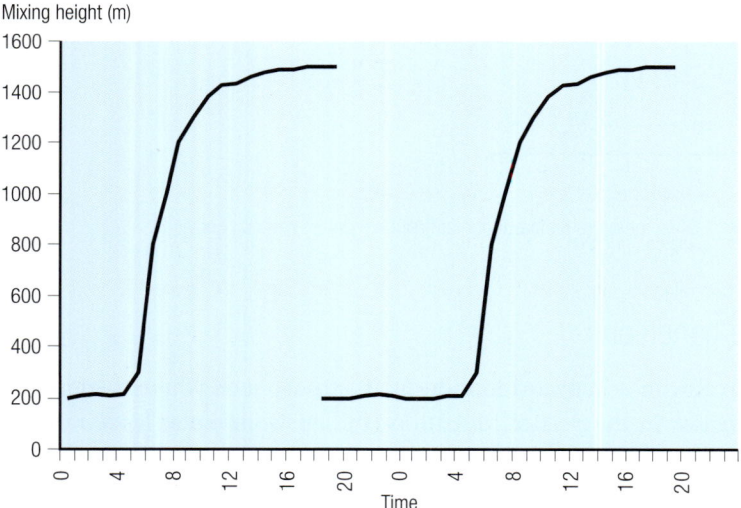

Figure 8.4 A typical diurnal variation in mixing height over two days during summer season. The mixing height grows when the sun rises (in this case about 6 am) and starts to heat the Earth surface. The warm surface heats the overlaying air and the buoyant lifting of this air mass creates convective turbulence and growth of the mixing height (unstable conditions). In the afternoon the mixing height collapses and a new low mixing height is formed (stable conditions).

Stable conditions

The situation is opposite when an air parcel moves upwards in an atmosphere that has a vertical temperature profile (the environmental lapse rate) decreasing less than the dry adiabatic lapse rate or in the more extreme case even increases with height. The air parcel will in this case be colder than the temperature it meets in the surroundings, and the air parcel will therefore decelerate. The motion of the air parcel will be suppressed and subsequently cease. The conditions in the atmosphere are in this case *stable* (Figure 8.5). Such a situation may appear when the surface is cooled e.g. as a result of radiation loss during a cloud free night, in this case the temperature in the layer near the surface may increase with height; this is termed a ground-based *inversion*. Other types of inversion may take place; e.g. a frontal inversion can occur when two air masses meet, having very different temperatures, humidity and pressure, and the warm air overrides the cold air. Furthermore there is the advective inversion that appears when a warm air mass passes over a much colder surface.

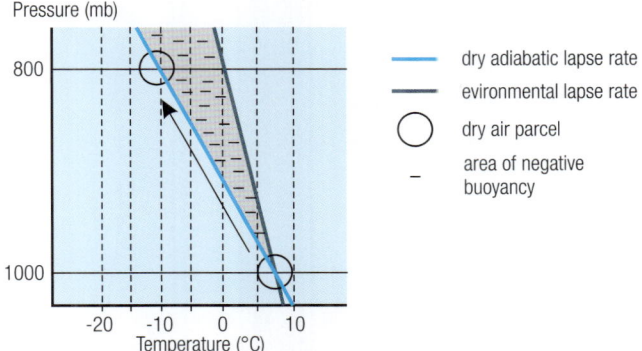

Figure 8.5 Vertical temperature profiles in the atmosphere during stable conditions.

Inversions suppres and may even prevent vertical mixing of air and thus restrict the dispersion of air pollution. They are therefore one of the main causes of air pollution episodes in cities. The cold surface further promotes domestic heating which may thus aggravate the situation. This can be especially serious (as in the disaster in London 1952) if the pollution shades for the sunlight and thus prevents a heating that would have lifted the inversion during the day.

The practical aspects of the three stabilities are applied in Gaussian plume modelling as described in Chapter 14.

8.4 Horizontal motion

Short range transport of air and thus of air pollution e.g. in cities can be complex as a result of local topological features. The problems are described in more details in Chapter 14 on air pollution modelling. But in principle they are simple because they take place in a constant scenario where sometimes even chemical reactions can be ignored.

Many pollutants have a lifetime in the atmosphere of a few days, and the average wind velocity is normally some hundred kilometres per day. It means that the pollutants can be transported over long distances and often over national borders before appreciable transformation is happening.

Horizontal air motions over longer distances are mainly governed by pressure gradients, the Coriolis force and friction. The pressure gradient force is proportional to the pressure gradient, but oriented in the opposite direction. Furthermore this force is inversely proportional to the density of the air. The effect of the friction depends on the roughness of the surface (See Figure 8.1)

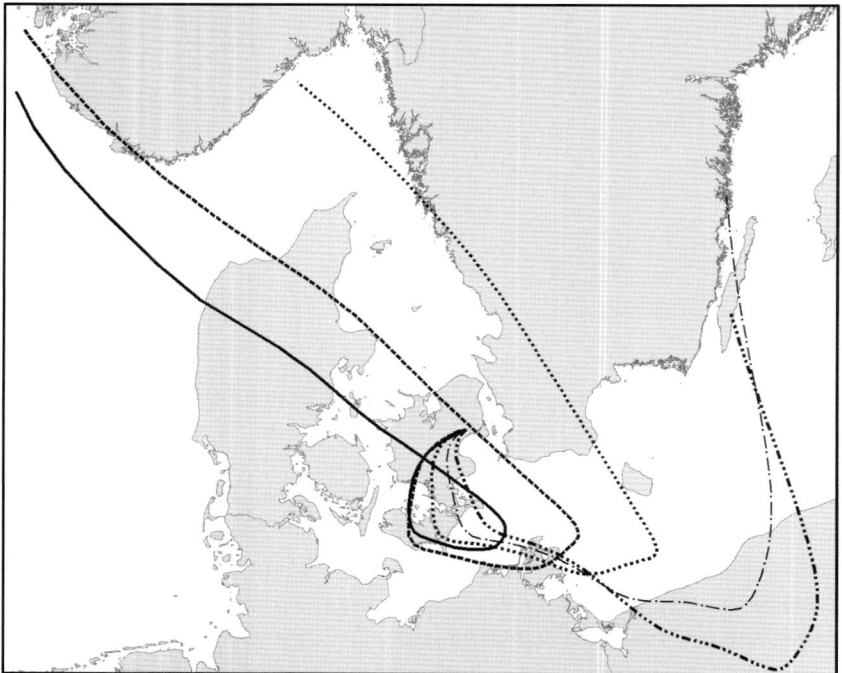

Air mass transport at different heights (m) above the surface

——— 200 ----- 400 ········ 600 ·—·— 800 ----- 1000

Figure 8.6 The path of an air parcel released at a height of 200, 400, 600, 800 and 1000 m. Calculated for March 3. 2007 at 12.00.

and results, as described in Chapter 1, in a turning of wind direction and a reduction of wind speed near the surface. A plume that is mixed up vertically will thus be transported at different speeds and in different directions at different heights. These factors vary along with the path of the air masses and thus along the path of the pollution. At the same time meteorological conditions, including precipitation, may vary along with the path of the pollution and will influence the chemical transformations of the pollution (Chapter 9). All these factors must be taken into account, when the long range transport is modelled (Chapter 14).

As an example is shown (Figure 8.6) the calculation of an air parcel released at different heights.

8.5 The general circulation

In Chapter 1 the general global circulation system is described. It has a profound impact on the distribution of air pollutants on larger scales.

The formation of circulation cells (Hadley and Ferrel cells) means that the two global hemispheres are by and large separated. Therefore air pollutants with a reasonable short lifetime (days) will not cross the Equator. Since there are significantly more emissions on the Northern hemisphere, the concentrations will be higher here. For long-lived pollutants like carbon dioxide the average concentration may be the same, but the yearly variations due to emission and absorption by vegetation are larger on the Northern hemisphere.

The spiralling air movements (Figure 1.10) give predominant wind directions in different regions of the Earth. Furthermore, in Northern Europe, the weather has a tendency to fluctuate rapidly due to the influence throughout the year of low pressure zones moving in from the Atlantic. This means that long range transport of pollutants at specific sites is episodic, as the day to day air mass transport from source to receptor may vary to a large degree.

Finally in Northern Europe the wind is predominantly from the West. This has a bearing on long range transport of air pollution, which is by no means fair. In England e.g. most of the air pollution is blown *out* of the country. Denmark has the further advantage that most of the cities are lying on the East coasts. The south of Sweden, on the other hand, is much more exposed to pollution from Denmark than the other way around.

8.6 Local features

Locally special features can have great impact on the dispersion and thus on the resulting pollution levels.

Characteristic local wind systems

Some areas have characteristic wind systems that can be due to special topological conditions. In mountainous areas ascending air can be formed over hot mountain sides during the day often causing cloud formation and precipitation. During the night, the process is reversed and the cold air lies in the valley and results in an atmospheric stability that can give pollution episodes.

If a larger air mass moves over a declining terrain a so-called *katabatic* (from Greek: to go down) wind is formed. Some of the strongest episodes appear in Norway, where the air is further cooled by passing snow or ice covered areas for eventually to pass a narrow canyon or fjord. The French *Mistral* is another example. Normally the pollution consequences are modest, but inconvenient natural dust can appear.

Föhn is a warm, dry wind that can develop on the leeward side of a mountain e.g. on the North side of the Alps. When the wind is forced over the mountain, it is cooled, giving off its humidity and will be warmer when it again moves downwards. This may form an inversion and reduce the possibilities for dispersion of pollution.

Urban climate and circulation

A city normally has a higher temperature than the surrounding countryside, typically a few degrees. There are several reasons: The city normally has a lower albedo and thus absorbs solar radiation better; more energy is used in the city predominantly for traffic and domestic heating or air conditioning; buildings and other constructions slow down the wind and thus give less cooling. As the buildings further act as a heat reservoir there will be less difference between day and night temperatures. The higher temperature and the fact that precipitation is largely led away in sewers mean that the air is dryer. Further the vegetation is modest, which means that the evaporation is restricted. All these factors influence the relation between emission and resulting concentration of air pollution.

This so-called urban heat island can produce a special circulation, where the air ascends over the warm city and is substituted with colder air (country wind) coming in from the surroundings (Figure 8.7). A similar phenomenon (sea breeze) can arise when the air over a city near the coast is heated, ascends and is substituted with air from the sea. In both cases previously emitted pollution can be returned to the source-area. These phenomena are not pronounced in northern Europe, but are well-known in southern Europe.

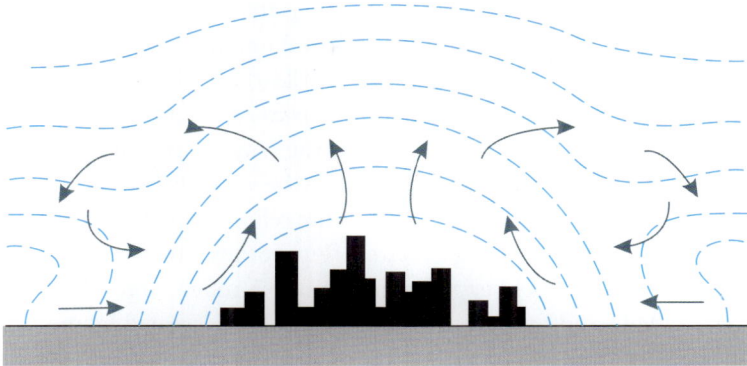

Figure 8.7 A city acts as a heat island with its own climate. In quiet weather a special urban circulation can appear. The lines are isotherms.

Street canyon effects

In relation to air pollution, urban areas are strongly inhomogeneous entities with significant hot spots. The air pollution level is the result of local emissions as well as contributions from pollution transport from both nearby sources inside and remote sources outside the city. A characteristic hot spot is the street canyon. In street canyons air pollutants are released from vehicle transport, local domestic heating and smaller industries and the emissions from these sources are not diluted as efficiently as it is usually the case for emissions released from tall sources. Thereby, these sources contribute significantly to the pollutant concentrations at ground level. The emission of pollutants from traffic usually follows a fixed pattern throughout the day and the week. The highest diurnal mean emissions are found for the working days, and the highest hourly mean emissions take place during the rush hour periods. Meteorological conditions are also very important for the pollutant levels inside the urban street canyons with the wind speed as the primary driving parameter. The concentration inside the street canyon may be considered as two contributions; one from the local traffic in the street itself and one from background pollution entering the street canyon from the roof level:

$$c = c_b + c_s \tag{8.5}$$

Where c is the concentration in the street, c_b is the urban background contribution, and c_s is the contribution from local traffic inside the considered street. The background contribution may then furthermore be considered as two contributions; one from nearby sources in the urban area (typically it will mainly be from traffic in other streets), and one from regional (from sources within a distance of a few hundreds of km) and from long range transport (from sources up to thousands of km away).

The most characteristic situation occurs if the wind at roof levels has a component perpendicular to the street. Thereby the wind direction at street level is opposite to the one above roof, due to the reflection of the wind flow on the facade of the buildings along the street (Figure 8.8). This phenomenon has been studied in a few field experiments, and to a much larger extent in a variety of detailed wind tunnel studies. The turbulence created by moving vehicles in the street is a relevant dilution mechanism especially in cases of low ambient wind speeds. This so-called traffic produced turbulence has to be included in advanced street pollution models. (Chapter 14)

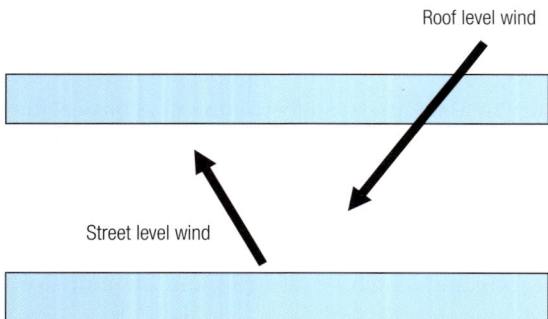

Figure 8.8 Illustration (seen from above) of the reflection of the wind inside a narrow street canyon. In this example the recirculation zone covers the entire street canyon. For very wide streets, the recirculation zone will only cover parts of the street and it will only be inside this zone that the wind direction is changed by reflection. Outside the recirculation zone the wind will in this case be similar to above the roof.

8.7 Water

As described in Chapter 1 water in all forms plays an important role in the global climate. It is also important both for the distribution and for the impacts of air pollution.

Precipitation

Precipitation is important for the wet deposition of pollution (Chapter 8), and it is e.g., as mentioned in Chapter 1, the reason for high depositions on the Norwegian west coast, which gets a high precipitation when the westerly wind hits the mountainous coastline.

Clouds and fog

Particles are necessary as condensation nuclei in the formation of clouds and fog and played an important role in the 1952 disaster in London (Chapter 2).

Water vapour may condensate on any available surface in the free atmosphere. The most efficient nuclei are aerosols with an aerodynamic diameter ranging between 0.1 and 1 μm. Non-hygroscopic aerosols (e.g. soil dust) can only initiate droplet formation at over-saturation of the air (i.e. relative humidity above 100%), whereas hygroscopic aerosols (aerosols that attract water i.e. sodium chloride and ammonium sulphate containing aerosol particles) may form droplets at a relative humidity of only 80%.

Impacts

A dirty surface may be constantly humid, which can be highly important for the degradation of materials that often takes place in a thin electrolytic layer on the surface of the material (Chapter 16). Therefore material damage may be more important in a humid climate than in a dry at the same concentration levels of pollution.

The impact of air pollution on vegetation may also depend on precipitation, since plants lacking water are generally more sensitive.

8.8 Literature

Graedel, T.E. and Crutzen, P.J. (1993): *Atmospheric Change. An Earth system Perspective.* W.H.Freemann, New York, USA.

Holton, J.R. (1992): *An Introduction to Dynamic Meteorology.* Academic Press, San Diego, USA.

Seinfeld, J.H. and Pandis, S. N. (1998): *Atmospheric Chemistry and Physics, from Air Pollution to Climate Change,* John Wiley & Sons, New York, USA

Wallace, J.M. and Hobbs, P. V. (1977): *Atmospheric Science. An introductory Survey.* Academic Press, San Diego, USA.

9 Atmospheric Chemistry

Matthew S. Johnson

The troposphere and stratosphere are the most important regions of the atmosphere from the perspective of human society and the biosphere. This chapter will describe the main features of the photochemical processes occurring in these two regions.

We will begin with a description of the chemical composition of the troposphere and stratosphere. Stable, long-lived gases are thoroughly mixed in the troposphere because of relatively rapid vertical transport; their respective molar fractions are constant throughout. For species with short atmospheric lifetimes on the other hand, including most air pollution, this is not the case. The concentrations of the unstable species are largest close to their sources and their concentrations change due to dilution as well as chemical reactions.

In the stratosphere species are characterized by complex vertical distributions, except for N_2 and Ar whose mole fractions remain constant and O_2 which is also practically constant. The most important species in the stratosphere, ozone (O_3), peaks several kilometres over the *tropopause*, the divide between the two spheres. Methane declines steadily with altitude throughout the stratosphere due to chemical processes, the most important being oxidation by the hydroxyl radical OH. The mole fractions for the CFC's and N_2O also decline, due to photolytic reactions.

9.1 Inventories

The composition of the atmosphere has been presented in detail in Chapter 1. Here we will only discuss a number of key substances that are relevant to atmospheric reactions and their kinetics. We also provide data concerning the UV-radiation in the troposphere and stratosphere.

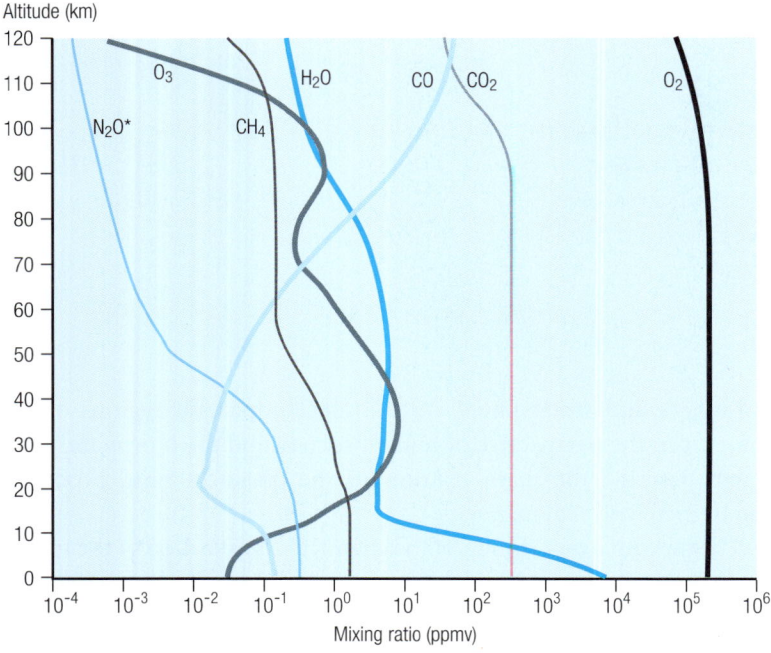

Figure 9.1 U.S. Standard Model Atmosphere Profiles from 0 to 120 km for the major radiating atmospheric gases (based on Anderson et al., 1986).

It is obvious that the molecules fall into different categories. The first category includes the completely stable molecules such as CO_2, N_2 and O_2, where O_2 is the basis of ozone chemistry. The amount of ozone is relatively small. As seen in Figure 9.1 its mixing ratio is only 10^{-5} compared with 0.21 for oxygen. The deflection of the O_2 profile is minor. N_2O and methane are stable in the troposphere, but decline in the stratosphere. Water shows an interesting behaviour. It simply freezes out at the tropopause – sometimes the word hygropause is seen in this connection. Its gradual rise in the stratosphere is due to the oxidation of methane and hydrogen. The case of ozone shall occupy the next section, and in addition Chapter 19 is devoted to ozone.

Figure 9.2 shows how the key species involved in the formation of tropospheric pollution are distributed in the modern atmosphere. The amount of surface-level ozone has doubled, approximately, in the northern hemisphere relative to the pre-industrial atmosphere. The amount of ozone increases near the stratosphere due to the subsidence of stratospheric air. Nitrogen oxides act as catalysts in the troposphere in the chain reactions (see below) that convert hydrocarbons into oxidized species, generating ozone and particles. Their concentrations show a dramatic peak in the northern hemisphere near the surface, due to the use of fossil fuels. In addition to their role in promoting atmospheric photochemistry nitrogen oxides are precursors for acid rain. The hydroxyl radical is the main reactive species of the troposphere and removes a wide variety of air pollution. It is formed by a combination of sunlight, ozone and water vapour, giving rise to the spatial distribution shown in the figure. Carbon monoxide is the main atmospheric reagent. It is emitted directly at the surface by combustion in oxygen poor conditions, and it is generated in the atmosphere through the oxidation of virtually all hydrocarbons. About ¾ of atmospheric OH will react with CO, producing 1/6 of the atmosphere's CO_2.

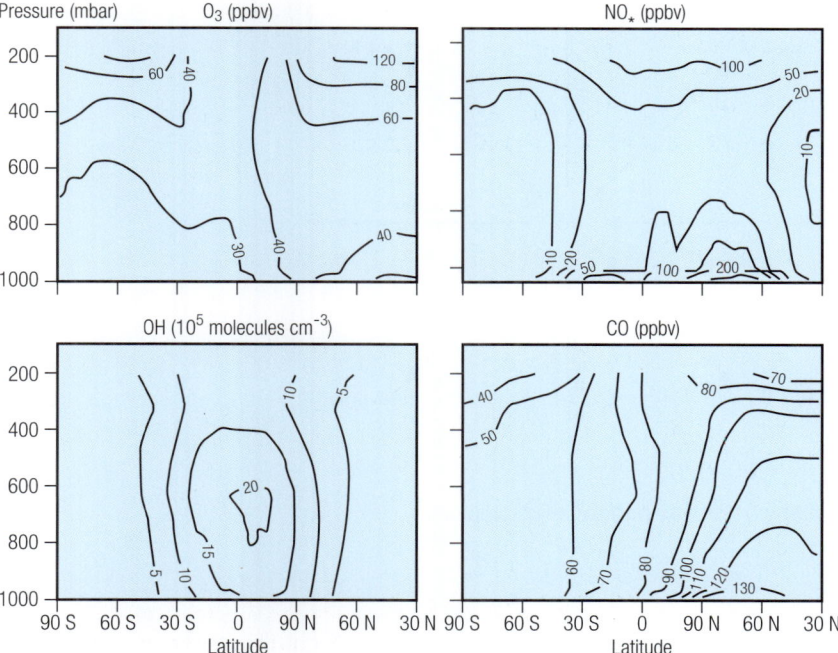

Figure 9.2 Distributions of key tropospheric species in the modern atmosphere (Adapted from Wang and Jacob, 1998).

9.2 Photolysis in the atmosphere

Photolysis is an example of a unimolecular reaction, one in which a single molecule breaks into two fragments. The usual means of obtaining the energy required to break a bond is through the absorption of light. In other cases the energy may be provided by the thermal bath, and there are many important examples of such reactions in the atmosphere. The energy may also come from the internal energy of the reagents themselves, or from the chemical energy released when a bond is formed between the two reagents. The product forms if the reagents are able to get rid of this energy via a collision with another molecule in a so-called three body reaction. The variety of possible means of forming, and fates of the excited state A* are summarised in Figure 9.3. Non-photolytic unimolecular reactions are discussed in Section 9.4 dealing with association reactions.

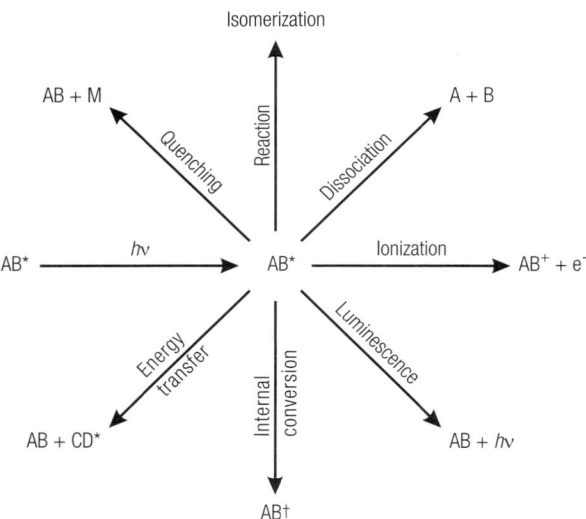

Figure 9.3 Summary of photochemical processes.

At the top of the atmosphere the actinic flux (light flux at those wavelengths capable of inducing photochemistry) resembles a blackbody emission curve at a temperature of 5780 K, the temperature of the surface of the sun. This light is absorbed by atmospheric molecules and the closer one approaches Earth's surface the lower the light flux, especially in the ultraviolet region. At the shortest wavelength dioxygen is the main absorber and the result of the absorption of 1 photon is the dissociation of at most 1 molecule:

$$O_2 + h\nu \ (\lambda < 170 \text{ nm}) \rightarrow 2O \tag{9.1}$$

The rate at which dioxygen dissociates due to radiation in a certain wavelength interval is proportional to the number density of dioxygen C_{O_2} and to the number of photons in the interval

$$-\left[dc_{O_2}\right]_\lambda = c_{O_2}\left[\sigma_{O_2;\lambda}\Phi_{O_2;\lambda}\right]q_\lambda d\lambda \tag{9.2}$$

The constant of proportionality is enclosed in parenthesis and is subdivided into two factors. The first is $\sigma_{O_2;\lambda}$ which is the *absorption cross section* for oxygen at the wavelength considered. It is a quantity with dimensions of area – hence the term cross section – that gives the probability of a photon being absorbed by the molecule. The second factor $\Phi_{O_2;\lambda}$, called the *photolysis quantum yield*, is a dimensionless quantity between 0 and 1 that gives the probability of the molecule breaking apart once the photon has been absorbed.

On integration over the wavelength range where photolysis takes place a normal first order rate equation is obtained:

$$-\frac{dc_{O_2}}{dt} = j_{O_2}c_{O_2}; \quad j_{O_2} = \int\left[\sigma_{O_2;\lambda}\Phi_{O_2;\lambda}\right]q_\lambda d\lambda \tag{9.3}$$

j_{O_2} is called the *photolysis rate constant* for dioxygen; the dimensionality is s^{-1}.

Atmospheric chemistry is driven by sunlight. Sunlight photolyses molecules, generating radicals that seed chain reactions that can involve millions of steps before the radicals are removed. The spectral distribution of sunlight as a function of altitude is shown in Figure 9.4.

A number of features of Figure 9.4 are important in atmospheric photochemistry. The first is that light at wavelengths shorter than ca. 290 nm does not reach the troposphere. In addition the absorption cross sections of O_2 and O_3 (cf. Figure 9.5) allow UV radiation at a wavelength of 200 to 210 nm to penetrate deep into the stratosphere. This UV window is critical to the photochemistry of the stratosphere. There are three main reasons for this. The first is that this radiation is responsible for photolysis of the CFC gases. In addition it photolyses N_2O, which is the main source of NO_x to the stratosphere. Finally it photolyses OCS, providing a source of sulphur to the stratospheric sulphate aerosol layer.

The excited molecule produced by the absorption of a photon does not always dissociate. For many systems the photolysis quantum yield is less than unity.

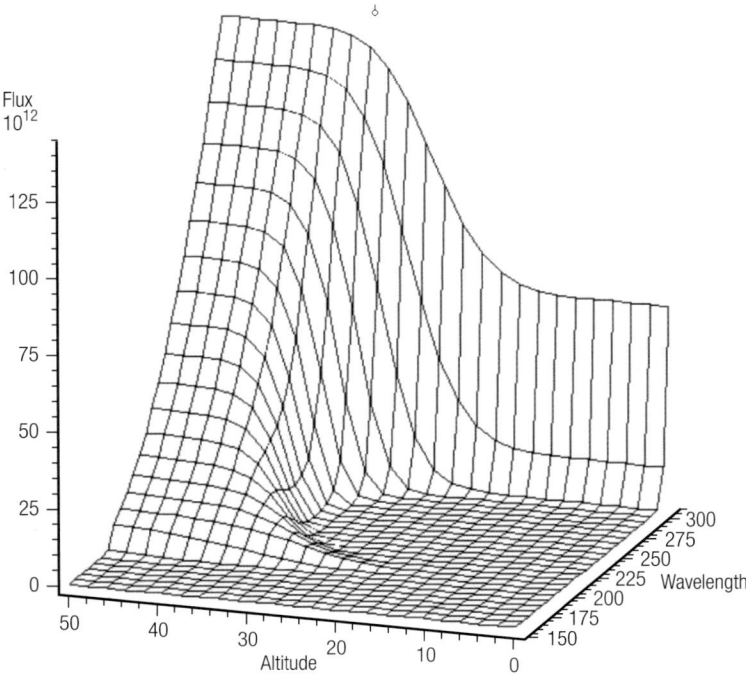

Figure 9.4 Actinic flux as a function of altitude. Solar actinic flux calculated using the sun's Planck spectrum, propagated through an atmosphere of molecular oxygen and an ozone layer, solar zenith angle 90°. Altitude / km, Wavelength / nm, Actinic Flux / photons cm^{-2} s^{-1} nm^{-1}.

Photolysis in the stratosphere

Oxygen and ozone

The photolysis of dioxygen is the most important photolysis reaction in the stratosphere since it is the source of *odd oxygen* ($O_x \equiv O + O_3$; $[O_x] \approx [O_3]$).

$$O_2(X^3\Sigma_g^-) + h\nu \ (\lambda < 170 \text{ nm}) \rightarrow 2O(^3P) \tag{9.4}$$

The electronic state labels $X^3\Sigma_g^-$ and ^3P for the ground states of the dioxygen molecule and the oxygen atom have been added. The fate of the ground state oxygen atom is predominantly ozone formation – a three-body recombination which is looked at closer in Section 9.4. Since the radiation intensity declines steadily towards Earth' surface while the oxygen concentration rises there must be a height at which ozone formation (and hence ozone concentrations) peaks. This is observed at around 25 km. A portion of the absorption spectrum of O_2 relevant to the stratosphere is shown in Figure 9.5.

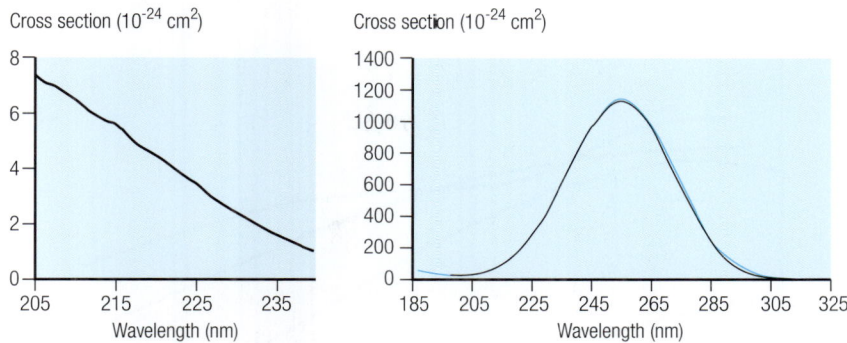

Figure 9.5 a A portion of the Herzberg continuum system of molecular oxygen. The photolysis of oxygen in the stratosphere occurs mainly between 2C0 and 220 nm. (b) Absorption spectrum of ozone from 185 to 325 nm. Figures based on JPL-06.

The consideration is the photolysis of ozone, which occurs by visible and UV radiation as follows:

$$O_3 + h\nu(\lambda < 1180 \text{ nm}) \rightarrow O(^3P) + O_2$$

$$O_3 + h\nu(\lambda < 305 \text{ nm}) \rightarrow O(^1D) + O_2 \quad\quad\quad (9.5)$$

The absorption spectrum is shown in Figure 9.5(b). For wavelengths shorter than 305 nm the recommended photolysis quantum yield is 0.9, independent of temperature. Both the oxygen atom and the dioxygen molecule are produced in excited states. The fate of both is quenching (collisional relaxation) to their respective ground states by collision with other molecules. However the singlet atom can also react with other species, in particular H_2O, CH_4, N_2O and H_2, see below in Section 9.3. The photolysis reaction is one reason why the ozone concentration does not continue to grow. There are other reasons which we shall consider later; see also Chapter 19. Absorption of UV-radiation by dioxygen and ozone in the stratosphere protects Earth's surface against harmful UV-radiation which can damage e.g. plants, plankton and people. Only radiation with a wavelength longer than 290 nm reaches the troposphere and subsequently Earth's surface.

Chlorine-containing compounds, CFC's and HCFC's

The anthropogenic CFC compounds are stable in the troposphere. Lovelock found that the atmospheric burden of the CFCs was approximately equal to the total amount that had been produced. Molina and Rowland were awarded the Nobel Prize for their discovery of what happened to the CFCs, namely

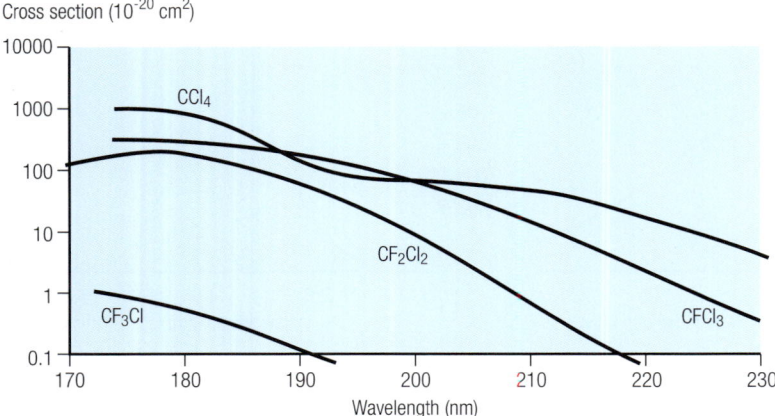

Figure 9.6 Absorption cross sections of chlorine and fluorine derivatives of methane, CCl_4, $CFCl_3$, CF_2Cl_2, CF_3Cl. The cross section of CF_4 is too small to be shown in this Figure. Based on data from JPL-06.

that the only significant process leading to their destruction was photolysis in the stratosphere. The chlorine released by this reaction destroys ozone in a catalytic process CFC-12 will serve as an example:

$$CF_2Cl_2 + h\nu(\lambda < 220\ nm) \rightarrow CF_2Cl + Cl \tag{9.6}$$

The absorption cross sections of three CFCs and CCl_4 are shown in Figure 9.6. Figure 9.7a shows the vertical profile of CFC-11 and CFC-12. Figure 9.7b shows the photolysis rates as a function of altitude for a number of CFC's and N_2O. The absorption spectra of CFCs and HCFCs are similar and peak in the vacuum ultraviolet. The common feature is a $n_{Cl} \rightarrow \sigma^*$ transition on the C-Cl chromophore. Variations in intensity arise mainly from the number of C-Cl bonds present, and the other substituents on a given molecule act to shift the position of the peak.

The fate of the radicals such as to CF_2Cl is discussed in Chapter 19. The fate of the chlorine atoms is to abstract H from methane or O from ozone. We return to these reactions in connection with abstraction reactions. HCFC's can react with OH in the troposphere, thereby greatly decreasing the relative amount of chlorine that reaches the stratosphere. The same is true for CH_3Cl. In addition a fraction of HCFCs are photolysed in the stratosphere, also releasing Cl. The absorption cross section of HCFC-141b, a common CFC substitute, is shown in Figure 9.8 for comparison to the CFCs.

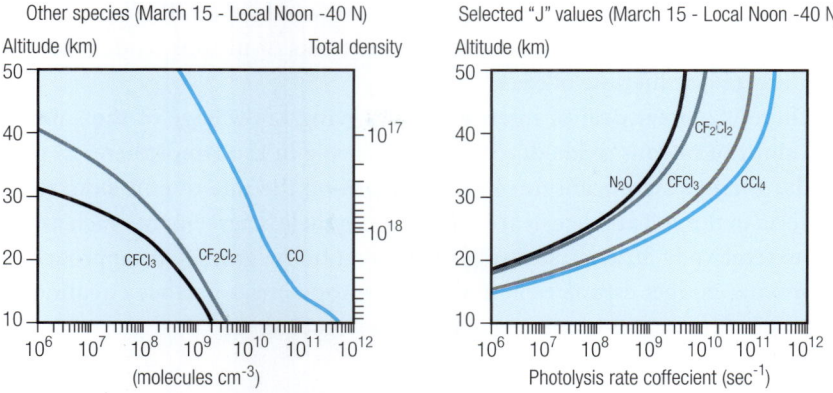

Figure 9.7 CFCs in the stratosphere.(a) Vertical profiles of $CFCl_3$, CF_2Cl_2 (and CO). (b) Vertical profiles of J values (photolysis rate coefficients (s^{-1}) for $CFCl_3$ and CF_2Cl_2 (and N_2O, CCl_4). (JPL-97).

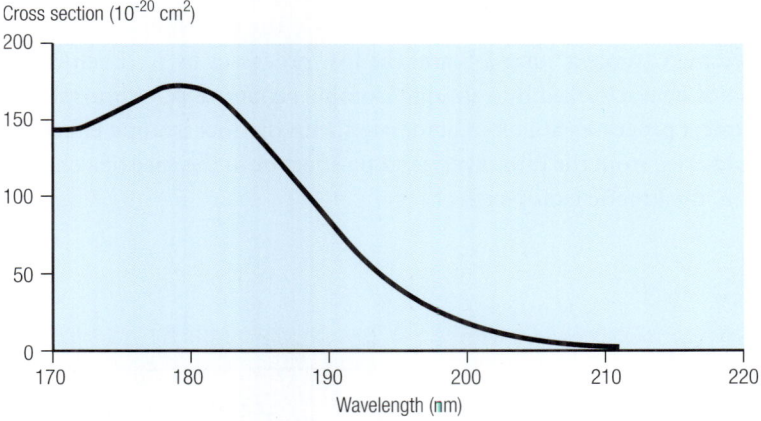

Figure 9.8 Absorption cross section of HCFC-141b. 1,1-dichloro-1-fluoroethane, CCl_2FCH_3, is a common CFC substitute. Based on JPL-06.

Nitrogen-containing compounds, N_2O, NO_x, NO_y

Nitrous oxide is produced by nitrifying and denitrifying bacteria in the soils and oceans. Its atmospheric concentration is rising because of increased use of fertiliser which is a concern because nitrous oxide is a long-lived green-house gas and the most important source of NO_x to the stratosphere. Around 90% of nitrous oxide is removed from the atmosphere by photolysis:

$$N_2O + hv \ (\lambda \approx 205 \text{ nm}) \rightarrow N_2 + O(^1D) \tag{9.7}$$

The photolysis occurs in the 'stratospheric UV window' (cf. Figure 9.4). The combined photolysis and chemical reaction sinks of nitrous oxide result in an atmospheric lifetime of ca. 120 years.

There is a great deal of interest in improving knowledge of the sources and sinks of nitrous oxide due to its central role in the atmosphere. In particular there is significant uncertainty regarding the rate of emission, since bacteria in the soil and ocean are diffuse. Nonetheless agreements such as the Kyoto treaty rely on accurate budgets for greenhouse gases. One approach to improving budget estimates is to use the isotopic mass balance equation to constrain the sources of nitrous oxide (Johnson et al., 2002).

$$\delta_{obs} = \frac{\sum_i \delta_i S_i}{\sum_i S_i} + \frac{\sum_i \varepsilon_i L_i}{\sum_i L_i} \tag{9.8}$$

In this equation δ_{obs} represents the enrichment (relative to an accepted standard) of a given isotope in a sample, e.g. $^{14}N^{15}N^{16}O$. The emissions sources are represented by S_i (mole a^{-1} or g a^{-1}) and the loss processes by L_i. Each emission source is characterized by a unique isotopic signature δ_i. Atmospheric photochemical processes are often associated with distinct isotope effects ε_i, which could arise from the photolysis isotope effect or in the case of a chemical reaction, the kinetic isotope effect.

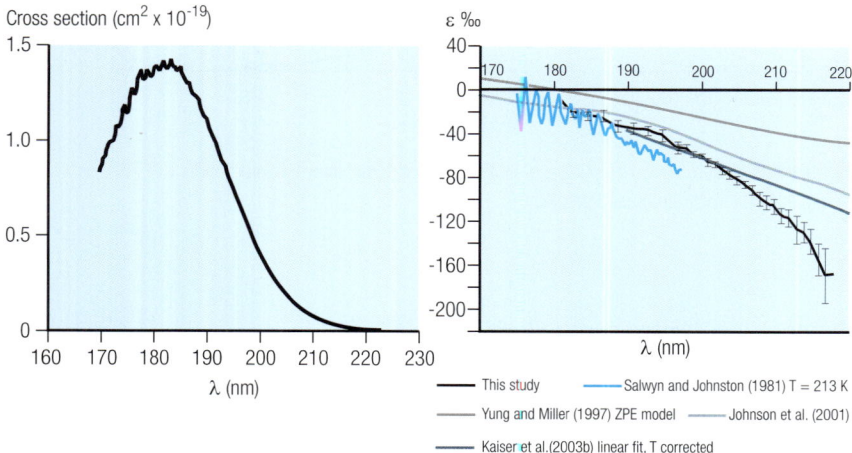

Figure 9.9 Photolysis of nitrous oxide. The left panel shows the absorption cross section of nitrous oxide (Yoshino et al., 1984). The right panel shows the photolytic fractionation constant for $^{14}N^{15}N^{16}O$ measured at a temperature of 233 K (von Hessberg et al., 2004 and references therin), compared to theoretical models and a broadband photolysis study.

Using available information (9.8) can be solved to constrain S_i, the strengths of the individual source terms. The value of δ_{obs} is relatively easy to determine, and several groups around the world have established reliable δ_i values. The loss processes L_i are straightforward: 90% photolysis and 10% reaction with $O(^1D)$. Isotopic substitution has been shown to have an important effect on the absorption cross sections of nitrous oxide isotopomers and isotopologues (von Hessberg et al., 2004). The photolysis isotopic fractionation constant ε_i for $^{14}N^{15}N^{16}O$ as a function of wavelength is shown in Figure 9.9b. It can be seen that at a wavelength of 205 nm, at a stratospheric temperature of 233 K, the photolysis cross section of $^{14}N^{15}N^{16}O$ is ca. 8% smaller than that of $^{14}N^{14}N^{16}O$. For nitrous oxide the heavy isotopologues and isotopomers are photolysed more slowly. The atmosphere is enriched in these species, and according to Equation (9.8) this enrichment is balanced by isotopically light biological sources, relative to the atmosphere's composition.

Nitrogen oxides, collectively called NO_x, are the most important radical family in terms of ozone depletion. They have a rich photochemistry. As discussed in the previous section the source of stratospheric NO_x is nitrous oxide

$$N_2O + O(^1D) \rightarrow 2NO \tag{9.9}$$

NO reacts with O_3 to produce NO_2 which is easily photolysed, cf. Figure 9.9:

$$NO + O_3 \rightarrow NO_2 + O_2 \tag{9.10}$$

$$NO_2 + h\nu \; (\lambda < 415 \text{ nm}) \rightarrow NO + O(^3P) \tag{9.11}$$

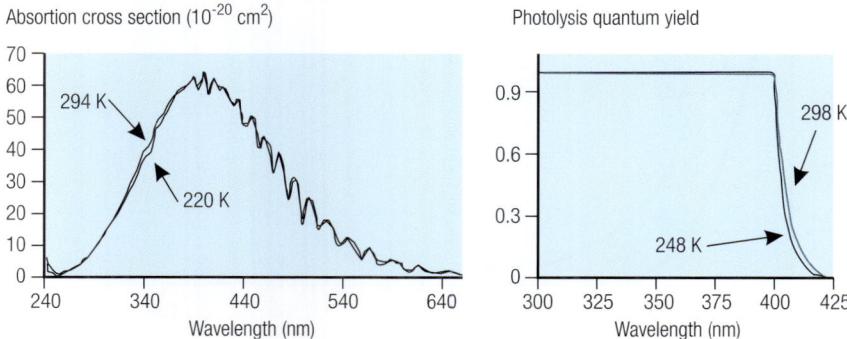

Figure 9.10. The visible absorption spectrum of the nitrogen dioxide molecule. (a) Absorption cross section. (b) Photolysis quantum yield for $NO_2 + h\nu \rightarrow NO + O$. Based on data from JPL-06

The oxygen atom reforms ozone resulting in a null cycle, but the photolysis of NO_2 is far more important than as a small source of heat.

The photolysis of NO_2 competes with the reaction:

$$NO_2 + O \rightarrow NO + O_2 \tag{9.12}$$

The combination of 9.10 and 9.11 is the main ozone destruction mechanism in the middle of the troposphere. At night, when NO_2 is not photolysed, it can react as follows:

$$NO_2 + O_3 \rightarrow NO_3 + O_2 \tag{9.13}$$

$$NO_2 + NO_3 + M \rightarrow N_2O_5 + M \tag{9.14}$$

Dinitrogen pentoxide (anhydrous nitric acid) serves as an overnight reservoir for NO_x radicals, i.e. NO and NO_2, since it is quickly photolysed when the sun comes up (cf. Figure 9.10). N_2O_5 can also be hydrolysed on stratospheric aerosols:

$$N_2O_5 + H_2O_{(s)} \rightarrow 2HNO_{3(s)} \tag{9.15}$$

Reactive NO_x is converted to inorganic NO_y. The other sink for NO_x radicals is reaction with OH:

$$NO_2 + OH + M \rightarrow HNO_3 + M \tag{9.16}$$

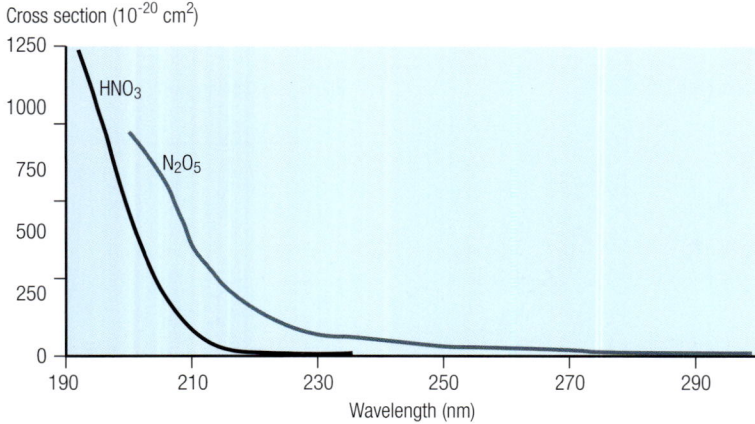

Figure 9.11 Absorption spectra of HNO_3 and N_2O_5. Based on data from JPL-06.

The fate of nitric acid is either deposition, reaction with OH or photolysis (cf. Figure 9.11):

$$HNO_3 + h\nu \ (\lambda < 220 \text{ nm}) \rightarrow OH + NO_2 \tag{9.17}$$

The photolysis cross section and photolysis quantum yields for NO_3 are shown in Figure 9.12. The nitrate radical illustrates several of the possible fates of the excited state:

$$NO_3 + h\nu \rightarrow NO_3^*$$

$$NO_3^* \rightarrow NO + O_2$$

$$NO_3^* \rightarrow NO_2 + O(^1D)$$

$$NO_3^* \rightarrow NO_3 + h\nu$$

$$NO_3^* + M \rightarrow NO_3 + M^*$$

The visible band extends from 400 to 700 nm, as shown in Figure 9.12. At wavelengths shorter than 585 nm the fate of the excited state is to produce NO_2 and ground state oxygen atoms. At longer wavelengths the NO + O_2 channel becomes competitive, but quantum yields decline long before the absorption cross section goes to zero. The fate of nitrate radicals excited at these longer wavelengths is either to re-radiate the light or to be collisionally quenched.

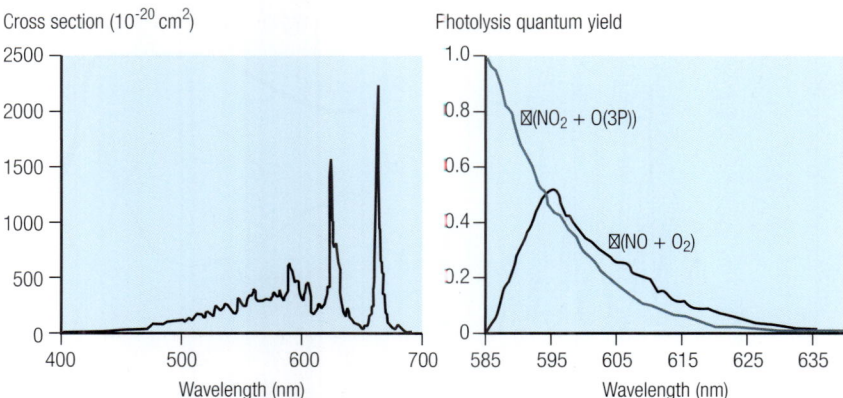

Figure 9.12 The visible absorption of the nitrate radical. (a) Absorption cross section. (b) Photolysis quantum yield (ϕ). Based on data from JPL-06.

Photolysis in the troposphere

Nitrogen dioxide, formaldehyde and acetaldehyde

During its passage through the stratosphere solar radiation has lost a great deal of its UV radiation (cf. Figure 9.4). Only wavelengths larger than 290 nm penetrate into the troposphere. Hence photochemistry is limited to a few species, most importantly O_3, NO_2 and some organic compounds, mainly aldehydes with formaldehyde (HCHO) and acetaldehyde (CH_3CHO) as the most important examples. The absorption cross section and photolysis quantum yield for NO_2 are summarised in Figure 9.10. The photolysis of O_3, discussed in the previous section, is the first source of OH in the troposphere.

There are two product channels for the photolysis of formaldehyde in the tropospheric ultraviolet region:

$$HCHO + h\nu \rightarrow H + HCO \qquad (\lambda \leq 330 \text{ nm}) \qquad\qquad (9.18)$$

$$HCHO + h\nu \rightarrow H_2 + CO \qquad (\lambda \leq 361 \text{ nm}) \qquad\qquad (9.19)$$

The fate of CHO is to be converted to CO and HO_2 by reaction with O_2; H also forms HO_2 by reaction with O_2. The absorption spectrum and photolysis quantum yields for formaldehyde are shown in Figure 9.13.

Dynamics appear to play a role in the branching ratios between the molecular and radical channels. The cross section and photolysis quantum yield for acetaldehyde are shown in Figure 9.14. Acetaldehyde is produced in the oxidation of many hydrocarbons and its photolysis is an important source of HO_x radicals.

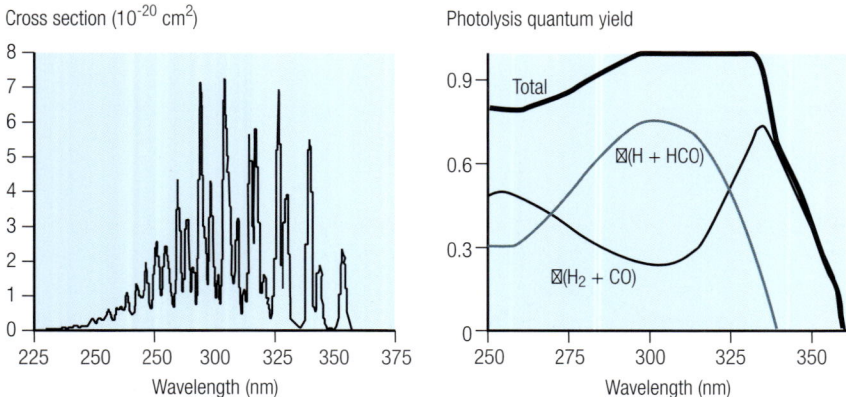

Figure 9.13 Absorption cross section and photolysis quantum yield for formaldehyde, HCHO at 298 K. Based on data from JPL-06.

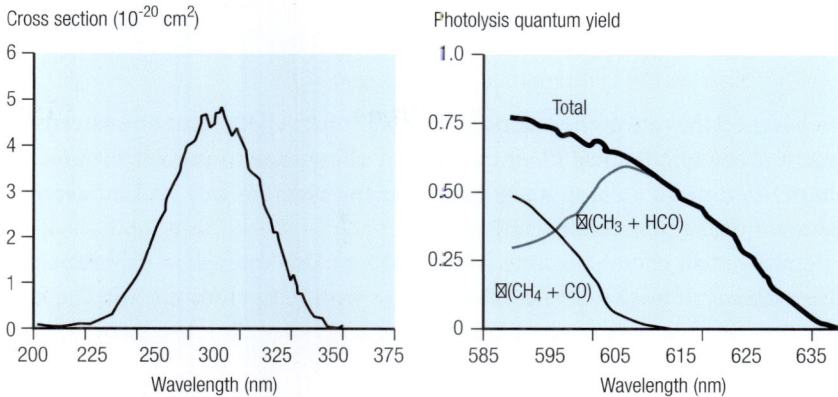

Figure 9.14 Absorption cross section and photolysis quantum yield for acetaldehyde, CH_3CHO at 298 K. Based on data from JPL-06.

9.3 Second order reactions

This type of reaction occurs frequently in the atmosphere between closed shell molecules and radicals. One bond is exchanged for another, resulting in a new radical and a new closed shell molecule, thus second order reactions serve to propagate atmospheric chain reactions. The general form is:

$$A + B\cdot \rightarrow X + Y\cdot \qquad (9.20)$$

For these reactions the rate coefficient k has dimensions of cm^3 s^{-1}. Units of cm^3 $molecule^{-1}$ s^{-1} are also encountered however 'molecule' is not a proper unit according to IUPAC and the International System of units.

One example of a second order reaction is the oxidation of methane:

$$CH_4 + OH \rightarrow H_2O + CH_3 \qquad (9.21)$$

Here a CH bond is broken and an OH bond formed. Thermodynamics will tell whether such a reaction is allowed, but equally important is the *rate* at which it occurs, the topic of chemical kinetics.

It is a good approximation that many atmospheric processes removing gases are first order in concentration. Consider (9.20) for the common case of [A] >> [B·]. For a well-mixed system the chemical lifetime can be defined for A with respect to removal by B·:

$$\tau = \frac{[A]}{r_L} = \frac{[A]}{k[A][B]} = \frac{1}{k[B]} \tag{9.22}$$

Here is used the rate of the reaction $r_L (cm^{-3}\, s^{-1}) = k[A][B]$. Thus one extremely practical use of chemical kinetics is that it allows one to predict the atmospheric lifetime of a compound based on the reaction rate and an average value for the radical concentration .

The Swedish chemist Svante Arrhenius was able to explain experimental observations of how reaction rates change with temperature using the rate equation that has been named in his honor:

$$k = Ae^{-E/RT} \tag{9.23}$$

The Arrhenius equation describes many experimental observations using only two variable parameters, a pre-exponential factor A and an energy E. E is often referred to as the activation energy of the reaction. The Arrhenius equation works because it describes two fundamental aspects of many chemical reactions, namely, there is some probability that the reagents will have the correct orientation (A), and there is a reaction barrier that must be overcome. The reaction barrier is overcome using internal energy, whether this is translational, rotational, vibrational or electronic. At thermal equilibrium the energies found in the different degrees of freedom will be described by a Boltzmann distribution. Thus the probability that a given collision can overcome the reaction barrier is an exponential function involving the opposite of $1/T$.

Table 9.1 summarizes the Arrhenius parameters for some of the important processes removing methane from the troposphere and stratosphere. The excited state of oxygen $O(^1D)$ reacts with no barrier, whereas $O(^3P)$ does not react. The OH and Cl reactions both proceed with significant reaction barriers and temperature coefficients.

	A-factor $(cm^3\, s^{-1})$	E/R (K)	k(298 K) $(cm^3\, s^{-1})$
$O^1D + CH_4$	1.5×10^{-10}	0	1.5×10^{-10}
$O^3P + CH_4$	No Reaction	No Reaction	No Reaction
$OH + CH_4$	2.45×10^{-12}	1775	6.3×10^{-15}
$Cl + CH_4$	7.3×10^{-12}	1280	1.0×10^{-13}

Table 9.1 Arrhenius parameters for some methane oxidation reactions.

Second order reactions in the stratosphere

A variety of radicals are found in the stratosphere, including OH, Cl and O. The concentration of water vapour in the stratosphere has increased dramatically. By one estimate there is 50% more water vapour in the stratosphere now than there was in 1950 (Rosenlof et al., 2003). One cause may be more powerful convective storms in the tropics that transport water across the tropopause through convective lofting (circumventing the 'cold trap'), but another is chemical. The atmosphere contains more H_2 and CH_4 than it did only a short time ago. These molecules are oxidized via second order reactions in the stratosphere and their hydrogen ends up in the form of water vapour.

$$CH_4 + O(^1D) \rightarrow CH_3 + OH \tag{9.24}$$

$$CH_4 + OH \rightarrow CH_3 + H_2O \tag{9.25}$$

$$H_2 + O(^1D) \rightarrow H + OH \tag{9.26}$$

$$H_2 + OH \rightarrow H + H_2O \tag{9.27}$$

Increased stratospheric water vapour leads to an increased amount of stratospheric ice particles. These ice surfaces make the stratosphere more sensitive to chlorine-catalysed ozone depletion than it would otherwise be; this is of particular concern at mid-latitudes. This issue is discussed further in Section 9.5.

Second order reactions in the troposphere

The production of air pollution in the troposphere is fueled by hydrocarbons. Catalytic reactions based on the HO_x and NO_x radical families produce ozone and oxidized hydrocarbons that condense to form particles. The combination of ozone and particles is known as smog. NO_x is found in the troposphere due to fossil fuel combustion, cf. Figure 9.2. The formation of HO_x is initiated by the photolysis of O_3:

$$O_3 + h\nu(\lambda < 305 \text{ nm}) \rightarrow O(^1D) + O_2(^1\Delta_g) \tag{9.5}$$

$$O(^1D) + H_2O \rightarrow 2OH \tag{9.28}$$

$$O(^1D) + M \rightarrow O(^3P) + M \tag{9.29}$$

Hydroxyl radicals are able to abstract a hydrogen atom from virtually any atmospheric pollutant containing one, for example CH_4, HNO_3, C_6H_6, isoprene, etc.

$$RH + OH \rightarrow R + H_2O \tag{9.30}$$

$$R + O_2 + M \rightarrow RO_2 + M \tag{9.31}$$

$$RO_2 + NO \rightarrow RO + NO_2 \tag{9.32}$$

Consider the last carbon atom on the alkoxy radical RO, that is, let RO = $R'CH_2O$:

$$R'CH_2O + O_2 \rightarrow R'CHO + HO_2 \tag{9.33}$$

$$HO_2 + NO \rightarrow OH + NO_2 \tag{9.34}$$

The cycle begins with the photolysis of ozone to produce an excited oxygen atom. Usually this atom will be quenched (9.29) and reform ozone, but a small fraction of the time it will react with water (9.28) to form two hydroxyl radicals. This reaction is an important sink for ozone in the clean troposphere.

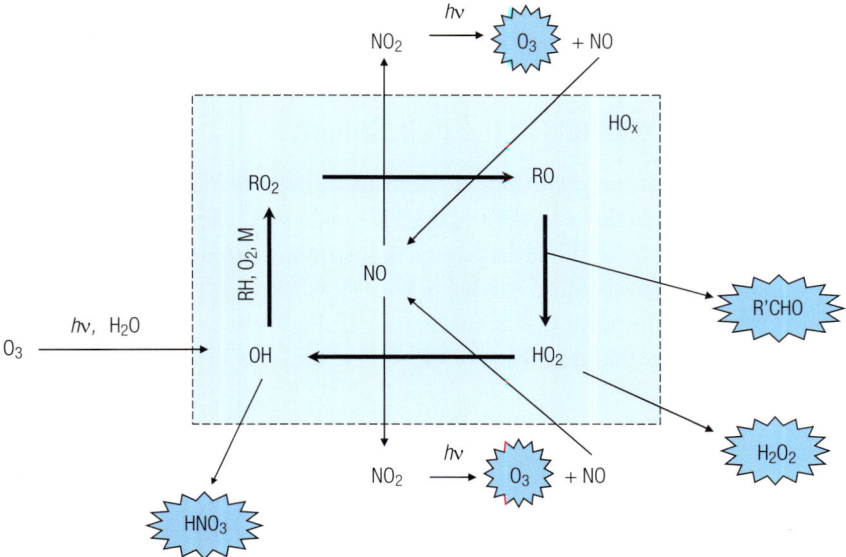

Figure 9.15 The photochemical cycle of the troposphere. The reactions are explained in the text.

The hydroxyl radical will react with a hydrocarbon to produce an alkyl radical, R, which quickly reacts with molecular oxygen. Molecular oxygen is unusual in having a triplet ground state and these two unpaired electrons make it an easy reaction partner for atmospheric radicals. In high NO_x conditions the alkyl peroxyl radical will react with NO to produce NO_2 and an alkoxy radical, 9.32. As shown in 9.33 this radical will form an aldehyde and an HO_2 radical, which is converted to OH via 9.34. In the polluted atmosphere reaction 9.34 is a more important source of OH than is the photolysis of ozone. The reaction sequence is summarized in Figure 9.15.

RO and RO_2 are considered to be part of the HO_x radical family in the troposphere because they are so closely linked to OH and HO_2. As shown in the Figure, there are two main mechanisms removing HO_x in the troposphere. The first is the self-reaction of HO_2 radicals:

$$HO_2 + HO_2 \rightarrow H_2O_2 + O_2 \tag{9.35}$$

And the second is the production of HNO_3 from OH and NO_2. This is an association reaction and will be discussed in the following section. If the first of these two chain termination reactions is most important then the conditions are said to be low- NO_x. Atmospheric chemistry is then limited by the conversion of HO_2 into OH via NO. In the second case the atmosphere is said to be high NO_x, and the catalytic cycle is limited by the reaction of OH with RH.

The reaction sequence generates two NO_2 molecules each time the cycle goes around. These will be photolysed according to 9.11 and the oxygen atom thus produced will form ozone, an association reaction discussed in the next section. Figure 9.15 shows how photochemical smog is produced in the troposphere: ozone is produced via NO_2, and oxygenated species such as aldehydes are formed that are able to condense into particles.

Hydroxyl

Hydroxyl is central to tropospheric chemistry, determining the atmospheric lifetime of many pollutants. It has a high reactivity and can abstract hydrogen atoms from carbon, nitrogen, chlorine, etc. In addition it can react by addition to double bonds. The initial HO· is due to the UV photolysis of ozone:

$$O_3 + h\nu(\lambda < 305 \text{ nm}) \rightarrow O(^1D) + O_2 \tag{9.5}$$

followed by:

$$O(^1D) + H_2O \rightarrow 2OH \tag{9.28}$$

in competition with:

$$O(^1D) + M \rightarrow O(^3P) + M \tag{9.29}$$

However, once OH and HO_x are present, more HO· is formed by the cycle shown in Figure 9.14, in particular:

$$HO_2 + NO \rightarrow OH + NO_2 \tag{9.34}$$

The mean tropospheric concentration of OH has been estimated by considering the atmospheric budget of methyl chloroform, CH_3CCl_3, using the mass balance equation $dm/dt = r_p - r_L$. The change in atmospheric mass with time depends on the difference between the rate of production r_p and the rate of loss r_L. The concentration of methyl chloroform in the atmosphere is easily measured, and from this one can determine the time derivative, dm/dt. Methyl chloroform has only anthropogenic sources and chemical industry production statistics allow an accurate determination of the rate of production, r_p. Methyl chloroform is removed from the atmosphere through reaction with OH:

$$CH_3CCl_3 + OH \rightarrow CH_2CCl_3 + H_2O \tag{9.36}$$

The rate coefficient of this reaction k has been determined in the laboratory. The rate of removal is therefore $r_L = k\ [CH_3CCl_3]\ [OH]$. Every term in the mass balance equation can be identified allowing the determination of the tropospheric global average concentration of OH, $1.2 \cdot 10^6$ cm^{-3}. One advantage of the method is that methyl chloroform is well-mixed in the troposphere so this method results in a true global average. A disadvantage is that the production of methyl chloroform was banned by the Copenhagen Amendment to the Montreal Protocol. Within a few years the atmospheric concentration of CH_3CCl_3 will be too small to detect.

9.4 Association reactions

Association reactions are especially important in atmospheric chemistry because they terminate chain reactions, and thus determine the efficiency of catalytic cycles. This occurs when two open shell radicals combine to form a stable product. In addition some bimolecular reactions occur through a three-body type process, in which a collision partner is needed to remove excess energy from a reaction intermediate. The general form is as follows:

$$A + B + M \rightarrow AB + M \quad k_3 \text{ with dimensions } cm^6 \, s^{-1} \tag{9.37}$$

This is a composite reaction made of three elementary steps:

$$A + B \rightarrow AB^* \qquad \text{association, } k_a \tag{9.38}$$

$$AB^* \rightarrow A + B \qquad \text{dissociation, } k_d \tag{9.39}$$

$$AB^* + M \rightarrow AB + M \qquad \text{quenching. } k_q \tag{9.40}$$

AB^* is a short-lived intermediate, and so we can use the steady state approximation:

$$\frac{d[AB^*]}{dt} = k_a[A][B] - k_d[AB^*] - k_q[AB^*][M] = 0 \tag{9.41}$$

Solving this for $[AB^*]$:

$$[AB^*] = \frac{k_a[A][B]}{k_d + k_q[M]} \tag{9.42}$$

We can now describe the rate of the overall process in terms of the three elementary processes:

$$\frac{d[AB]}{dt} = k_q[AB^*][M] = \frac{k_q k_a[A][B][M]}{k_d + k_q[M]} \tag{9.43}$$

One characteristic is that without the third body M, reaction is impossible. Another is that k_d increases with increasing temperature, meaning that three body reactions tend to get faster as temperature decreases, the opposite of what is predicted by the Arrhenius equation.

Association reactions in the stratosphere: Ozone and the catalytic cycles

It is clear that the presence of ozone is the defining characteristic of the stratosphere. The ozone layer was first described using the Chapman reactions:

$$O_2 + hv \rightarrow 2O \tag{9.44}$$

$$O + O_2 + M \rightarrow O_3 + M \tag{9.45}$$

$$O_3 + hv \rightarrow O + O_2 \tag{9.46}$$

$$O + O_3 \rightarrow 2O_2 \tag{9.47}$$

The second and third of these reactions convert solar UV radiation (Figure 9.4) into heat, giving the stratosphere its characteristic temperature inversion. The absorption of UV radiation in the stratosphere protects the surface from high energy radiation, making life outside of the oceans possible. Ozone depletion is a grave public health and environmental concern.

Ozone is formed through an association reaction, 9.45 and is removed by the fourth reaction in the Chapman scheme, but this reaction is only a fraction of the total stratospheric loss rate for ozone. There are many catalytic cycles of the form:

$$O_3 + X \rightarrow O_2 + XO \tag{9.48}$$

$$XO + O \rightarrow X + O_2 \tag{9.49}$$

$$\text{Sum: } O + O_3 \rightarrow 2O_2 \tag{9.50}$$

Examples of X include OH, NO, Cl and Br. The efficiencies of the radical cycles are determined by the processes that remove the radicals. In addition the radical families are linked to one another through association reactions. The OH + NO_2 reaction described in the next section is a good example of such a link, since it removes both HO_x and NO_x radicals from the stratosphere. Another example is the formation of chlorine nitrate from the chlorine monoxide radical and nitrogen dioxide:

$$ClO + NO_2 + M \rightarrow ClONO_2 + M \tag{9.51}$$

The rate of this reaction is shown in Figure 9.16.

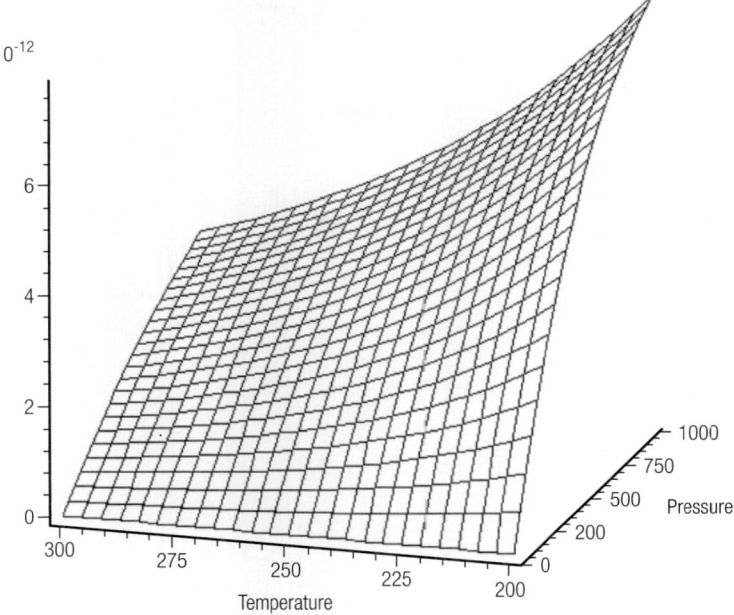

Figure 9.16. The rate of association (k, cm^3 s^{-1}) of ClO and NO$_2$ at pressures (p, mbar) and temperatures (T, K) relevant to the troposphere and stratosphere.

Troposphere: The formation of nitric and sulphuric acids

The main anthropogenic components of acid rain are nitric and sulfuric acids. Both are formed by three-body association reactions. Figure 9.17 shows the overall, second-order rate coefficient of the association reaction

$$OH + NO_2 + M \rightarrow HNO_3 + M \qquad (9.52)$$

as a function of temperature and pressure. As shown in the Figure the rate decreases with increasing temperature. This temperature dependence is typical for a three body reaction, since the lifetime of the intermediate complex AB* decreases with increasing temperature. Also typical for an association reaction, the rate increases with pressure. This reaction is in the fall-off region at all atmospheric conditions.

About half of anthropogenic acid rain is nitric acid generated from NO$_x$ emissions via 9.52. The other half is sulphuric acid; the most important precursor is SO$_2$ produced by coal combustion. Sulphur dioxide in the troposphere reacts in the gas phase and in droplets, in addition to wet and dry deposition. The gas phase lifetime is about a week. The oxidation is initiated

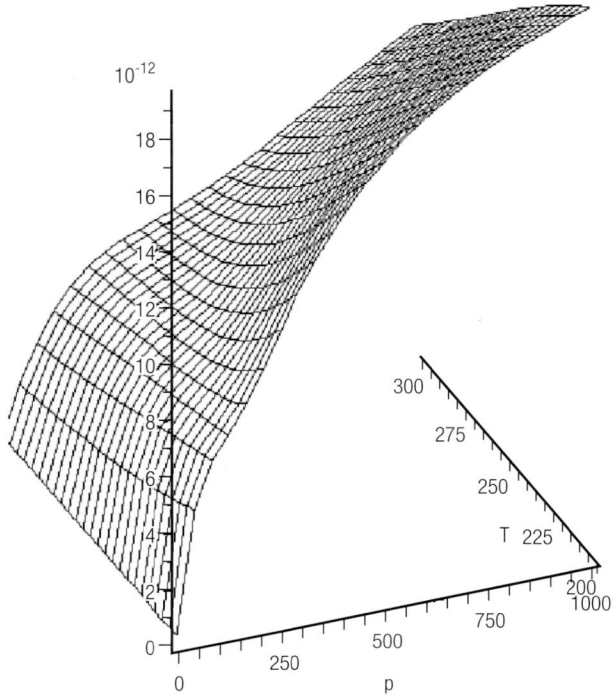

Figure 9.17 The rate of association (k, cm³ s⁻¹) of OH and NO_2 at pressures (p, mbar) and temperatures (T, K) relevant to the troposphere and stratosphere.

by the hydroxyl radical:

$$SO_2 + OH + M \rightarrow HOSO_2 + M \qquad (9.53)$$

The HO_x radical is regenerated in the following reaction:

$$HOSO_2 + O_2 \rightarrow HO_2 + SO_3 \qquad (9.54)$$

Sulphur trioxide reacts rapidly with water to produce sulphuric acid:

$$SO_3 + H_2O + M \rightarrow H_2SO_4 + M \qquad (9.55)$$

The rate-limiting step is 9.53. The rate coefficient for this reaction as a function of temperature and pressure for atmospheric conditions is shown in Figure 9.18.

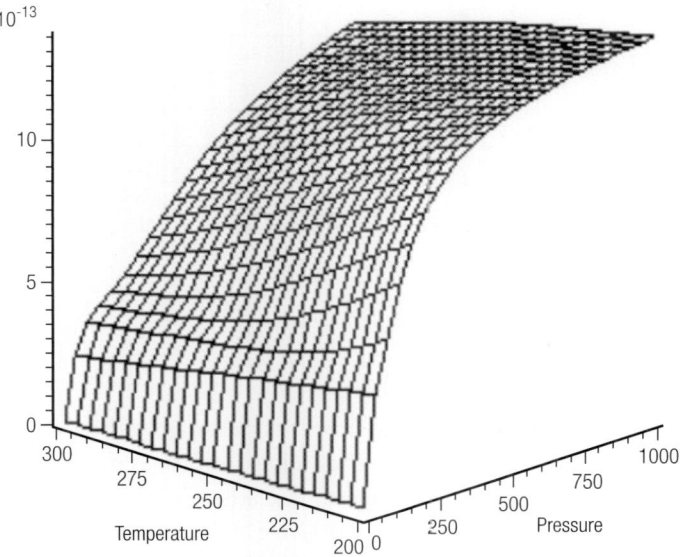

Figure 9.18. The rate of association (k, cm³ s⁻¹) of OH and SO₂ at pressures (p, mbar) and temperatures (T, K) relevant to the troposphere and stratosphere.

Table 9.2 shows the three-body rate coefficient for a series of association reactions between the oxygen molecule and several small radicals. These reactions occur on similar potential energy surfaces; the energy of the bond formed and the activation barrier is very similar. However, the rate of reaction varies by over a factor of a million. This is because when the reagents come together they form a new bond, and the bond energy is released into the association complex AB*. By definition, it is enough energy to break the bond; the activated complex is metastable. However, as described by RRKM theory, the location of the energy is a matter of probability. In a system such as O_3^* with few vibrational degrees of freedom, there is a high probability that the energy will be localized in the O_2-O bond, thereby breaking it. Because the dissociation rate of the intermediate complex is so fast there is not adequate time for collisional quenching and overall, the association reaction is slow. In the case of $C_2H_5O_2^*$, in contrast, there are 21 vibrational modes. Statistically there is a much lower probability that the energy will be localized in the C_2H_5-O_2 bond. The intermediate has a longer intrinsic 'unimolecular' lifetime, allowing adequate time for collisional quenching and a dramatically faster overall rate constant.

A	B	X	k_0(300K; cm^6 s^{-1})
O	O_2	O_3	6.0×10^{-34}
CH_3	O_2	CH_3O_2	4.5×10^{-31}
CF_3	O_2	CF_3O_2	3.0×10^{-29}
C_2H_5	O_2	$C_2H_5O_2$	1.5×10^{-28}

Table 9.2 Comparison of rates for some thermolecular reactions A + B + M → X + M

9.5 Heterogeneous Chemistry

Atmospheric particles are a matter of significant interest due to their effects on human health and climate. In addition they allow new types of chemical reactions occur that cause important changes in the atmosphere.

Small aerosol particles are able to pass directly through the nose and deposit in the lungs, introducing foreign material into the blood stream. The effects of inhaling particulate matter include asthma, lung cancer, cardiovascular disease and death (Chapter 15).

The condensed phase has a profound impact on the chemistry of the atmosphere (Martin et al., 2003). For example the Antarctic ozone hole would not exist were it not for heterogeneous chemistry, and clouds have an effect on climate that is as important as that of the greenhouse gases. The chemical composition of the atmosphere changes when clouds are present. The introduction to this section will describe fundamental phenomenon such as deposition and condensation, and specific examples from the stratosphere and troposphere will be given below.

A few basic concepts determine the physical and chemical distribution of aerosols in the atmosphere. The first is the deposition time. There is a competition between gravitational settling of a particle and mixing (both Brownian motion and turbulence), meaning that particles above a certain size will not remain in the atmosphere for a significant length of time. All atmospheric particles achieve their steady-state settling velocities within a fraction of a second. The settling velocity is given by the formula (Seinfeld and Pandis, 2006):

$$v_t = \frac{\mu \, \mathrm{Re}}{\rho D_p} \qquad\qquad (9.56)$$

Here v_t is the settling velocity, μ is the gas viscosity, Re is the Reynolds number, ρ is the density of the particle and D_p its diameter. For example, a water droplet with a diameter of 200 µm has a terminal velocity of 75 cm s^{-1}.

An important point is that particles with diameters larger than roughly 2 µm have relatively short residence times, since the effects of eddy diffusion and turbulence are not enough to keep them aloft. Examples include wind-blown dust, sea spray and plant particles.

One way to form new particles is through the condensation of vapours. Small particles (so-called Aitkin nuclei) in the size range of 10 to 100 nm then grow through coagulation and condensation. The intermediate particle range, from 100 nm to 1 µm, is known as the accumulation mode. This size range is too small to fall out of the atmosphere. Roughly speaking Aitkin nuclei represent the greatest number of particles in the atmosphere, accumulation mode particles represent the greatest surface area, and large sedimentation class particles represent the greatest volume.

Vapour pressure increases with surface curvature, and this means that if a large particle and a small particle are placed next to each other, the small particle will shrink and the large particle will grow. There is a characteristic water supersaturation necessary for a given particle to become activated, that is, to start to grow. The point of activation is a chemical property, linked to the hygroscopicity of the molecules in the particle. A given particle core may go through several clouds before it is removed from the atmosphere by deposition, each cloud episode representing a cycle of supersaturation by water vapour, growth and dehydration. Cloud formation in large portions of the atmosphere is limited by the availability of cloud condensation nuclei and thus anthropogenic pollution has an important effect on climate. Industrial activity can be an important regional source of cloud condensation nuclei, in the form of sulphates generated by coal combustion, soot and mechanically generated dust.

The most important natural sources of particles are dust, volcanoes, forest fires and sea spray. Anthropogenic sources are mainly linked to combustion through transportation and power generation, in addition to processes that generate dust such as construction, farming and deforestation. In addition natural and anthropogenic emissions include precursors that can form secondary aerosol particles when they are oxidized in the atmosphere.

Mineral dust reflects the composition of the crust, including iron and aluminium oxides, calcium carbonate and silicates. Sea salt particles formed from sea spray have a large component of sodium chloride. Secondary aerosols can involve nitrates, sulphates and organic acids in addition to ammonium. Carbon in aerosols can be reduced or oxidized, or in the form of black carbon soot, so-called elemental carbon.

Heterogeneous Chemistry in the Stratosphere

Christian Junge was the first to discover a layer of sulphate aerosols found in the stratosphere, using balloon-borne mass spectrometers. The layer, concentrated in the lower stratosphere, consists mainly of an aqueous sulphuric acid solution at 60 to 80%, at temperatures of -80 to -45 °C respectively. Volcanoes inject significant quantities of sulphate into the stratosphere, resulting in beautiful sunsets and cooler temperatures for a year or two after the eruption. Significant volcanic eruptions in the 20th century include Agung (1963), El Chichón (1982) and Pinatubo (1991). Stratospheric sulphate scatters incoming sunlight, cooling the planet, and leads to mid-latitude ozone depletion since it makes the ozone layer more vulnerable to chlorine-containing gases, as discussed below. The source of sulphate in non-volcanic periods is a matter of debate. The first source to be identified (Crutzen, 1976) was the photolysis of OCS, the most abundant sulphur containing species in the atmosphere. However, recent models have indicated that lofting of tropospheric SO_2 and sulphate through the tropical tropopause may inject even more sulphur.

There were several barriers to understanding the formation of the Antarctic ozone hole. One of the most significant was explaining how chlorine monoxide radicals could be present at such high concentrations. Normally, over 90% of the inorganic chlorine in the stratosphere is found in HCl and $ClNO_3$. (Inorganic chlorine is denoted ClO_y, as opposed to chlorine radicals, Cl and ClO, denoted ClO_x.). The ClO concentrations observed using microwave limb spectrometry on a NASA ER-2 mission into the ozone hole showed that over 90% of chlorine was present as ClO_x, not ClO_y. The chemical mechanism was first explained by Susan Solomon and co-workers (Solomon et al., 1986):

$$HCl + ClNO_3 \rightarrow Cl_{2(g)} + HNO_{3(s)} \tag{9.57}$$

This reaction occurs on polar stratospheric clouds, a special kind of ice particle made of sulphuric and nitric acids and water. The PSCs only form at extremely low temperatures, and the temperatures in the Antarctic polar vortex are the lowest found anywhere in the atmosphere. During the Antarctic winter the temperature drops, forming PSCs at the dry conditions found in the stratosphere. ClO_y is processed to form chlorine gas, and other similar heterogeneous reactions form HOCl. When spring arrives the sun comes out:

$$Cl_2 + h\nu \rightarrow 2Cl \tag{9.58}$$

$$HOCl + h\nu \rightarrow OH + Cl \tag{9.59}$$

$$Cl + O_3 \rightarrow ClO + O_2 \qquad\qquad (9.60)$$

$$\tfrac{1}{2}(ClO + ClO \rightarrow Cl_2O_2 \rightarrow Cl + ClO_2 \rightarrow 2Cl + O_2) \qquad\qquad (9.61)$$

$$\text{Net: } O_3 \rightarrow 3/2\ O_2$$

Concentrations of chlorine monoxide radicals are so anomalously high in the Antarctic that they are able to react with themselves via 9.61.

There are a number of reasons why the Antarctic ozone hole, once it forms, does not heal easily. As shown in 9.57 the nitric acid produced by the reaction stays in the condensed phase. This conversion of NO_x to NO_y (and subsequent loss of NO_y through sedimentation) means that NO_2 is not available to sequester ClO through the association reaction 9.51. A key property of stratospheric ozone is that it absorbs solar UV radiation, producing heat that gives the stratosphere its characteristic temperature distribution. Without ozone temperatures in the Antarctic vortex remain low, and therefore favorable for PSCs. The remaining mechanism for removing ClO_x from the vortex is the reaction of chlorine atoms with methane:

$$CH_4 + Cl \rightarrow CH_3 + HCl \qquad\qquad (9.62)$$

The concentration of ozone from 60°S to 60°N latitude has dropped by 6% since the early 1980s (WMO, 2002), a matter of major public health and environmental concern. The mechanism behind this involves several anthropogenic changes resulting from air pollution. The concentration of water vapour in the stratosphere appears to have increased by 50% since the second world war (Rosenlof, 2003). This is thought to be due to a combination of increased concentrations of water precursors CH_4 and H_2 in the stratosphere, and increased convective activity near the equator that lofts more water vapour across the tropopause. In addition, climate change has lead to not only an increase in the surface temperature, but also to a decrease in stratospheric temperature. More water vapour and lower temperature have lead to more stratospheric aerosol particles. These particles are able to hydrolyse N_2O_5 that otherwise would only be a temporary reservoir for NO_x during the night:

$$NO + O_3 \rightarrow NO_2 + O_2 \qquad\qquad (9.10)$$

$$NO_2 + O_3 \rightarrow NO_3 + O_2 \qquad\qquad (9.13)$$

$$NO_2 + NO_3 + M \rightarrow N_2O_5 + M \qquad\qquad (9.14)$$

$$N_2O_5 + H_2O_{(s)} \rightarrow 2HNO_{3(s)} \qquad\qquad (9.15)$$

The hydrolysis of dinitrogen pentoxide leads to the denitrification of the stratosphere. With less NO_x available, there is less NO_2 to convert HO_x and ClO_x into inactive forms:

$$ClO + NO_2 + M \rightarrow ClONO_2 + M \qquad\qquad (9.51)$$

$$OH + NO_2 + M \rightarrow HNO_3 + M \qquad\qquad (9.52)$$

The net impact is to increase the efficiency of the HO_x and ClO_x ozone-removing cycles, leading to ozone depletion.

Heterogeneous Chemistry in the Troposphere

Particulate pollution in the troposphere can be of two kinds, primary and secondary. Primary particles are generated at the surface and include emissions from industry (soot, oven exhaust) and agriculture, dust from construction and transportation (tires, brakes, roads), and so on. Secondary particles condense within the atmosphere from chemical precursors. Examples include terpenes and other aromatics that become oxidized, in addition to conversion of SO_2 and NO_2 into H_2SO_4 and HNO_3.

There is a significant thermodynamic barrier to the nucleation of new particles in the atmosphere. The development of a successful first-principles theory for nucleation is a significant challenge. There are many possible mechanisms of nucleation, including (Seinfeld and Pandis, 2006):

Sulphuric acid-water binary nucleation
Sulphuric acid-ammonia-water ternary nucleation
Nucleation of organic vapours
Ion-induced nucleation
Halogen oxide nucleation

The nucleation rate in the atmosphere cannot be measured directly since the finest particle sizers currently available have a threshold of 3 nm. The formation rate of particles at this size ranges from 0.01 to 10 cm^{-3} s^{-1} in the boundary layer and up to 100 cm^{-3} s^{-1} in urban areas. New particles quickly form larger particles due to coagulation. The main species fuelling aerosol growth is sulphuric acid, especially in polluted areas, but organic species and heterogeneous chemical reactions also play an important role.

Tropospheric particles have many effects, one of the most important being the correlation between particle concentrations and mortality (Chapter 15). In addition particles have important effects on climate and atmospheric chemistry (Chapter 20).

The effects on climate are generally divided into two categories, direct and indirect. The direct effect arises because particles scatter light, increasing the Albedo of the planet. (Many particles, such as soot, also absorb sunlight and can warm the surface, but this effect is not as important as scattering). The distribution of scattered light depends on the relationship between the wavelength of light and the particle size. Rayleigh scattering results from particles smaller than the wavelength of light, Mie scattering if the particle and wavelength are of similar dimensions and geometric scattering for particles large compared to the wavelength. For visible light, particles with sizes less than about 100 nm give rise to Rayleigh scattering; here the amount of forward and backscattered light is equal. The intensity of the scattered light scales as the inverse fourth power of wavelength, thus blue light scatters much more easily than red, explaining the intensely red sunsets after a major volcanic eruption, and in part explaining the colour of the sky in polluted urban areas.

At the other extreme are the large water droplets found in clouds and falling as rain, which give rise to geometric scattering. Rainbows are one example of this phenomenon.

Clouds have a complicated effect on climate, since they cool the surface by scattering visible sunlight, and can warm the surface by absorbing infrared heat. Roughly speaking high clouds give a positive radiative forcing, warming the surface. This is because they are relatively diffuse and not efficient at scattering incoming sunlight. In addition the upper troposphere is quite cold and so these clouds are not effective at radiating infrared energy to space; they are able to absorb some of the earth's infrared heat before it escapes the atmosphere and re-radiate it to the surface. Low clouds cool the surface by reflecting sunlight to space.

Atmospheric particles also have an indirect effect on climate, distinct from the direct effect (absorption and scattering of light). The cooling due to the indirect effect is possibly as large as the warming effect of anthropogenic CO_2; the indirect aerosol climate effect is the subject of intense research. The indirect effect can be understood by considering that the presence of more cloud condensation nuclei in a cloud would mean that it would have more and smaller water droplets, for the same total amount of condensed water. Thus pollution, by increasing the concentration of cloud condensation nuclei and changing the distribution of water particle sizes within a cloud, can make clouds whiter. The geographical distribution of climate change shows that warming has not been as rapid over the industrial regions of Europe, North America and Asia.

A recent study has modelled the photochemical effects of aerosols on tropospheric oxidants (Martin et al., 2003). The driving forces were scattering and absorption of UV radiation by aerosols and the reactive uptake of HO_2, NO_2 and NO_3. It was found that aerosols decrease the rate of production of OH from O_3 by 5 to 20% at the surface, throughout the northern hemisphere, and by a factor of two in biomass burning regions, largely due to black carbon. Aerosol uptake accounts for from 10 to 40% of HO_x radical loss in the boundary layer over polluted regions, and for more than 70% of the HO_x loss where there is biomass burning. Overall the global mean OH concentration is decreased by 9% by the presence of aerosols, which translates directly into a longer lifetime for the greenhouse gas CH_4 and the tropospheric pollutant CO.

9.6 Literature

Anderson, G.P. et al. (2006): *AFGL Atmospheric Constituent Profiles 0 to 120 km*, Air Force Geophysics Laboratory, Hanscombe Air Force Base, Massachussets, 1986. http://handle.dtic.mil/100.2/ADA175173.

Bingham, G.E. et al. (1997): *Cryogenic infrared radiance instrumentation for shuttle (Cirris 1A) Earth limb spectral measurements, calibration and atmospheric O-3, HNO3, CFC-12 and CFC-11 profile retrieval.* Geophys. Res.-Atmos. *102(D3)*, 3547-3558.

Crutzen, P.J., (1976): *Possible importance of CSO for sulfate layer of stratosphere*, GRL *3(2)*, 73-76.

Grage, M.M.-L. et al. (2006): *HCl and DCl: A Case Study of Different Approaches for Determining Photo Fractionation Constants*, Physical Chemistry and Chemical Physics, 8, 4798-4804.

IUPAC-05: Atkinson, R., et al. (2005): *Summary of Evaluated Kinetic and Photochemical Data for Atmospheric Chemistry*, IUPAC Subcommittee on Gas Kinetic Data Evaluation for Atmospheric Chemistry, http://www.iupac-kinetic.ch.cam.ac.uk/.

Johnson, M.S. et al. (2002): *Isotope effects in atmospheric processes*, Chemical Society Reviews, *31*, 313-323.

JPL-06: Sander, S.P. et al. (2006): *Chemical Kinetics and Photochemical Data for Use in Atmospheric Studies,* Evaluation Number 15, JPL Publication 06-2.

Martin, R.V. et al. (2003): *Global and regional decreases in tropospheric oxidants from photochemical effects of aerosols,* Journal of Geophysical Research, Atmospheres *D3,* 4097.

Pope, C. Arden, et al. (2002): *Cancer, cardiopulmonary mortality, and long-term exposure to fine particulate air pollution.* Journal of the American Medical Association. *287,* 1132-1141.

Rosenlof, K.H. et al. (2001): *Stratospheric water vapor increases over the past half-century,* Geophysical Research Letters *28(7),* 1195 – 1198.

Seinfeld, J.H and Pandis, S.N. (2006): *Atmospheric Chemistry and Physics, From Air Pollution to Climate Change,* second edition, Wiley,

Solomon, S. et al. (1986) *On the Depletion of Antarctic Ozone,* Nature *321,* 755 – 758.

von Hessberg, P. et al. (2004): *Ultra-violet absorption cross sections of isotopically substituted nitrous oxide species: $^{14}N^{14}NO$, $^{15}N^{14}NO$, $^{14}N^{15}NO$ and $^{15}N^{15}NO$,* Atmospheric Chemistry and Physics 4, 1237-1253.

Yoshino, K. et al. (1984): *High-resolution absorption cross section measurements of N_2O at 295 K-299 K in the wavelength region 170-222 nm,* Planetary and Space Science, *32(10),* 1219-1222.

Wang, Y. and Jacob, D.J. (1998): *Anthropogenic forcing on tropospheric ozone and OH since preindustrial times,* J. Geophys. Res.-Atmos., *103(D23),* 31123-31135.

WHO (World Health Organization) **(2000):** *Air Pollution, Fact Sheet No. 187*

WMO (World Meteorological Organization) **(2002):** *Scientific Assessment of Ozone Depletion: 2002,* Global Ozone Research and Monitoring Project Report No. 47.

10 Deposition

Ole Hertel and Lise Marie Frohn

The atmosphere is a transport medium for both gaseous and particulate pollutants of anthropogenic as well as natural origin. The scavenging of these pollutants from the atmosphere is termed atmospheric deposition. Quantification of the atmospheric deposition is of crucial importance in the assessment of the acidification and eutrophication of marine and terrestrial ecosystems, and thereby also for the development of environmental strategies, i.e. for the protection of a high biodiversity in the ecosystems. The two forms of atmospheric deposition – dry and wet – represent the final path of the pollutants in the atmosphere. Dry deposition covers the direct scavenging to the ground (or other) surfaces by turbulent transport and/or gravitational settling, whereas wet deposition is the uptake of pollutants into cloud, rain, hail and snow flakes and subsequent transport together with the wet media by turbulent transport and gravitational settling down to the surface. In the following we will look further into the dry and wet deposition processes and discuss how these processes may be expressed mathematically e.g. in transport-chemistry models. However, to a start with we will look into some examples of practical assessments of deposition of gases and particles from the atmosphere to the surface.

10.1 Parameterisation of dry deposition

Dry deposition is basically the removal of gases and particles from the atmosphere to the surface in the absence of precipitation. Dry deposition is driven by turbulent transport and gravitational settling. It is an efficient removal process for the more soluble and reactive gases and generally a slow process for fine fraction particles. However, before dry deposition can take place, the pollutants need to be transported down to and subsequently absorbed onto the surface. The process depends on the meteorological conditions, the chemical characteristics of the specie as well as the physical, biological and chemical properties of the surface. We will elaborate more on these processes in the following. Dry deposition of gases and fine mode particles (diameter < 1µm) is often described as a "transport" over several steps. A parameterisation based on this methodology was developed by Wesely and Hicks (1977). This parameterisation makes use of an analogy to electricity and is therefore called the resistance method. The dry deposition velocity is here computed as the inverse value of the total resistance towards transport to and absorption onto the surface. The total resistance for gaseous compounds covers three different steps in the dry deposition process; each of these steps is expressed as a resistance towards dry deposition taking place:

1. Transport from the bulk atmosphere down to the surface by turbulent transport,
2. Transport through a quasi-laminar sub-layer just above the surface,
3. Absorption of the pollutant onto the surface itself.

If one or more of the resistances are large, then the dry deposition velocity will be small. The dry deposition velocity for a gaseous compound is in this parameterisation expressed mathematically as:

$$V_d = \frac{1}{R_t} = \frac{1}{R_a + R_b + R_c} \tag{10.1}$$

where V_d is the dry deposition velocity (cm s^{-1}) of the gaseous compound, R_t (s cm^{-1}) is the total resistance, R_a (s cm^{-1}) is the aerodynamic resistance, R_b (s cm^{-1}) is the laminar sub-layer resistance, and R_c (s cm^{-1}) is the surface resistance.

These resistances each cover one of the three steps in the above listing (Figure 10.1), and they are explained in more detail in the following.

For the dry deposition of particles in the atmosphere the gravitational settling may also play a significant role; and this is accounted for in the following expression for the dry deposition of particles:

$$V_d = \frac{1}{R_a + R_b + R_a \times R_b \times V_s} + V_s \tag{10.2}$$

where V_s is the gravitational settling velocity of the particle, which depends on parameters like the size and mass of the article and may be calculated as:

$$V_s = \frac{D_p^2 \rho_p C_c g}{18\mu} \tag{10.3}$$

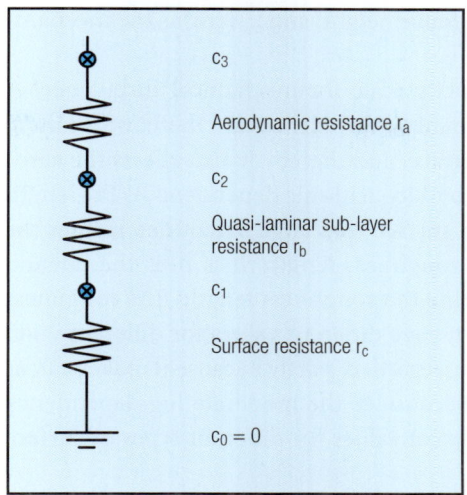

Figure 10.1 Principle of the resistance method by Wesely and Hicks (1977).

where D_p is the diameter of the particle (cm), ρ_p is the density of the particle, μ is the aerodynamic viscosity (equal to $v \times \rho$, where v is the kinematic viscosity of air and ρ is the density of air), and C_c is the so-called slip corrector factor (or Cunninghan corrector factor):

$$C_c = 1 + \frac{2\lambda}{D_p}(1.257 + 0.4\mathrm{Exp}(-0.55D_p / \lambda)) \qquad (10.4)$$

where λ is the mean free path of air molecules; the mean free path varies approximately inversely with the density of the air and is about 0.06µm for typical ambient conditions at sea level.

Aerodynamic resistance

Disregarding gravitational settling that is important only for the larger particles, turbulent transport is the process that brings various material including pollutants from the bulk atmosphere and down to the surface. The first step in the resistance method is thus the aerodynamic resistance, which is the resistance towards transport by turbulent movements down to the surface. This resistance may for neutral conditions be expressed as:

$$R_a = \frac{1}{\kappa \times u_*}\log\left(\frac{z_{ref}}{z_0}\right) \qquad (10.5)$$

where κ is the dimensionless von Karman's constant (≈ 0.4), u_* (cm s^{-1}) is the friction velocity, z_{ref} (cm) is the reference height, and z_0 (cm) is the mechanical roughness length.

The friction velocity is the velocity scale for mechanical turbulence. A rough surface generates more mechanical turbulence and thereby also more mixing of the air than a smooth surface, and thereby it also affects the aerodynamic resistance (although only by a logarithmic dependency). The length of the obstacles that forms a given surface to a large extent determines the associated downwind mechanical roughness length. It is thus the upwind surface characteristics that determine the roughness length. The roughness length will therefore in reality often have different values for different wind directions. This is, however, often disregarded; partly because it makes calculations more complex and partly because of the moderate log-dependency on the aerodynamic resistance. Typical values for different types of surface are given in Table 10.1.

Surface	Mechanical roughness length z_0 (cm)
Water (lake, fjord, ice etc)	0.01 – 0.1
Lawn	1
Uncut grass	5
Fully grown root crops	10
Areas with disperse trees and grass	25
Tree covered surface	100
Low density residential area	200

Table 10.1 Typical values for the mechanical roughness length over different surfaces.

From these values it can seen that i.e. an urban area has a considerably larger roughness length than a smooth water surface. Over a water surface the mechanical roughness length will vary with the wind conditions due to the formation of breaking waves; a dependency that according to experimental studies may be expressed as:

$$z_0 = \frac{0.13v}{u_*} + \frac{0.0144u_*^2}{g} \qquad (10.6)$$

where v is the kinematic viscosity for air (0.15 cm^2 s^{-1} at 20 ^{0}C), and g is the gravitational acceleration (981 cm s^{-2}).

It is common to account for stability conditions in the calculation of the aerodynamic resistance by adding a similarity function to the expression in Equation 10.5. The reference height is here the height at which the dry deposition velocity is calculated. It is important to note that u_* depends on the mechanical roughness length. The larger the roughness, the larger is also u_*.

Thereby the aerodynamic resistance decreases and the dry deposition velocity will tend also to increase (depending on which of the resistances that dominate).

In general during daytime the aerodynamic resistance tends to be low, and may in practice be disregarded. Exceptions are the highly reactive gases like HNO_3, NH_3 and HCl that deposit on any surface they get into contact with.

Quasi-laminar sub-layer resistance

The resistance method operates with a quasi-laminar sub-layer close to the surface; a layer through which transport takes place only by molecular (for gases) or Brownian (for particles) diffusion. However, the quasi-laminar sub-layer is not a real single physical layer close to the surface. The introduction of the sub-layer is instead a way to account for many viscous layers, that are present adjacent to the often numerous obstacles comprising the effective surface in the real world. Assuming that we have steady-state conditions, the flux through this sort of artificial sub-layer may be expressed as:

$$F_b = B \times u_* (c_2 - c_1) \tag{10.7}$$

where c_1 (i.e. in $\mu g\ m^{-3}$ or ppbv) is the concentration at the surface; and B is a dimensionless transfer coefficient, which by convention has been scaled by the friction velocity u_*. The laminar sub-layer resistance is then:

$$R_b = \frac{1}{B \times u_*} \tag{10.8}$$

As already stated, the transport through this laminar layer takes place by molecular diffusion for gases and for particles by Brownian motion. The dimensionless Schmidt number, Sc is a way to account for this diffusivity of the gases

$$Sc = \frac{v}{D} \tag{10.9}$$

where D is the molecular diffusivity of the gas in air (values for a number of selected gases are given in Table 10.2), and v again is the kinematic viscosity (Equation 10.6).

Gaseous specie	Diffusivity (10^{-6} m² s⁻¹)
SO_2	13.6
NO	19.2
NO_2	15.5
HNO_3	11.8
NH_3	23.4
O_3	16.4
HCHO	19.2

Table 10.2 Molecular diffusivity for selected gases at 25 °C and 1 atmosphere. The diffusivity may be computed at a different temperature and pressure using the expression:
$D(P;T) = (T / 298.15)^{1.75} \times D(1 ; 298.15) / P$, where P is in atmospheres and T in Kelvin.

Experimental studies have shown that the quasi-laminar sub-layer resistance is relatively insensitive to the mechanical roughness length, and this leads to a useful expression for the quasi-laminar sub-layer resistance for gaseous compounds (Seinfeld and Pandis, 1998):

$$R_b = \frac{5 Sc^{2/3}}{u_*} \tag{10.10}$$

Particles are transported by the Brownian motions, where the Brownian diffusivity depends on the particle size. For the smallest particles the transfer is rapid and hence the resistance is small. The slowest transfer through the quasi-laminar sub-layer is found for the particle range between 0.1 and 1 µm; this range and especially the range close to 1 µm is important since a significant part of the long-range transport of e.g. sulphate, nitrate and ammonium is associated with particles in this size range. For particles the quasi-laminar sub-layer resistance may be expressed as:

$$R_b = \frac{1}{u_* \left(Sc^{-2/3} + 10^{-3/St} \right)} \tag{10.11}$$

where St is the dimensionless Stokes number, which is sometime also referred to as the inertial parameter; it is an index of the impactability of the particle, which is defined as:

$$St = \frac{u_*^2 \times V_s}{g \times v} \tag{10.12}$$

where V_s is the gravitational settling velocity (cm s⁻¹) for the particle which is expressed in Equation 10.3.

Surface resistance

For many gases, the surface resistance is the limiting factor for the dry deposition process. The surface resistance is at the same time the most complex variable in the resistance model, and depends both on the characteristics of the surface and of the gaseous compound in question. Several expressions have been derived for the surface resistance; one version was originally developed by Wesely (1989). The basic idea is here that i.e. in a forest the deposition may take place onto the leaves or through the stomata in the top of the canopy, but it may also deposit directly to the soil or other surfaces. This leads to a number of additional resistances of which some are working in parallel to each other and others in series:

$$R_c = \left(\frac{1}{R_{cl} + R_m} + \frac{1}{R_{lu}} + \frac{1}{R_{dc} + R_{cl}} + \frac{1}{R_{ac} + R_{gs}} \right)^{-1} \qquad (10.13)$$

where the first term includes the leaf stomata resistance (R_{cl}) and the mesophyll resistance (R_m), the second term is the outer surface resistance in the upper canopy (R_{lu}) and include the cuticular resistance in healthy vegetation and other outer surface resistances, the third term covers resistances in the lower canopy and include resistance due to buoyant convection (R_{dc}) and resistance to uptake by leaves, twigs and other surfaces, the fourth term is the resistance at ground and include processes depending on the height of the canopy (R_{ac}) and a resistance for uptake in soil, litter etc at the ground surface (R_{gs}).

The above formulation (Equation 10.13) is applied to all types of surfaces, but in many cases like i.e. for an urban area some of these resistances will naturally not be relevant. The various resistances are functions of global radiation, ambient air temperature, diffusivity in air and solubility (Henry's law coefficients). Wesely (1989) derived expressions for each of these resistances for two gaseous compounds – sulphur dioxide and ozone. The resistances for other gaseous compounds are derived through a sort of interpolation between the values for ozone and sulphur dioxide using an index for solubility and reactivity of the gaseous compound in question. In the recent dry deposition parameterisation in the EMEP model this has been extended to include values for ammonia and performing interpolation between three sets of values to drive resistances for other compounds (Simpson et al., 2003).

The surface resistance has a strong seasonal variation, but varies also between day and night. In reality parameters like the humidity and type of soil are also of importance for the surface resistance. Values for such parameters are, however, generally difficult to obtain for transport-chemistry modelling, but may be accounted for in process studies e.g. in connection with analysis of experimental data.

For some species special conditions play a role in the dry deposition process; one example is ammonia. Experimental studies have indicated that the deposition of ammonia may depend on the ambient air concentration in such a way that the deposition decreases with increasing concentration; a kind of saturation of the surface or the ecosystem is apparent. In this connection it is common to talk about a so-called compensation point at which an ecosystem or a given plant does not receive or release ammonia. Only for ambient concentrations above the compensation point, the ecosystem or the plant will receive ammonia from the atmosphere, for lower concentrations the ecosystem will emit ammonia. Over sea ammonia may be emitted from the surface in case of high ammonium concentrations in the water. The process depends among other factors on the pH of the water. Air-sea exchange is important for a number of other gaseous compounds like i.e. CO_2.

10.2 Parameterisation of wet deposition

Wet deposition may basically take place through two different processes: in-cloud and below-cloud scavenging. In-cloud scavenging is the uptake in cloud droplets that later rains out and fall to the ground as rain, snow or hail. Below-cloud scavenging is the uptake in rain, snow or hail that fall to the ground. In-cloud scavenging is generally the most efficient of the two processes due to the longer residence time of cloud droplets (in rain clouds typically about 20 minutes) compared to rain, snow and hail (seconds to a few minutes).

Most clouds do not generate precipitation and a large part of the compounds that have been taken up by cloud droplets are released back to atmosphere, when the cloud evaporates. Model studies have shown that in average a cloud is formed about 7 times before leading to precipitation. However, inside the cloud droplets fast chemical transformations take place, and the clouds therefore act as a kind of catalysts for the chemical transformations in the atmosphere; i.e. about half of the atmospheric conversion of SO_2 to sulphate takes place inside cloud droplets.

The uptake of gases in cloud or rain droplets is a function of the size of the droplet, the diffusivity in air and solubility in water of the gas, how fast the gas in brought into contact with the droplet due to the movement of the droplet, and the rate of chemical conversion of the gas inside the droplet. The uptake of a gas in a cloud or rain droplet may be expressed as:

$$\frac{dC_w}{dt} = \frac{3F_g D_g}{r^2}\left[C_g - \frac{C_{w,i}}{H^*}\right] - kC_w C_{w,0} \qquad (10.14)$$

where C_w is the concentration inside the droplet (mol l^{-1}), F_g is a dimensionless ventilation coefficient for the gas (the relation between uptake in a non-moving and a falling droplet), D_g is the diffusivity of the gas in air (cm^2 s^{-1}), r is the radius of the droplet (cm), $C_{w,i}$ is the concentration of the gas on the surface of the droplet (mol l^{-1}), H* is the dimensionless Henry's law coefficient, k is the reaction coefficient for the conversion of the gas inside the droplet (s^{-1}), $C_{w,0}$ is the concentration of a specie inside the droplet (mol l^{-1}) by which the considered gas reacts. Equation 10.14 is derived in Seinfeld and Pandis (1998).

The velocity of falling droplets generates a circulation inside the droplet leading to an increase uptake, which is expressed in the ventilation rate. The ventilation rate is close to one for cloud droplets, whereas it may be up to 10 for falling rain droplets. The diffusivities for a few selected gases are shown in Table 10.2.

Collision efficiency

Uptake of aerosol particles in a falling rain droplet requires a collision between the two. It is a function of the velocity of the rain droplet, the diam-

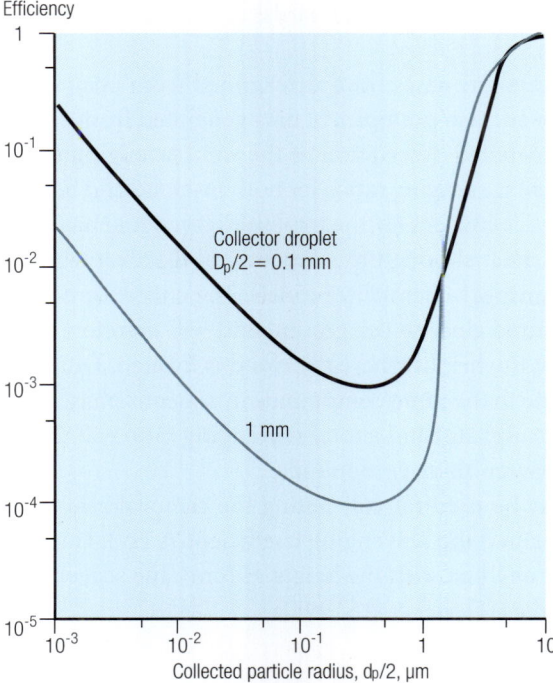

Figure 10.2 The collection efficiency E for two droplet sizes (Slinn, 1983) as function of the collected particles size. Unit density of the collected particle has been assumed. (From Seinfeld and Pandis, 1998).

eter, density and viscosity of both rain droplet and aerosol particle. The collection efficiency is the product of the collision efficiency and the coalescence efficiency. The coalescence efficiency depends on the size of the collector droplet and is close to unity for droplets smaller than 100μm. The collection efficiency from a semi-empirical study by Slinn (1983) is shown as function of the collected aerosol particle size in Figure 10.2 for two different sizes of collector droplets.

Scavenging ratios

Scavenging ratios is a common approach in describing the wet deposition of species in transport-chemistry models. The basic idea is that a certain ratio between the concentration in the wet media and the surrounding ambient air is persistent. This ratio is called the scavenging ratio S:

$$S = \frac{Caq}{C} \tag{10.15}$$

where C_{aq} is the concentration inside the droplet (aqueous phase) and C is the concentration in the surrounding ambient air before the wet scavenging takes place (it may be considerably lower afterwards; in some cases close to zero).

Scavenging ratios have in many cases been determined from analysis of experimental data, but there are also examples of data generated from calculations with models with a detailed description of the wet scavenging processes. It is possible to calculate scavenging ratios for both in-cloud and below-cloud scavenging, but often a coefficient for the total wet scavenging has been derived from experimental data, although this has several disadvantages in relation to application in transport-chemistry models. The concentration in ambient air is usually measured close to the ground and will therefore usually be considerably lower at the height where the cloud is formed. Different gases and particles contribute to the same compounds in aqueous phase (like i.e. ammonium, nitrate and sulphate) and a total scavenging ratio will therefore depend on the ratio between these compounds.

The scavenging ratio may be used for calculating the removal rate from the atmosphere, which is termed the scavenging coefficient Λ (s^{-1}). Considering the lower atmosphere as a box with the height H (cm), the scavenging coefficient may be expressed as (Hertel et al., 1995):

$$\Lambda = \frac{S \times I}{H} \tag{10.16}$$

where I is the precipitation intensity (cm s^{-1}).

Disregarding any other chemical or physical process that is leading to removal or formation of the gas, the change in ambient air concentration (C) as function of time may then be expressed as:

$$C = C_0 \times Exp(-\Lambda \times t) \tag{10.17}$$

where C_0 is the initial concentration of the species in the surrounding ambient air.

It is possible to distinguish between in-cloud and below-cloud scavenging coefficients. It this case H has to be the height over which the considered scavenging process takes place. For in-cloud scavenging this is the depth of the cloud, and for below-cloud scavenging it is the distance from cloud base and down to the ground. Table 10.3 provide examples of in-cloud and below-cloud scavenging coefficient for selected species.

Specie	Scavenging coefficient (s^{-1})	
	In-cloud	Below-cloud
SO_2	1.5×10^{-5}	0.0
NO	1.9×10^{-11}	0.0
NO_2	1.4×10^{-10}	0.0
HNO_3	1.4×10^{-3}	6.2×10^{-5}
NH_3	1.4×10^{-3}	9.5×10^{-5}
O_3	1.4×10^{-10}	0.0
Particles	1.3×10^{-3}	5.0×10^{-6}

Table 10.3 Examples of in-cloud and below-cloud scavenging coefficients for selected species at a precipitation intensity of 1 mm per hour and a mixing height of 1 km.

In many transport-chemistry models scavenging coefficients like the above have been applied in order to limit the calculation time. However, increasing computer power now makes it possible to apply more complex descriptions that take into account chemistry, dynamics and physics of the clouds.

The scavenging coefficients in Table 10.3 show that in-cloud scavenging is a more efficient removal process than below-cloud scavenging; In-cloud scavenging generally accounts for more than 80% of the wet deposition.

Wet deposition is the most important removal of fine fraction particles in the atmosphere. When wet deposition takes place it is for many air pollutants in general a much faster process than dry deposition, but it typically only rains about 5 to 10% of the time, and dry deposition may therefore in the long term be of the same order of magnitude as wet deposition. This is, however, depending on the considered species and the distance to the source regions.

10.3 Literature

Frohn, L.M. and Hertel, O. (2004): *Atmospheric deposition of nutrients.* In: Wassmann, P. and Olli, K. (eds): Drainage basin nutrient inputs and eutrophication: an integrated approach. University of Trosmø, Norwegian College of Fishery Sciences and Tartu University, Department of Botany and Ecology. Pp. 2-24. Available at: http://lepo.it.da.ut.ee./~olli/eutr/Eutrophication.pdf.

Hertel, O. et al. (1995): *Development and Testing of a new Variable Scale Air Pollution Model – ACDEP.* Atmospheric Environment, *29, 11,* 1267-1290. doi:10.1016/1352-2310(95)00067-9

Seinfeld, J.H., and Pandis, S.N., (1998): *Atmospheric Chemistry and Physics, From Air Pollution to Climate Change,* John Wiley & Sons Inc., New York, USA.

Simpson, D. et al., (2003): *EMEP Status Report 1/03 Part I Transboundary acidification and eutrophication and ground level ozone in Europe: Unified EMEP model description.* EMEP/MSC-W Report. Norwegian Meteorological Institute.

Slinn, W.G.N., (1983): *Precipitation scavenging,* In: Atmospheric Science and Power Production 1979, Chapter 11. Division of Biomedical Environmental Research, U.S. Department of Energy, Washington D.C., USA.

Wesely, M.L., and Hicks, B.B., (1977): *Some factors that affect the deposition rates of sulphur dioxide and similar gases on vegetation,* J. Air Poll. Control Ass., *27,* 1110-1116.

Wesely, M.L., (1989): *Parameterizations of surface resistance to gaseous dry deposition in regional-scale, numerical models,* Atmospheric Environment, *23,* 1293-1304.

IV

Techniques

Determination of air pollution emissions can in principle be carried out two ways: either by individual measurements (Chapter 11) or by emission inventories (Chapter 12). Individual measurements are normally used for large plants or in basic studies, but are impractica in normal management of a larger number of smaller sources. Here emissions are determined in a few characteristic cases and then expressed as emission per used amount of fuel, per driven kilometre, per unit produced in a factory or whatever. The total emission is then calculated as the product of activities and the so-called *emission factors*.

Qualitative observations and more or less primitive measurements were for many years the only means for describing the resulting air pollution. But in recent decades the measuring techniques have greatly improved (Chapter 13). In addition dispersion modelling (Chapter 14) on large computers, which can rank results of measurements and verify theories, and also treat hypothetical situations by means of scenario studies, has appeared. Today a combination of advanced measurements and dispersion modelling has become an important tool both in research and in air quality management. In combination with meteorological forecasts also pollution prognoses are possible.

11 Emission measuring techniques

Lars Kristian Gram

Air emissions are primarily emitted through exhaust pipes, stacks, ventilation ducts etc. but they can also come from diffuse sources such as pipe leaks, open windows, storage, spillage etc. This chapter describes measuring techniques for stationary sources such as power plants and industry where the emission comes from a well-defined source. Emissions from diffuse sources are normally determined based on an established mass balance.

Emissions are measured for a number of reasons:

- Emission limit value compliance assessment
- Emission inventories
- Calibration and function testing of automatic measuring systems (AMS)
- Establishing emission factors
- Establishing mass balances
- Process related measurements (regulation purposes, plant operation)
- Acceptance tests (proof of guarantee)

11.1 General principles for emission measurements

Emission measurements are distinguished from air quality measurements in that concentrations, temperatures and humidities can be much higher and that the gas has a velocity (flowing through a pipe).

Environmental requirements for maximum concentrations in the emitted gas will always be based on reference conditions (e.g. dry gas, 101.3 kPa and 0 °C). Energy producing plants will have a further reference parameter, namely a certain oxygen content (e.g. set at 10 vol %). Correction to reference conditions is performed by means of the law of ideal gases. Emission measurements in a flowing gas stream can either be performed by extractive or in situ measurements.

Extractive measurements

Extractive measurements are performed on a representative partial gas flow, which is extracted through a sample system to a monitor or a manual sampling train.

All components of the sample system should be chemically inert towards the gases being sampled. Condensation, deposition or adsorption of the pollutants to the surface of the sample system must be avoided. A sample system can consist of a gas conditioning system (filtration of particles, removal of humidity etc.), and/or be heated to above the dew point of the condensable vapours in the gas sample. The sample system must be leak-tight in order to avoid mixing with fresh air.

In situ measurements

A representative cross section of the stack is utilized as a measuring cell, or the sensor is placed directly in the gas flow. These systems do not have the disadvantages of extractive measurements with potential loss of substance and leaks. Contamination of optical windows and pollution of sensors are on the other hand some of the problems in situ instruments have to deal with.

Automatic measuring systems, AMS

Automatic measuring systems are defined as permanently installed monitors (in situ or extractive) that monitor concentration, flow or temperature when the plant is in operation.

AMS can be installed in order to regulate process parameters or to gain other relevant process information. The most common use of AMS is due to environmental requirements:

- As a replacement for regular compliance assessment measurements
- Demands in national environmental regulation
- Specified in EU directives (European Parliament and the Council, 2000, 2001)

11.2 Measuring principles

Extractive monitors

Monitors are characterized by their continuous operation. Most monitors have a gas stream flowing through a monitoring cell and give an analogue signal proportional to the concentration. Some monitor types are semi continuous and takes in a sample that is analyzed in minutes or seconds, and only then takes in another sample. A series of typical detection principles are listed in Table 11.1.

Extractive monitors are available for many gases and even particles under some conditions. Detection limits, interferences and problems with warm and humid gases are some of the challenges to extractive monitoring.

The sample system should take in a representative partial gas stream from the stack, condition it and transport it to the monitor in a state where gas concentration can be monitored without error. The sample system is an essential part of the measurement, and severe errors can occur if proper precautions are not taken. Dilution, filtration, condensation, drying, heating, etc. are the means of gas conditioning that should be chosen in consideration to the parameter, stack conditions interferences, monitor principle, monitor range etc.

Parameter	Typical detection principle
O_2	Para-magnetism
CO_2	Infrared detection (IR)
TOC	Flame-ionisation detection (FID)
SO_2	Ultraviolet detection (UV)
NO_x	Chemiluminiscence
CO	Infra red detection (IR)
HCl	Fourier transform infrared detection (FTIR)
Hg	Cold vapour atomic fluorescence analyzer

Table 11.1 Examples of air emission parameters and relevant detection principles.

Manual sampling

Particles (incl. fine and ultra fine particles)

Particles can vary in size, density and shape etc. Furthermore, particle size distribution and particle concentration can vary over the cross section of a sampling plane and even vary over time. In order to take out a representative sample from a stack under these conditions, adequate precautions should be taken:

- The sampling cross section should be placed far downstream from the last disturbance (minimum 5 times the inner diameter of the stack) and also upstream disturbances (minimum 2 times the inner diameter). This will in most cases ensure a uniform flow profile and an even distribution of particles over the sampling cross section. When minimum demands are not met an increase in the number of traverse points should be applied.
- The sampling should be performed as a traverse, with sampling in equal time periods in a number of sampling points over the sampling cross section. Number and placement of sampling points is dependant on the design of the stack
- Isokinetic sampling is extremely important in order to ensure representative sampling. Aside from stack design and number of traverse points the sample flow through the nozzle shall have exactly the same velocity and direction as the gas in the stack. Measuring errors from non-iso kinetic sampling are shown in Figure 11.1.
    - Too high sampling velocity means "over sampling", where smaller particles are "sucked" in to the nozzle and the measured value will be too high.
    - Too low velocity will lead to "under sampling", where some of the smaller particles are pressed outside the nozzle. Results will be too low
    - The error from "over sampling" is smaller than the error form "under sampling" which is also reflected in the requirements in the standard EN 13284-1 (European Standard, 2001) that the isokinetic sampling rate should be between 95 and 115%. Outside this interval samples should be rejected

An example of "instack" filter sampling system is shown in Figure 11.2. "Out-stack" sampling systems are also accepted in the standard even though deposition of particles is a risk and a rinsing procedure can be necessary. Furthermore the sample probe and filter holder should be heated to at least 20 °C above the gas temperature in the stack or 20 °C above the dew point of the stack gas. "In-stack" systems do not need heating, but can be preheated

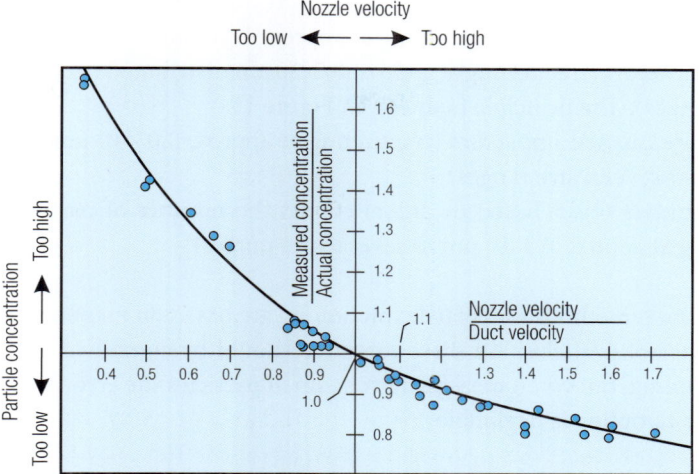

Figure 11.1 Example of measuring errors at non-isokinetic sampling (velocity in nozzle not equal to velocity in duct).

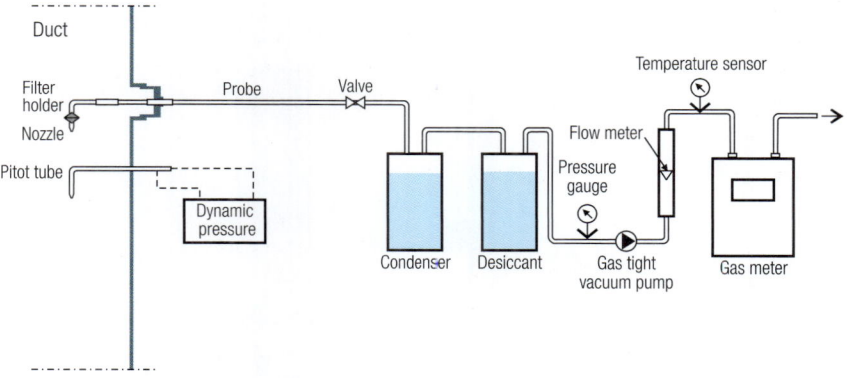

Figure 11.2 Example of "in-stack" filter sampling system (European Standard, 2001).

before sampling to avoid possible condensation on the filter, when a cold filter holder and probe enters a warm and humid gas close to the dew point.

Semi-volatiles are thermic unstable particles (e.g. inorganic salts or organic substances) that can evaporate or decompose during sampling or conditioning of filters.

Semi-volatiles are typically seen in biomass incineration and under certain scrubber conditions. When semivolatiles are present in the stack it is important to choose sampling temperature and conditioning temperature thoroughly.

Particle fractioning methods:
- Cyclones or pre separators (fractioning at approx. 1-20 μm)
- Cascade impactors (fractioning at approx. 0.2-20 μm distributed over several size ranges). The principle is shown in Figure 11.3.
- Low pressure cascade impactors (fractioning at approx. 0.03-10 μm distributed over several size ranges)
- Particle counters (laser based or other) counts the number of particles (fractioning at approx. 0.1-10 μm in several size ranges)

The most common methods for particle fractioning are based on mass rather than on number of particles. All these methods should be combined with isokinetic sampling, but when measuring very small particles the error from non-isokinetic sampling is negligible.

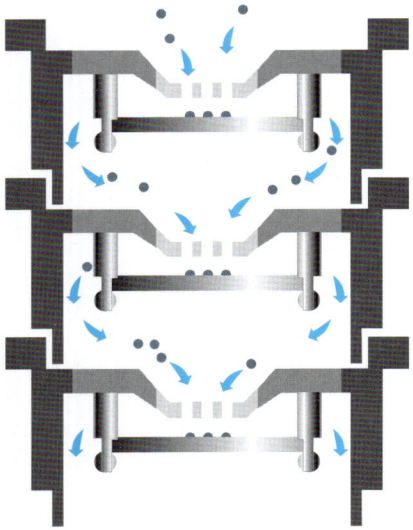

Figure 11.3 The cascade impactor collects particles size selectively. An impactor can have several stages, and on each stage the velocity of the gas stream is increased meaning that increasingly smaller particles are deposited on the plates. The mass of the deposited particles is determined by differential weighing.

Gases

Manual sampling of gases is performed either as collection of matter by absorption or adsorption or in a container followed by a suitable method of analysis (HPLC, GC/FID, GC/MS, ICP/MS etc.). In some cases isokinetic sampling of gases is necessary e.g. when droplets or particle bound gases are present. Trace metals are sampled as a combination of particle sampling and absorption, and both filter and absorption liquid are analysed.

Collection on liquid absorbents:

A representative partial gas stream is filtered and heated (to avoid condensation on filter or in the probe) and led through 2-3 cooled impingers containing absorption liquid. Downstream from the impingers the gas is led through a drying unit, a pump, a flow meter and a calibrated gas meter. When isokinetic sampling is required, the system can be combined with a isokinetic sample system and a split-flow as shown in Figure 11.4. A non-split-flow system requires rather large impingers in order to maintain good absorption efficiency because the isokinetic flow rate is rather large.

Absorption liquids can be water, alkaline solutions or acid. Absorption can be disolvation or a chemical reaction.

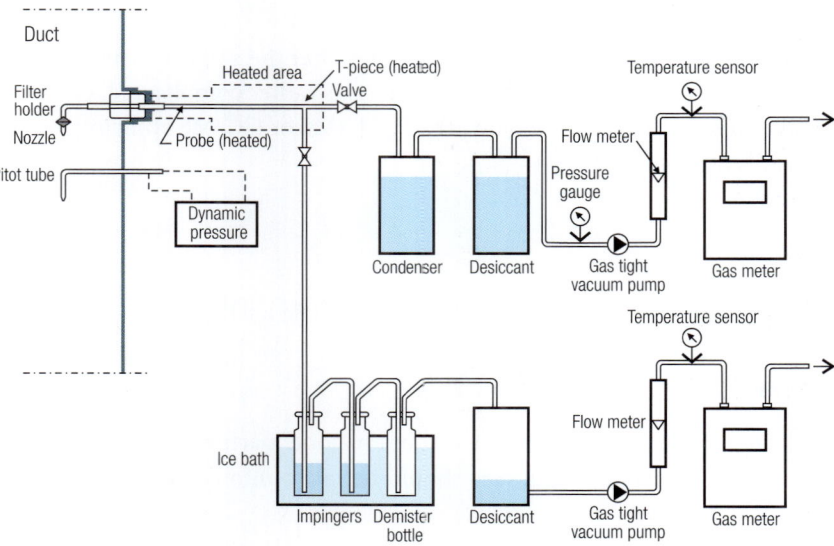

Figure 11.4 Example of absorption system for HCl with isokinetic sampling (European Standard, 1998).

Adsorption on solid sorbents

A representative partial gas stream is led through a solid sorbent in a column or in a glass tube. Two or more sections should be applied to test for breakthrough. Sample volume is determined with a gas meter or a calibrated constant flow pump. Solid sorbents can be activated carbon with or without coating, silica gel, organic polymers (XAD, tenax, PUR) etc. Solid sorbent glass tubes are commercially available in numerous combinations. The method is not applicable to warm and humid gases without special precautions:

- Dilution to below the dew point
- Condensation of water vapour and cooling of the gas. The condensate should be collected and analysed with the solid sorbent

Organic micro pollutants (PAH, dioxin and PCB)

Organic micro pollutants are sampled through a combination of several methods:

- Isokinetic sampling on filters
- Condensation and collection of water
- Adsorption on solid sorbents, organic polymers

Some methods allow condensation of water after the solid sorbent. All three phases are extracted and analysed. Analysis is rather complicated, especially for dioxins, and is performed on HPLC, GC/MS or GC/MS applying the isotope dilution method.

11.3 Quality in emission measurements

Planning

Thorough planning is essential to obtain good performance in emission measurements. Following issues should be taken into consideration (European Standard, 2007):

- Scope
- Measurement parameters, and supporting parameters
- Expected pollution concentration
- Emission limit values
- Limit of detection (should be less than 10% of the emission limit value)
- Possible interferences
- Uncertainty

- Design of measurement sites
  - safe access
  - number and placements of sampling ports
- Process
- Selection of measuring method
- Measurement strategy
  - number and duration of single measurements
  - surveillance with monitors
  - coordination with process factors (e.g. 80% load, over a batch)

Accreditation

Emission laboratories that hold an accreditation are assessed with respect to requirements in international standards and are thereby competent to carry out specific conformity assessment tasks, e.g. compliance with emission limit values. The accreditation means that an emission laboratory can demonstrate to clients, public authorities and others that their products or services fulfil demands in standards, rules and regulations or other specified demands on safety, health, quality or environment.

In some countries all compliance measurements (in emission) has to be performed by accredited laboratories.

Quality assurance for automatic measuring systems, AMS

A system for quality assurance for AMS is required to ensure valid results. The EU directives (European Parliament and the Council, 2000, 2001) stipulate demands for the quality of AMS, and such a quality assurance system is described in the standards EN ISO 14956 (European Standard, 2002) and EN 14181 (European Standard, 2004).

This quality assurance system described in the standards is commonly referred to as QAL 1-3 (quality assurance level) and AST (annual surveillance test)(Table 11.2).

Standard	Level	Description	Usually performed by
EN ISO 14956	QAL 1	Quality check of the monitor	Monitor producer
	QAL 2	Calibration and validation of the AMS	Accredited laboratory
EN 14181	QAL 3	Ongoing quality assurance during operation	Installation owner
	AST	Annual surveillance test	Accredited laboratory

Table 11.2 The four quality levels.

QAL 1 Quality check of the monitor

Test of the monitors suitability to a certain defined measuring task. The test comprises functional and qualitative controls based on well known statistical methods, and should be given as a written guarantee to the plant owner, stating that the monitor matches the demands in the standard.

QAL 2 Calibration and validation of the AMS

This quality level covers the on-site installation of the monitor. The calibration consists of minimum 15 parallel measurements with standard reference methods (SRM). The validation is performed with commonly used statistical methods. QAL 2 is performed every 3 or 5 years or when changes in the AMS point of installation or operation principle.

QAL 3 Ongoing quality assurance during operation

The installation owners self-checking of the AMS with e.g. calibration gases. Results from the self-check and the AMS performance characteristics (from QAL 1) are utilized in a calculation of the AMS stability (drift and precision). The method is based on statistical control charts.

AST Annual surveillance test

Verification of monitor performance. The test consists of minimum 5 parallel measurements with standard reference methods (SRM) and a monitor check with calibration gases.

11.4 Literature

European Parliament and the Council (2000): *Directive 2000/76/EC of 4 December 2000 on the incineration of waste.*

European Parliament and the Council (2001): *Directive 2001/80/EC of 23 October 2001 on the limitation of emissions of certain pollutants into the air from large combustion plants.*

European Standard (1998): EN-1911-1: *Stationary source emissions – Manual method of determination of HCl*
 a. *Part 1: Sampling of gases*
 b. *Part 2: Gaseous compounds absorption*
 c. *Part 3: Absorption solutions analysis and calculation*

European Standard (2001): EN 13284-1: *Stationary source emissions – Determination of low range mass concentration of dust – Part 1: Manual gravimetric method.*

European Standard (2002): EN ISO 14956: *Air quality – Evaluation of the suitability of a measurement procedure by comparison with a required measurement uncertainty.*

European Standard (2004): EN 14181: *Stationary source emissions – Quality assurance of automated measuring systems.*

European Standard (2007): EN 15259. *Air quality – Measurement of stationary source emissions – Measurement strategy, measurement planning, reporting and design of measurement sites.*

12 Emission inventories

Jytte Boll Illerup

Accounts of the amount of air pollutants emitted, so-called emission inventories, have many forms and are used for many different purposes:

- to evaluate whether reduction targets are fulfilled and emission ceilings are met both nationally and internationally
- to advise national authorities in decisions concerning regulation of air emissions from sectors or individual sources
- to create input to dispersion and deposition modelling on scales from local to global.

Depending upon the scope the inventories have a spatial resolution from global scale down to individual sources and, when necessary, a time resolution from years down to minutes.

This chapter gives an overview of the international obligations the European countries have with regard to reporting the emission inventories, as well as examples of the models and methods used to estimate the history, projections and geographic distribution of the emissions.

12.1 International obligations

Global and regional air pollution is regulated by several conventions and protocols with the objective of controlling and reducing the emission of pollutants affecting the global and regional environment. Furthermore, regional air pollution in Europe is regulated by EU directives. Most European countries are party to the conventions and the directives and are, therefore, obliged to prepare and report annual air emissions inventories on a national basis.

The two main conventions are the United Nations Framework Convention on Climate Change (UNFCCC) and the UNECE-Convention on Long-Range Transboundary Air Pollution (LRTAP).

Regional

Regional air pollution is regulated by a number of protocols, so far eight, under the LRTAP convention. The objective of the newest and most comprehensive protocol – the Gothenburg Protocol – is to control and reduce the emissions of SO_2, NO_x, NMVOC and NH_3 in order to reduce exceedance of the critical loads with regard to acidification, eutrophication and the effect of photochemical air (ozone) pollution.

Due to the increasing accumulation in nature of heavy metals (HMs) and persistent organic compounds (POPs), two protocols under the convention now control emissions of HMs and POPs. The HM protocol targets three particularly harmful metals: cadmium, lead and mercury. The POP protocol focuses on a list of 16 substances, which comprises eleven pesticides, two industrial chemicals and three by-products/contaminants.

Emission inventories are also reported under various EU directives, including the National Emission Ceilings (NEC) directive, the Large Combustion Plants (LCP) directive and the Integrated Pollution Prevention Control (IPPC) directive. The European countries strive to attain the goals set in the protocols and directives through national actions plans.

Global

One of the provisions of the UNFCCC (the Climate Convention) was to stabilise the greenhouse gas emission from industrialised nations by the end of 2000. The parties to the convention realised that the stabilisation goal was inadequate and a legally binding agreement was reached, committing the industrialised countries to reduce six greenhouse gases by 5.2% up to 2008-2012 compared to the 1990 level. In 2005 the Kyoto Protocol came into force. The EU is a party to the Kyoto Protocol and is therefore under a legal commitment to meet its requirements. The European Union as a whole must reduce emissions of greenhouse gases by 8%. However, within the EU, Member States have made a political agreement – the Burden Sharing Agreement – on the contribution of each state to the overall EU reduction level of 8%. See Chapter 20 for more details.

12.2 Emissions in Europe

Reporting obligations

In Table 12.1 the pollutants that the countries are required to report to the LRTAP Convention, the Climate Convention and the EU are listed.

Report	Pollutants
LRTAP Convention	SO_2, NO_x, CO, NMVOC, NH_3, As, Cd, Cr, Cu, Hg, Ni, Pb, Se, Zn, Dioxins, PAH and hexachlorobenzene (HCB)
UNFCCC and EU	CO_2, CH_4, N_2O, SO_2, NO_x, CO, NMVOC, HFCs, PFCs, SF_6

Table 12.1 Emissions inventory reporting obligations to conventions.

The emission inventories reported to the LRTAP Convention and to the Climate Convention are organised in 6 main source categories:1 Energy, 2 Industrial processes, 3 Solvent and other product use, 4 Agriculture, 5 Land-Use Change and Forestry, 6 Waste. Each source category has a number of subcategories.

For many European countries emissions from the incineration of municipal waste in power plants or district heating plants are included in the source category "Energy", rather than in the source category "Waste" because the energy is used for the production of heat and electricity.

Acidifying gases

Acid deposition of sulphur and nitrogen compounds mainly derives from emissions of SO_2, NO_x and NH_3. By means of weighting the individual substances according to their acidification effect, total emissions in terms of acid equivalents can be calculated by Equation 12.1:

$$Acidification\ index = \frac{m_{SO_2}}{M_{SO_2}} \cdot 2 + \frac{m_{NO_x}}{M_{NO_x}} + \frac{m_{NH_3}}{M_{NH_3}} = \frac{m_{SO_2}}{64} \cdot 2 + \frac{m_{NO_x}}{46} + \frac{m_{NH_3}}{17} \quad (12.1)$$

where m_i is the emission of pollutant i in tons and M_i is the molecular weight [g mol^{-1}] of pollutant i.

For the EU-15 in 2004 the relative weighted contribution of sulphur emissions was 28%, with 37% for NO_x emissions and 35% for NH_3 emissions (Figure 12.1). The most significant emission sources in 2004 were agriculture (34%), the energy industries (24%), road transport (18%) and energy use in

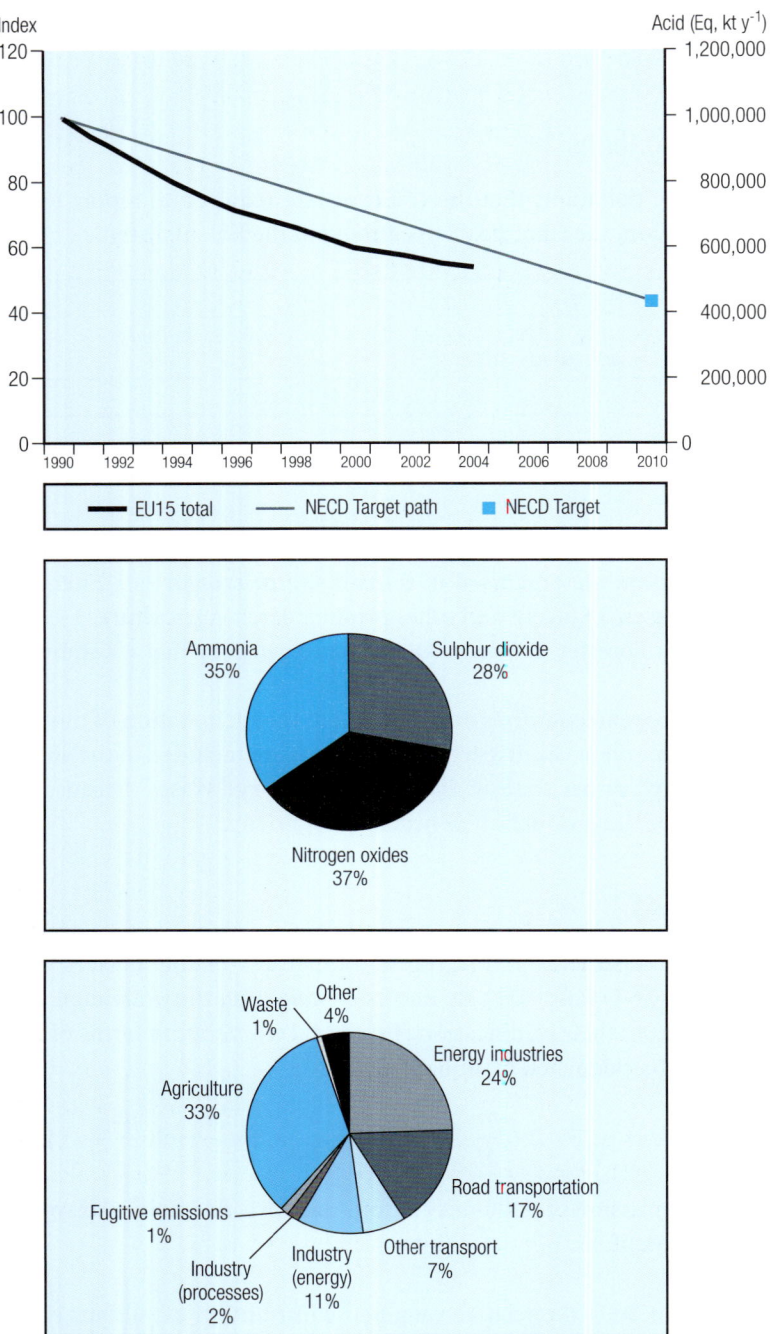

Figure 12.1 Emissions of acidifying gases for EU-15. Top graph: development since 1990. In the middle: distribution after compound in 2004. Bottom: distribution after source in 2004. (EEA, 2006a).

industry (10%). Since 1990 the largest decrease in emissions has occurred in the energy industries sector, where emissions have been reduced by around 50%. This significant reduction is mainly due to increased implementation of pollution abatement equipment e.g. flue gas desulphurisation, together with the use of low-sulphur fuels in power plants. The emissions of nitrogen oxide have fallen since 1990 due to abatement measures in road transport and large combustion plants, but these have to some extent been offset by increased road traffic. Ammonia emissions in the EU-15 are stabilising, although agriculture emissions, the major source, are associated with a high degree of uncertainty and are difficult to control.

Tropospheric ozone precursors

Tropospheric ozone is a secondary pollutant that is formed in the atmosphere from the precursors NO_x, NMVOC, CO and CH_4. It is thus not possible to indicate the emission of ozone directly. The indicator of aggregated ozone precursor emissions is the weighted sum of the emissions of NO_x, VOC, CO and CH_4 as in equation 12.2, where m_i is the emission of pollutant i in tons. (Leeuw, 2002):

Ozone precursors index =

$$1.22 \cdot m_{NO_x} + 1 \cdot m_{NMVOC} + 0.11 \cdot m_{CO} + 0.014 \cdot m_{CH_4}$$

(12.2)

Emissions of non-methane volatile organic compounds (44% of total weighted emissions) and nitrogen oxides (36% of total weighted emissions) contributed the most to the formation of tropospheric ozone in 2004 (Figure 12.2). Carbon monoxide and methane contributed 20% and 1% respectively. The emissions of NO_x and NMVOC were reduced significantly between 1990 and 2004, and the total emission of ozone precursors was reduced by 39% across the EU-15 countries between 1990 and 2004. Emission reductions are due mainly to the continued introduction of catalytic converters in cars and the increased penetration of diesel engines, but also as a result of the implementation of the solvents directive in industrial processes. Emissions from the energy and transport sectors have contributed 9% and 64% respectively to the total reduction in weighted ozone precursor emissions.

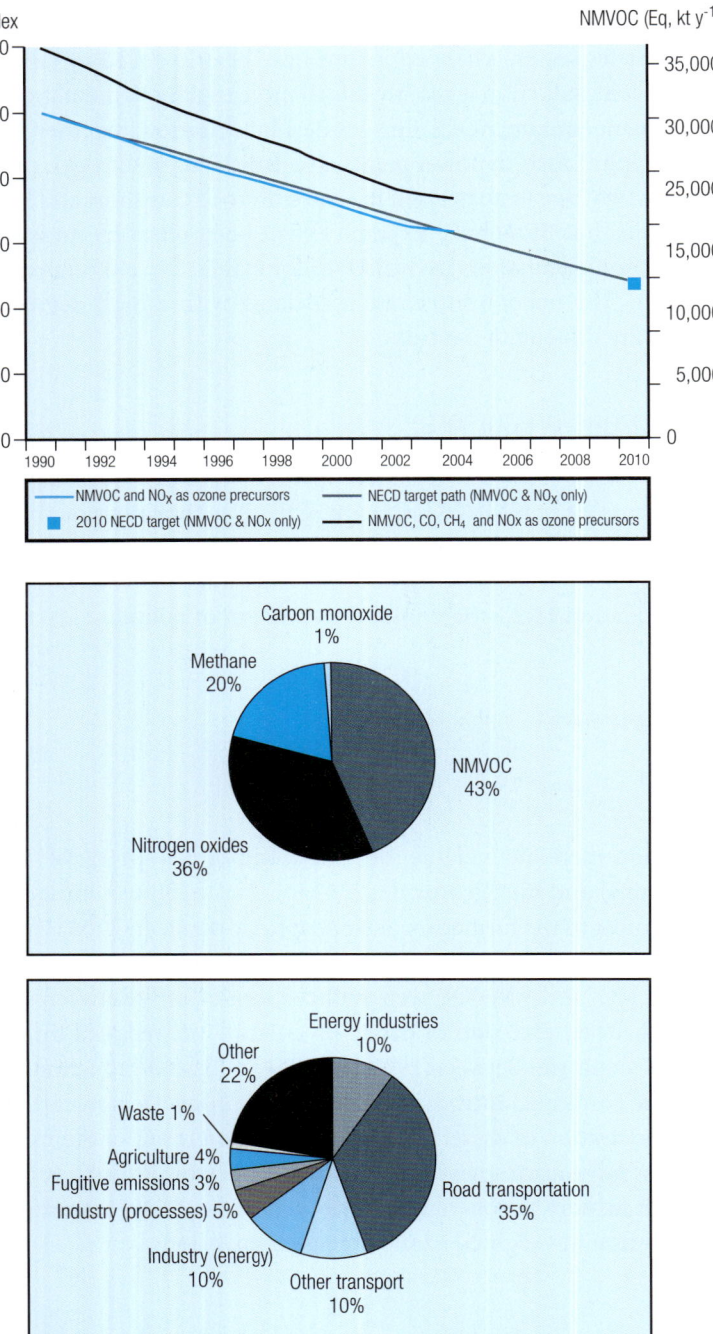

Figure 12.2 Emissions of ozone precursors for EU-15. Top graph: development since 1990. In the middle: distribution after compound in 2004. Bottom: distribution after source in 2004. (EEA, 2006a).

Greenhouse gases

The greenhouse gases reported under the Climate Convention are carbon dioxide (CO_2), methane (CH_4), nitrous oxide (N_2O), hydrofluorocarbons (HFCs), perfluorocarbons (PFCs), and sulphur hexafluoride (SF_6). They are very different in their impact on the global heat balance, and for CO_2, CH_4 and N_2O the so-called global warming potentials (GWP).

Based on weight and a 100-year period, methane is about 25 times more powerful a greenhouse gas than CO_2, and N_2O is about 300 times more powerful. The other greenhouse gases (hydrofluorocarbons, perfluorocarbons and sulphur hexafluoride) have considerably higher global warming potential values. For example, sulphur hexafluoride has a global warming potential of about 23,000. An aggregated greenhouse gas emission in CO_2 equivalents can be calculated by means of equation 12.3.

$$GHG_{eqv} =$$

$$GWP_{CO_2} \cdot m_{CO_2} + GWP_{CH_4} \cdot m_{CH_4} + GWP_{N_2O} \cdot m_{N_2O} + GWP_{F_{gases}} \cdot m_{F_{gases}}$$

(12.3)

For the EU-15 countries the most important sector is 'Energy (excluding transport)', which accounts for 59% of total emissions in 2004. The second largest sector is 'Transport' (21%), followed by 'Agriculture' (9%) and 'Industrial processes' (8%). The most important GHG by far is CO_2, accounting for 82% of total EU-15 emissions in 2004 (EEA, 2006b). CH_4 and N_2O are responsible for 8% each, and fluorinated gas emissions account for 1.6% of total GHG emissions in 2004 (Figure 12.3).

EU-15 GHG emissions have shown a small decrease of 0.9% between the base year and 2004. However, transport CO_2 emissions (with 20% of total EU-15 GHG emissions) increased by 26% due to road transport growth in almost all EU-15 member states. CO_2 emissions from energy industries increased by 4% due to increasing fossil fuel consumption in public electricity and district heating plants. Most member states experienced increases overall in CO_2 emissions between 1990 and 2004, whereas the large member states, Germany and the United Kingdom, reduced their emissions. N_2O emissions from industrial processes decreased by 78%, mainly due to specific measures at adipic acid production plants in the UK, Germany and France. Also N_2O emissions from agricultural soils fell by 11% between 1990 and 2004, due to a decline in fertiliser and manure use.

As illustrated in Figures 12.1 to 12.3, the most important sources for the emission of GHGs and other air pollutants are energy industries, transport, agriculture and use of solvents. Section 12.4 gives an overview of the methods and typical data used for estimation of the emissions from these sources.

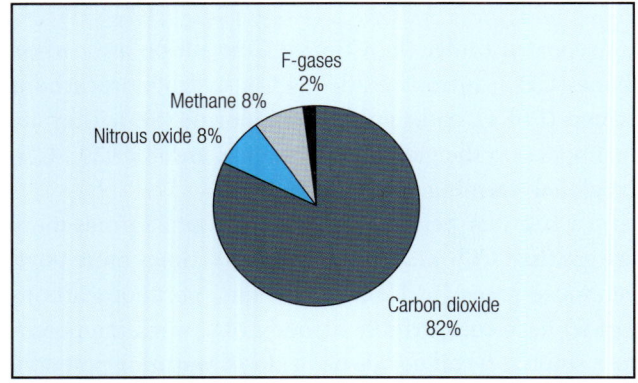

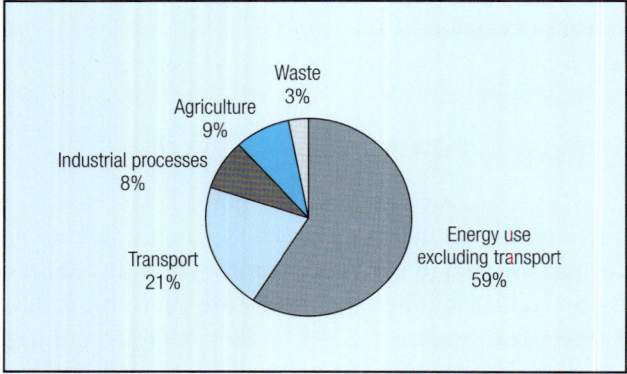

Figure 12.3 GHG emissions in CO_2 equivalents for EU-15 distributed after compound and source type.

12.3 Emission estimation methods

General

The European air emission inventories typically follow the international guidelines for estimating and reporting national air emissions. The most important guidelines are the Revised 1996 Intergovernmental Panel on Climate Change (IPCC) Guidelines for National Greenhouse Gas Inventories (IPCC, 1997), the Good Practice Guidance and Uncertainty Management in National Greenhouse Gas Inventories (IPCC, 2000), and the EMEP/CORINAIR Guidebook (EMEP/CORINAIR, 2005).

Since it is impossible to measure the emissions continuously from all relevant sources, the 'Emission Factor Method' is used, whereby activity rates for a given source are multiplied by an emission factor. Activities are numbers referring to a specific process generating emissions, while an emission factor is the mass of emissions per unit activity. Information on activities is mainly based on official statistics, and emission factors are either national values or default factors proposed by the international guidelines.

The emission as a function of time for a given pollutant can be expressed as:

$$E = \sum_s A_s(t) \cdot \overline{EF}_s(t) \qquad\qquad (12.4)$$

where A_s is the activity for sector s for the year t and EFs(t) is the aggregated emission factor for sector s.

In order to model the emission development as a consequence of changes in technology and legislation, the activity rates and emission factors of the emission source should be aggregated on an appropriate level at which relevant parameters such as process type, reduction targets and installation type can be taken into account.

If detailed knowledge and information of the technologies and processes are available, the aggregated emission factor for a given pollutant and sector can be estimated from the weighted emission factors for relevant technologies as in Equation 12.5, where P is the activity share of a given technology sector, $EF_{s,k}$ is the emission factor for a given technology and k is the type of technology.

$$\overline{EF}_s(t) = \sum P_{s,k}(t) \cdot EF_{s,k}(t) \qquad\qquad (12.5)$$

Methodological choice for individual source categories is important in managing overall inventory uncertainty. Generally, inventory uncertainty is lower when emissions are estimated using the most rigorous methods, but due to finite resources this may not be feasible for every source category. It is good practice (IPCC, 2000) to identify those source categories that make the greatest contribution to overall inventory uncertainty in order to make the most efficient use of available resources.

Energy

Emissions from the energy sector cover all emissions arising from combustion of fossil fuels but also fugitive emissions from other activities than combustion. Stationary combustion plants include very different types of technology – ranging from that installed in large power plants to small-scale residential stoves and boilers. Overall, stationary combustion plants cover plants in electricity and heat production, petroleum refining, industry, commercial, institutional and households. Transport and other mobile sources cover the important sectors: road transportation, national sea traffic, fishing, agriculture machinery, navigation and railways.

Stationary combustion plants

The inventory is typically based on activity rates from national energy statistics and on emission factors for different fuels, plants and sectors. Some emission factors refer to the international guidelines and some are country specific and refer to legislation, research reports or measurements from various combustion plants. Large plants such as power plants and municipal waste incinerators are typically registered individually as large point sources and emission data from the actual plants are used. This enables use of plant-specific emission factors that are based on emission measurements stated in annual environmental reports, etc.

On a European level, greenhouse gas emissions from stationary combustion represent the main source of GHGs. For stationary combustion plants CO_2 is the most important greenhouse gas and accounts for about 98% of the greenhouse gas emissions in CO_2 equivalents. The emissions mainly result from fossil fuel combustion, where the most important fuels are coal, fuel oil and natural gas. The CO_2 emission factors for these fuels are shown in Table 12.2.

Emission factors	Coal	Gas/diesel oil	Natural gas	Wood
CO_2 all sectors (kg GJ^{-1})	95[1]	78[1]	57[1]	0[1]
NO_x (g GJ^{-1})				
Combined heat and power plants > 50 MW	131[2]	131[2]	168[2]	69[2]
Residential plants	110[3]	68[3]	57[3]	75[3]
$PM_{2.5}$ (g GJ^{-1})				
Combined heat and power plants > 50 MW	2[2]	5[2]	0,1[2]	1[2]
Residential plants	400[3]	4[3]	0,5[3]	800
NMVOC (g GJ^{-1})				
Combined heat and power plants	15[2]	1,5[2]	2[2]	48[2]
Residential plants	485[3]	16[3]	11[3]	700

Table 12.2 Emission factors for CO_2, NO_x, PM and NMVOC for various fuels and appliances.
1: Illerup et al. 2006b, 2: Illerup et al. 2006a, 3: EMEP/CORINAIR, 2005.

Public electricity and heat production are the main sources for CO_2, SO_2 and NO_x emissions, while other sectors contribute more to the emissions of other pollutants. For instance, small-scale combustion plants – i.e. wood stoves and boilers in the residential sector are important sources for emissions of NMVOC, CO, particulate matter, as well as dioxins and PAH. Also waste incineration is one of the main sources of the dioxin emission, but due to increasing use of dioxin filters the dioxin emission from these plants has been decreasing.

For most substances the emission factors depend on fuel type as well as combustion conditions. Coal has the highest CO_2 emission factor and wood, which is a renewable fuel, is considered to have an effective emission factor of zero. The emission factors for nitrogen oxides, particulate matter and volatile organic are very much dependent on combustion conditions; high emissions factors for PM and NMVOC are seen at lower combustion temperatures, while the NO_x emission factors increase with increasing temperature.

Transport and off-road machines

Road transportation is the second largest source of GHG emissions and, as in electricity and heat generation, CO_2 is by far the most important greenhouse gas. Road transportation also represents a large source for the emissions of NO_x, NMVOC and CO.

Various national and European emission models are used to calculate the emissions for road traffic, e.g. the European model COPERT (Ntziachristos & Samaras, 2000). The models calculate the emissions for operationally hot engines, during cold start and fuel evaporation. The model also includes the emission effect of catalyst wear. Input data for vehicle stock and mileage are used in the models and are grouped according to average fuel consumption and emission behaviour. For each group, the emissions are estimated by combining vehicle and annual mileage numbers with hot emission factors, cold:hot ratios and evaporation factors.

Off-road working machines and equipment are grouped according to the following sectors: inland waterways, agriculture, forestry, industry, and household and gardening. In general, the emissions are calculated by combining information on the number of different machine types and their respective load factors, engine sizes, annual working hours and emission factors.

Emission factors	Gasoline	Diesel oil
NO_x (g GJ^{-1})		
Passenger cars, rural driving, conventional	1163	251
Passenger cars, rural driving, catalyst	176	-
Agriculture	86	879
Households	78	-
NMVOC (g GJ^{-1})		
Passenger cars, rural driving, conventional	452	19
Passenger cars, rural driving, catalyst	30	-
Agriculture	1032	100
Households	2141	-

Table 12.3 Examples of emission factors for NO_x, and NMVOC for transport Reference: Illerup et al, 2006a.

As for stationary combustion the emission factors depend on the combustion conditions. For passenger cars, a large difference is apparent for cars with and without catalysts.

Agriculture

The emission from the agricultural sector is mainly related to livestock production, land use and crop yield. Some of the most important input data to agriculture emission models are feed consumption, animal type, age and weight classes together with stable type and manure type.

Activity data for livestock, land use and crop yield are obtained from the official agriculture statistics. Data concerning feed consumption and nitrogen excretion are typically based on results from research activities in national agricultural research institutes. The emission factors are typically estimated from international guidelines or from national nitrogen mass balances, Table 12.4.

Substance	Dairy cattle	Pigs	Sheep
NH_3 [1]	26	3	1
CH_4 [2]	126	1	17

Table12.4 Examples of emission factors for livestock (kg head^{-1} year^{-1})
1: Illerup et al. 2006a, 2: Illerup et al. 2006 b.

The major part of the agricultural CH_4 emission originates from digestive processes and is mainly related to ruminants, e.g. cattle and sheep.

Solvents

Use of solvents and other organic compounds in industrial processes and households are important sources of evaporation of non-methane volatile hydrocarbons (NMVOC), and are related to the sources: paint application, degreasing, dry cleaning, and the manufacture and processing of chemical products. Generally there are two different ways to estimate the emissions of NMVOC from use of solvents and other products. The emission estimate can either be based on the levels of the activities or on single chemicals. The latter approach leads to a clearer picture of the influence from each specific chemical, which will enable a more detailed differentiation for products and the influence of product use on emissions. The procedure is to quantify the use of the chemicals and estimate the fraction of the chemicals emitted as a consequence of use. Mass balances are simple and functional methods for calculating the use and emissions of chemicals:

use = production + import - export - destruction/disposal - hold up
emission = use · emission factor   (12.6)

where 'hold up' is the difference in the amount in stock at the beginning and at the end of the year of inventory.

A mass balance can be made for single substances or groups of substances, and the total amount of chemical emitted is obtained by summing the individual contributions. Production, import and export figures can be extracted from national statistics and, for instance, information for industrial use categories and products specified for individual chemicals is available in the Nordic SPIN database (Substances in Preparations in Nordic Countries). This information is used to distribute the used amounts of individual chemicals in relation to specific products and activities.

12.4 Geographic distribution of emissions

Emission of pollutants to the atmosphere results in more or less regional or local air pollution. In order to be able to model the impact of the emissions, it is not sufficient to know the total amount emitted; it is also necessary to know the location and altitude of the emission.

On a European level, the EMEP programme (Co-operative Programme for Monitoring and Evaluation of the Long-Range Transmission of Air Pollutants in Europe) under the UNECE Convention assesses the transboundary transport of regional air pollution. The programme is carried out in collaboration with national emission experts, who report national emission data based on grids (50 km x 50 km). From these national data it is possible to model atmospheric transport, and to map emission loads, air quality and deposition (Chapter 14).

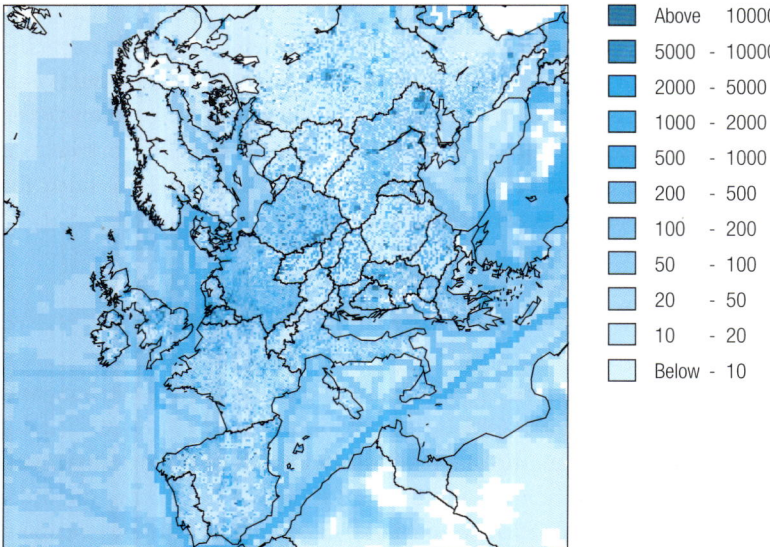

Figure 12.4. SO_2 emissions for Europe in the year 2004. Unit: kg y^{-1} km^{-2}.

Emissions of ammonia from agricultural activities and particulate matter (PM) from residential wood burning are examples of local air pollution, and various models have been developed to estimate the local emissions with a resolution as high as e.g. 100 x 100 m.

Emissions of NH_3 may cause both acidification and eutrophication, and models to estimate the local emission contain information on the number and location of farms, together with e.g. stable, crop, and animal types, as well as number of animals for each individual farm.

The effects of PM emissions from the residential use of wood stoves and small-scale wood boilers are related to health effects including bronchial infection. The basic data used in the models are data on e.g. housing area and heating type, obtained from dwelling and building registers. Address level registrations by local chimney sweeps, such as the number and type of stove or boiler, are also important input data. Through combining these data it is possible to estimate the potential wood consumption and emission for

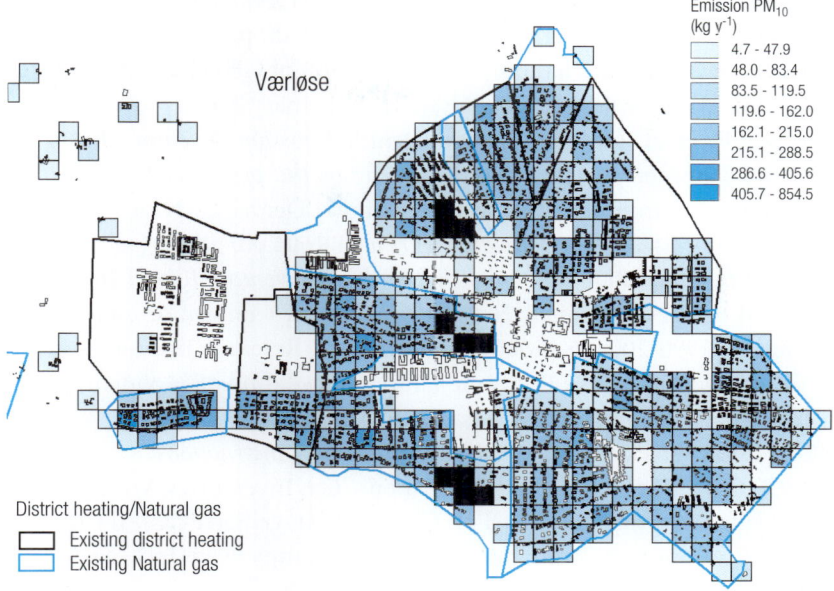

Figure 12.5 Example showing the distribution of PM_{10} emissions in a small Danish town, Værløse. Unit: kg y^{-1} per grid square (100 x 100 m).

each address. Figure 12.5 shows the distribution of PM_{10} emissions in a small Danish town.

For use in investigations on traffic pollution emission inventories based on traffic statistics can be made for individual streets and with a time resolution of maybe a quarter of an hour.

12.5 Literature

EEA (2006a): *EEA Core Set of Indicators.* http://themes.eea.europa.eu/IMS/CSI.

EEA (2006b): *Annual European Community greenhouse gas inventory 1990–2004 and inventory report 2006.* http://reports.eea.europa.eu/technical_report_2006_6/en/EC-GHG-Inventory-2006.pdf.

EMEP/CORINAIR (2005): *EMEP/CORINAIR Emission Inventory Guidebook 3rd Edition December 2005 Update*, Technical Report no 20, European Environmental Agency, Copenhagen. http://reports.eea.eu.int/EMEPCORINAIR4/en.

Gyldenkærne, S. et al. (2004): *Danish Integrated Emission Model for Agriculture (DIEMA).* Poster presented at the Workshop on End user requirements on spatial and temporal disaggregation of GHG budget organised by CARBOEUROPE, Spello, Italy, p 26-27 January 2004.

Illerup, J.B. et al. (2006a): *Annual Danish Emission Inventory Report to UNECE. Inventories from the base year of the protocols to year 2004.* National Environmental Research Institute, Denmark. NERI Technical Report No. 604. http://www.dmu.dk/Pub/FR604.pdf

Illerup, J.B. et al. (2006b): *Denmark's National Inventory Report 2006.* Submitted under the United Nations Framework Convention on Climate Change, 1990-2004. National Environmental Research Institute. – NERI Technical Report 589 (electronic). http://www2.dmu.dk/1_viden/2_Publikationer/3_fagrapporter/rapporter/FR589.pdf

IPCC (1997): *Greenhouse Gas Inventory Reporting Instructions. Revised 1996 IPCC* Guidelines for National Greenhouse Gas Inventories, Vol 1, 2 and 3. The Intergovernmental Panel on Climate Change 55(IPCC), IPCC WGI Technical Support Unit, United Kingdom. http://www.ipcc-nggip.iges.or.jp/public/gl/invs1.htm

IPCC (2000): *IPCC Good Practice Guidance and Uncertainty Management in National Greenhouse Gas Inventories.* http://www.ipcc-nggip.iges.or.jp/public/gp/english/

Leeuw, f.A.A.M. de (2002): *A set of emission indicators for long-range transboundary air pollution.* Environmental Science & Policy, 5, 135-145.

Ntziachristos, L. and Samaras, Z. (2000): *COPERT III Computer Programme to Calculate Emissions from Road Transport – Methodology and Emission Factors (Version 2.1).* Technical report No 49. European Environment Agency, November 2000, Copenhagen. Available at: http://reports.eea.eu.int/Technical_report_No_49/en (June 13, 2003).

13 Measurement of air quality and deposition

Finn Palmgren

Air quality measurements are under constant development due to increasing demand for protection of the human health and the environment in general. Air quality measurement methods have been developed and used for several hundred years. The first methods were generally based on absorption of the gaseous pollutants in liquids or on special surfaces and subsequent chemical analysis. Other methods are soot (black smoke) or soiling for determination of particulates. The measurement methods have been refined with respect to detection limits and specificity in relation to properties of the air pollutants and the technological development. Many of the old methods are still in use, but are refined in order to obtain the necessary sensitivity and specificity. However, the general trend has been in favour of automatic instrumentation in order to get actual data without too much manual work. A review of the most applied methods at present and in the near future is given in this chapter. More details have to be found in the literature.

General principles for air quality measurements

The pollutants must be evaluated according to their spatial and temporal distribution in order to establish efficient control strategies. This can be achieved through monitoring networks or campaigns based on well-selected and developed measuring facilities. These include instruments for the sampling and for the direct measurements of atmospheric pollutants of interest in environmental studies. The criteria for selection of methods could be technical, scientific, practical, economical etc. The measurements are often parts of an international or national measurement or monitoring programme, which prescribes a certain measuring method, i.e. reference or standard method.

The effects of air pollution on human health or the environment can be related to the total long term or acute exposure, which means that it is necessary to obtain long term averages – typically annual averages – as well as short term data, e.g. one hour averages. High time resolution is also useful in analysis of the collected data in relation to the sources and the meteorology in order to assess the contribution from the different sources and thereby to select efficient measures for reduction of the air pollution.

The main groups of air pollutants are gases and particulates, which requires different types of measurement methods. They are described separately in the following paragraphs, but it has to be taken into account that interactions occur between gases and particles in the air and also in connection with sampling and measurements.

13.1 Gaseous pollutants

Nitrogen compounds

The nitrogen oxides, generally called NO_x are determined by chemiluminescence with ozone. In a chemiluminescence analyser air is sampled through a filter and fed at a constant flow rate into the reaction chamber of the analyser, where it is mixed with an excess of ozone for the determination of nitrogen monoxide only. The emitted radiation (chemiluminescence) is proportional to the number of nitrogen monoxide molecules in the detection volume and thus proportional to the concentration of nitrogen monoxide. The emitted radiation is filtered by a selective optical filter and converted into an electric signal by a photomultiplier tube or a photodiode. For the determination of nitrogen dioxide, the sampled air is fed through a converter where the nitrogen dioxide is reduced to nitrogen monoxide and analysed in the same way as above. The electrical signal obtained from the photomultiplier tube or photodiode is proportional to the sum of concentrations of nitrogen dioxide and nitrogen monoxide, often defined as NO_x. The amount of nitrogen dioxide is calculated from the difference between this concentration and that obtained for nitrogen monoxide only (when the sampled air has not passed through the converter) (Figure 13.1).

This chemiluminescence method is based on the reaction:

$$NO + O_3 \rightarrow NO_2^* + O_2 \qquad\qquad (13.1)$$

$$NO_2^* \rightarrow NO_2 + hv \qquad\qquad (13.2)$$

Excited nitrogen dioxide (NO_2^*) emits radiation in the near infrared region (600 nm to 3000 nm) with a maximum centred around 1200 nm.

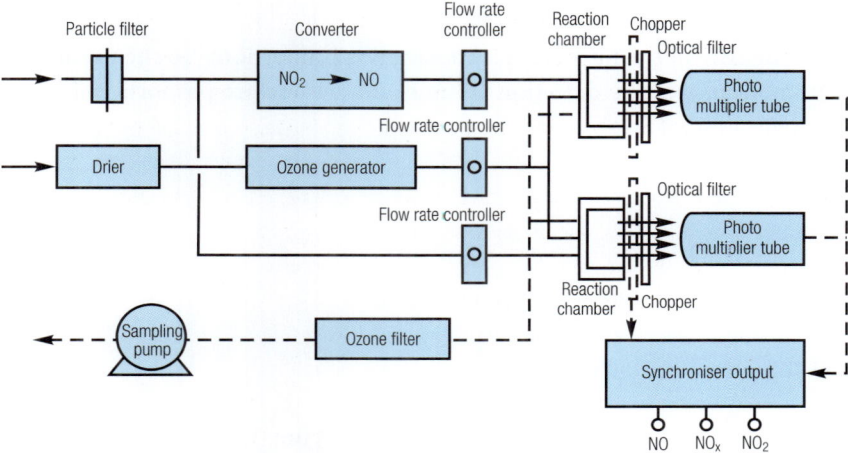

Figure 13.1 The principle of a chemiluminescence NO/NO_x monitor (CEN, 2004a).

The main disadvantage of the method is that several other nitrogen species are converted into nitrogen monoxide in the catalytic converter. These species include HNO_3, HONO, peroxyacetyl nitrate (PAN) and alkyl nitrates, which therefore potentially contribute to the apparent NO_2 signal.

Other examples of NO_2 measurement methods are: 1) diffusion tubes, where NO_2 gets into contact with impregnated surface by diffusion followed by chemical analysis, 2) denuders, where NO_2 is sucked through a tube with walls covered by an absorbent followed by chemical analysis, 3) sampling on a impregnated filter and followed by chemical analysis and 4) Differential Optical Absorption Spectroscopy (DOAS), where NO_2 is measured by optical absorption (the fine structure) in an open light path.

Similar methods are available for other nitrogen compounds.

Sulphur dioxide

SO_2 is measured by UV (ultraviolet) fluorescence. It is a result of emission of light from SO_2 molecules excited by UV radiation when they return to their ground state. The first reaction step is:

$$SO_2 + hv \rightarrow SO_2^* \tag{13.3}$$

In the second step the excited SO_2^* molecule returns to its ground state, emitting an energy hv' according to the reaction:

$$SO_2^* \rightarrow SO_2 + hv' \text{ (UV)} \tag{13.4}$$

The intensity of the fluorescence radiation is proportional to the number of SO_2 molecules in the detection volume and is therefore proportional to the concentration of SO_2.

Therefore: $F = k \times c$ (13.5)

where
F is the intensity of fluorescence radiation;
k is the factor of proportionality;
c is the concentration of SO_2.

Before entering the fluorescence analyser (Figure 13.2) the air sample is passed through a filter in order to exclude interferences caused by contamination with particles. The sampled air is scrubbed to remove any interference from aromatic hydrocarbons that may be present.

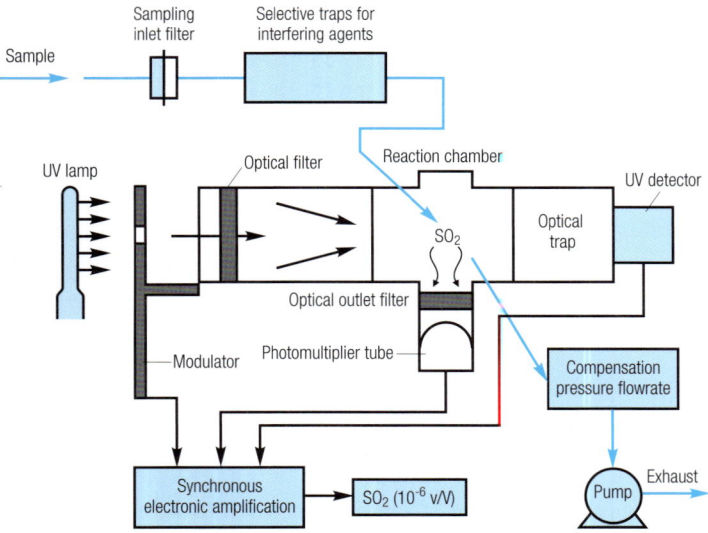

Figure 13.2 The principle of an UV fluorescence SO_2 monitor (CEN, 2004b).

The sampled air is then introduced into a reaction chamber, where it is irradiated by UV light in the wavelength range between 200 nm and 220 nm. The UV fluorescent light emitted in the wavelength range of 240 nm to 420 nm, is optically filtered and then converted to an electrical signal by a UV detector, for example, a photomultiplier tube.

The response of the analyser is proportional to the number of SO_2 molecules in the reaction chamber. Therefore, temperature and pressure either have to be kept constant, or if variation of these parameters occurs, the measured values have to be corrected.

The concentration of sulphur dioxide is directly measured in volume/ volume units if the analyser is calibrated using a volume/volume standard. The final results for reporting are expressed in $\mu g\ m^{-3}$ using standard conversion factors.

Ozone and other photochemical oxidants

The ozone measurement is based on UV absorption (Figure 13.3). Sampled air is drawn continuously through an optical absorption cell where it is irradiated by monochromatic radiation, centred on 253.7 nm, from a mercury discharge lamp. The UV radiation, which passes through the absorption cell, is measured by a sensitive photodiode or photomultiplier detector and converted to an electrical signal. Absorption of this radiation by the sampled air within the absorption cell is a measure of the ozone concentration in the ambient air.

Two different systems for the measurement of the ultraviolet absorption by ozone are usually employed. In one system the ultraviolet absorption

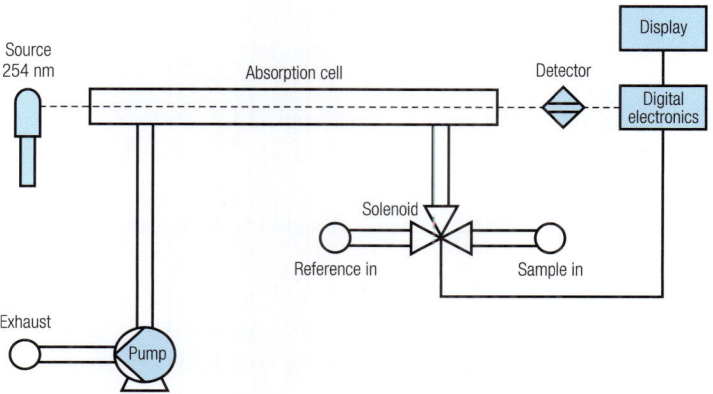

Figure 13.3 The principle of an ozone monitor based on UV absorption (CEN, 2004c).

by ozone is determined by means of the difference in ultraviolet absorption between a sample cell and a reference cell (dual-cell type). In the other system only a single sample cell is employed. The ultraviolet absorption of ozone is determined by alternately supplying sampled air containing ozone to the absorption cell and ozone-free sampled air. Ozone-free sampled air is obtained by passing the sampled air through an ozone catalytic converter which remove the ozone.

Most ozone analysers measure the temperature and pressure of the sampled air in the absorption cell. Using these data an internal microprocessor automatically calculates the measured ozone concentration relative to some chosen reference conditions. For analysers without this automated pressure and temperature compensation, the concentrations need to be corrected manually to the chosen reference conditions.

The concentration of ozone is measured in volume/volume units (if the analyser is calibrated using a volume/volume standard). The final results for reporting are expressed in $\mu g \ m^{-3}$ using standard conversion factors. The photochemical oxidants PAN can be measured by gas chromatography with electron capture detector or by chemiluminescence.

Carbon monoxide and other carbon compounds

Ambient CO concentration is measured by non-dispersive infrared (NDIR) methods (Figure 13.4). The attenuation of infrared light passing through a sample cell is a measure of the concentration of CO in the cell, according to the Lambert-Beer law.

Not only CO but also other molecules will absorb infrared light, in particular water and CO_2 have broad bands that can interfere with the measurement of CO. Different technical solutions have been developed to suppress cross-sensitivity, instability and drift in order to design continuous monitoring systems with acceptable properties. For instance:

- Measuring IR absorption of a specific wavelength (4.7 μm for CO)
- Dual-cell monitors, using a reference cell filled with clean air (compensation for drift)
- Gas filter correlation, "measuring" over a range of wavelengths

The concentration of carbon monoxide is measured in volume/volume units (if the analyser is calibrated using a volume/volume standard). The final results for reporting are expressed in $\mu g \ m^{-3}$ using standard conversion factors. CO_2 can also be measured by NDIR, but at another wavelength.

A large number of volatile organic compounds can be measured by gas

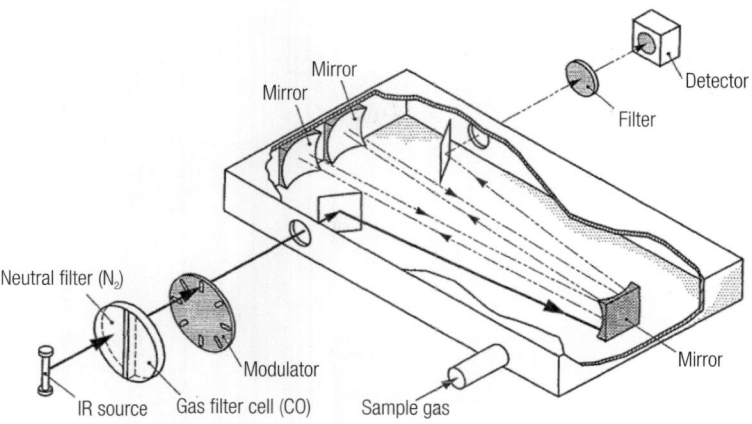

Figure 13.4 The principle of a CO monitor based on gas filter correlation (CEN, 2004d).

chromatography after collection in steel canisters or absorbed in tubes filled with different type of absorbents, e.g. active carbon.

13.2 Particles

Particles are often referred to as a single compound, but this is too simple. Particle pollution is a common denominator for various compounds, considerably differing in physical and chemical properties (Figure 13.5).

Consequently, each measurement method will have its artefacts and partly defy its measuring objectives. The measurement of concentrations of particulate matter in ambient air is therefore not straightforward. There is a variety of techniques available to measure mass concentrations, but due to the complex nature of particulate matter, the method chosen can significantly influence the result. Particles have very different shapes and sizes ranging from spheres over crystal shapes to needles, and the sizes range between a few nanometers to more than 100 µm.

Figure 13.6 shows typical size distributions in urban air as mass and as number distributions. Typical source contributions are indicated. Mass as well as number measurement methods are used in order to cover the wide range of particles sizes. The very small particles have nearly no mass and the number of the largest particles are very few. As an illustration, one 10 µm particles has the same mass as one million 0.1 µm particles. The figure shows often used particles parameters, e.g. PM_{10} and $PM_{2.5}$ (the mass of particles with diameter below 10 and 2.5 µm respectively).

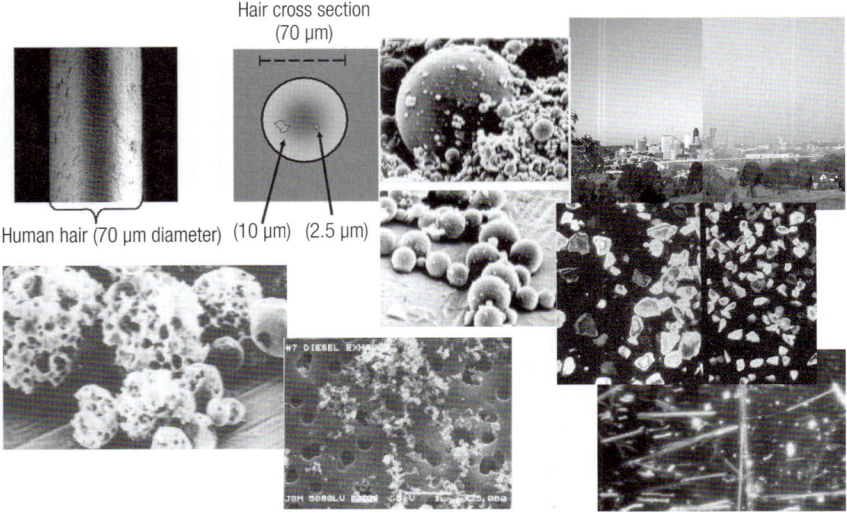

Figure 13.5 Examples of particles (from left to right): Compared to human hair: fly ash, photochemical particles (haze), oil coke, diesel soot and asbestos needles. Particles in the electron microscope pictures are typically in the μm range.

With gaseous pollutants it is possible to express concentrations as an amount fraction, the ratio of pollutant molecules to the total number of air molecules, for example with units of nanomoles per mole (nmol mol^{-1}) or,

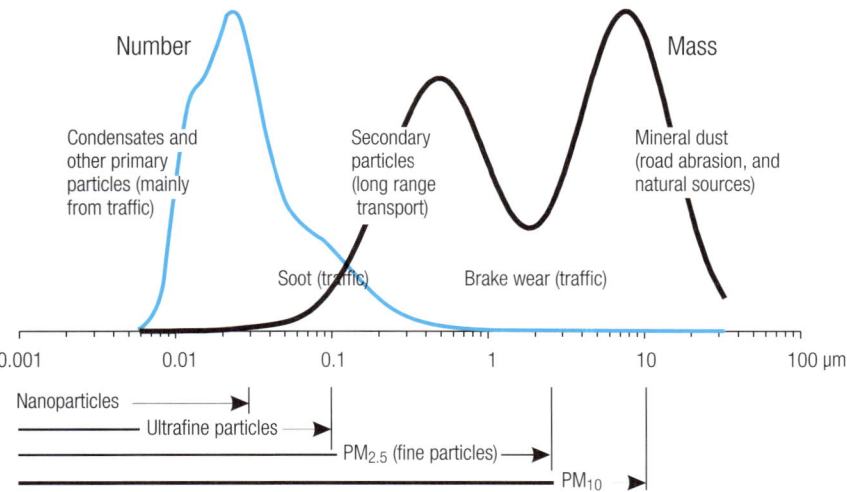

Figure 13.6 Mass and number distributions of urban particles. Typical source contributions are indicated.

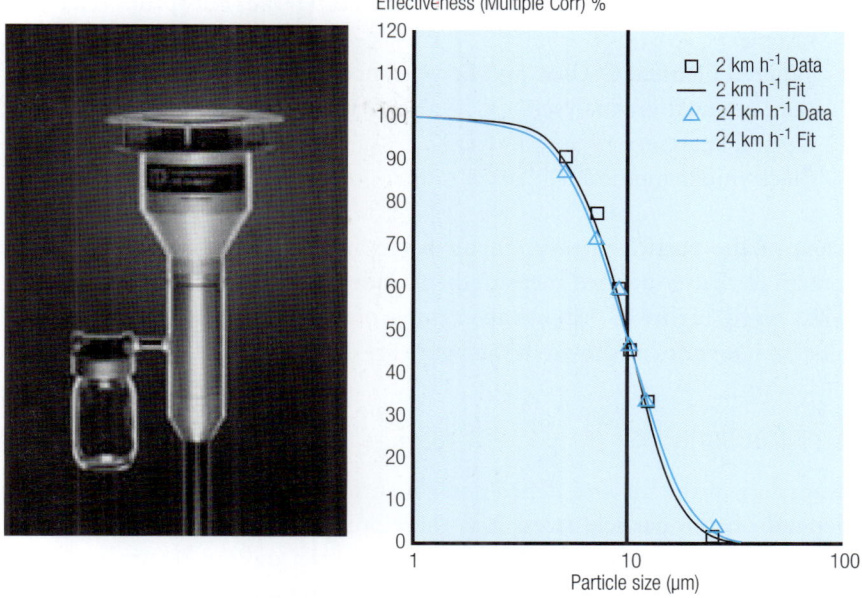

Figure 13.7 PM$_{10}$ sampling head (left) and the shape of the cut-off curve (right) shown for different wind speeds

more commonly, parts per billion volume (ppbv). This is not possible for particulate matter, and measurements are always given in units of particulate mass per unit volume of air (typically µg m^{-3}). When these units are used without specifying the temperature and pressure of the air, the same 'packet' of air will have a changing concentration as these air properties changes. The European legislation for particulate matter requires that the air volume used must refer to the ambient air temperature and pressure at the time of sampling. In practice this means that appropriate corrections need to be made if the flow rate used to calculate the sampled volume is not based on the actual volume of sampled air.

There is a variety of monitoring methods available for the measurement of mass concentrations of particulate matter in ambient air. These include both direct-reading instruments, which provide continuous measurements of particle concentrations, and filter-based gravimetric samplers, which collect the particulate material onto a filter, which subsequently must be weighed in a laboratory.

Commonly used methods for the mass measurement of particulate matter in ambient air include:

- Filter-based gravimetric samplers (including the European reference sampler)
- Tapered Element Oscillating Microbalance (TEOM) analysers
- Beta-attenuation analysers
- Optical analysers
- Black smoke method.

Most of the particle measurement instruments have an inlet size cut-off system for better defined measurement, independent of wind direction and wind speed. Figure 13.7 shows an example of a PM_{10} sampling head. The cut-off characteristics as shown in the figure are to some extent standardised

Size fractionation

Transport in the air and deposition on surfaces or in the human air-ways depend on the particle sizes. It is therefore crucial for assessment of the health effects, for understanding of the processes and for determination of transport and deposition to determine the number and/or the mass of particles for different particles sizes.

The most often used type is a filter sampler with a sampling head in the inlet, which cuts off the larger particles. Particle size fractions are defined as PM_{10} or $PM_{2.5}$. TSP (total particulate matter) is often particulate matter with a more or less well defined upper cut off.

A better controlled size fractionation can be obtained by a so-called impactor. An example, a cascade impactor, was shown in Figure 11.3 (p. 244). The nozzles in the impactor become smaller and smaller downstream resulting in larger and larger air velocities from stage to stage of the impactor, leading to smaller and smaller particles to be collected on the collection plates. The smallest particles are collected on a back-up filter.

Filter sampling

The most frequently used method for measurement of particles is collection on filters and subsequent analysis by traditional chemical or physical methods. The filter can be produced of fibres (paper, glass, Teflon etc.) or membranes produced of polymers (nylon, esters of cellulose, polycarbonate etc.) For all types of filters a so-called pore size can be determined. Collection efficiency can be defined as the percentage of particles, which are collected on the filter. The collection efficiency of a certain filter depends on the aerodynamic particle size, the air velocity through the filter and possible chemical reactions on the filter. The collection efficiency is a result of several mechanisms leading to collection on the surface of the filter material, e.g. impaction, interception, diffusion, sedimentation and electrostatic forces. The impaction and interception are the dominating mechanisms for large particles and/or at high air velocities. The diffusion is dominating for small particles and/or low air velocities. The same mechanisms determine the transport of particles into the lungs. It is important to select filters and air velocities for high collection efficiency in the whole range of particles sizes. A high efficiency for large and small particles often leads to low efficiency for particles in the range 0.2-0.5 μm, as shown in Figure 13.8.

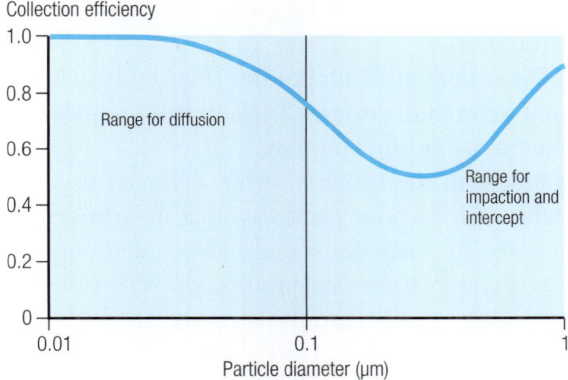

Figure 13.8 Filter collection efficiency versus particles size.

Filter samplers

Member States in the European Union must use standard measurement techniques and procedures in order to assess the air quality in a consistent way. They must follow the Directive 96/62/EC on ambient air quality assessment and management, and Directive 1999/30/EC (1st EC Air Quality Daughter Directive), which sets the parameters specific to the assessment of particulate matter (and also of sulphur dioxide, nitrogen dioxide and lead).

There are no traceable reference standards for $PM_{2.5}$ measurements. Therefore, the standard measurement method set out in EN 14907 (2005) effectively defines the measured quantity, specifically by the sampling inlet design and associated operational parameters covering the whole measurement process. The EU First Air Quality Daughter Directive (1999/30/EC) specifies that measurements of PM_{10} should be carried out using the reference method as defined in European Standard EN12341. This standard refers to three sampling devices that may be used:

- Superhigh volume sampler – the WRAC (Wide Range Aerosol Classifier)
- High volume sampler – the HVS PM_{10} sampler (68 m^3 hr^{-1})
- Low volume system – the LVS PM_{10} sampler (2.3 m^3 hr^{-1})

Each of these samplers consists of a PM_{10} sampling inlet that is directly connected to a filter substrate and a regulated flow controller. Following completion of the sampling period, the PM_{10} mass collected on the filter is determined gravimetrically. The filter is conditioned at 20 °C and 50% relative humidity prior to weighing.

The WRAC sampler is usually regarded as a 'primary standard' and is not suitable for deployment in the general ambient environment. The low volume sampler (often referred to as the 'kleinfiltergerät' or 'KFG') has been most widely used in many countries.

The reference method will be subject to both positive (for example due to an increase in particle-bound water) and negative artefacts (for example due to loss of semi-volatile compounds) during sampling.

A number of samplers based on the LVS have been designed to allow automatic filter changing after each 24-hour period, so that the instrument does not need to be visited daily. This includes systems which incorporate 8 separate PM_{10} sampling heads connected to a central pump via a solenoid switching system, and a sequential sampler that uses an automatic exchange mechanism.

Filter packs

Particle filters can be followed by one or more filters impregnated with material, which absorb gases, a so-called filter pack. The method has been widely used and is still in use for measurements of long-term averages in background areas, e.g. under the EMEP programme. The filters can be analysed by normal chemical analytical methods.

Automatic particle measurements

Examples of automatic particle measurement instruments are described in the following, but more types exist. Please, see the literature for further information.

The Tapered Element Oscillating Microbalance (TEOM) is a low volume air sampler with its filter mounted at the end of a tapered quartz tube, which acts like a tuning fork whose resonant frequency is very sensitive to small changes in the mass of the filter. The quartz tube is the frequency control element in an electronic oscillator. In this way it is possible continuously to monitor the filter mass via the oscillation frequency (Figure 13.9). A TEOM permits the measurements of airborne particulates time scales as short as ten minutes, providing information about the variation of particulate concentrations during the day and their correlation with potential sources.

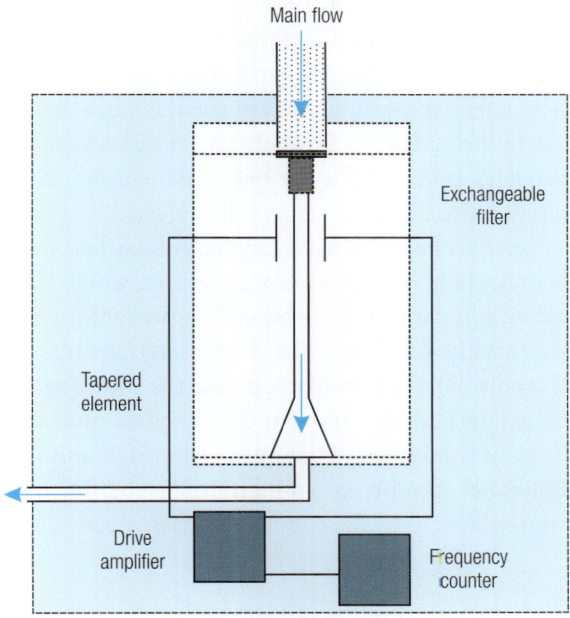

Figure 13.9 Working principles of a TEOM monitor.

To maintain the frequency stability of the instrument and prevent condensation of moisture on the filter, the TEOM element must be operated at a constant temperature above ambient temperature, e.g. 50 °C. This can lead to a significant loss of volatile components in the particulates as they "boil off" from the heated filter. This can result in an underestimation of particulate concentrations. The losses take place e.g. when monitoring secondary particles of ammonium nitrate or particles or when the sample contains fine droplets of condensed tars and other unburned fuel vapours. A correction for the losses can be made by a so-called FDMS system, which measures the mass of the volatiles, which afterwards can be added in order to give the total $PM_{10}/PM_{2.5}$ equivalent to the reference method (WRAC sampler).

The beta attenuation or beta gauge method operates by drawing air through a filter. Beta particles are passed through the particles deposited on the filter and the attenuation of these beta particles is measured in a sensor located on the other side of the filter. The attenuation is converted to an estimate of mass based on the absorption coefficient.

Although the absorption coefficient will vary with different particle compositions, the instrument is typically calibrated using the mass absorption coefficient for quartz. In practice the mass absorption coefficient may vary by up to 20%.

Advantages of the beta attenuation method include the provision of data at an hourly time resolution and the low labour and material costs associated with ongoing operation. Beta attenuation monitors are used in a number of ambient air quality monitoring programmes throughout Europe and other parts of the World. The particles collected for the beta attenuation can be analysed by different techniques, e.g. PIXE for elemental analysis.

Different type of *optical analysers* can be used for determination of the particle concentrations. One example is black smoke analysers, which have been used for several decades. A modern type is based on measurement of transmission of light through a white filter tape. The changes in transmission of light are a measure of the concentration of black smoke (or soot), which is a good indicator of combustion particles and can be related to mortality and morbidity. The sensitivity of modern instruments is very high and it is possible to measure the soot concentration with a high time resolution, e.g. down to 15 minutes, and thus related to traffic pattern in urban areas.

Particle counters

The particle mass is not a good measure for particles when they are in the nano meter range, as the mass is very low. Examples are particles from diesel

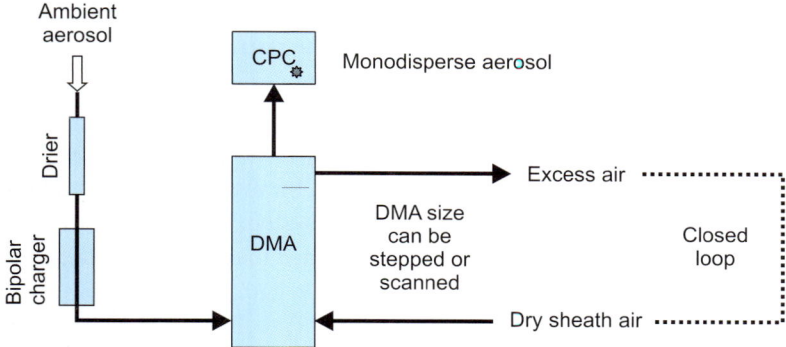

Figure 13.10 Principle of a DMPS (Differential Mobility Particle Sizer)

engines. It is then necessary to count the number of particles. The total number of particles can be counted by a condensations particle counter (CPC). The measurement principle is based on condensation of liquids (e.g. butanol or water) on the particles in order to increase their size so that they can be detected optically, i.e. particles larger than 0.3 μm.

The sizes of the particles are often an important parameter for assessment of the health effects, but also for determination of the origin of the particles. A DMPS (Differential Mobility Particles Sizer) or SPMS (Scanning Mobility Particles Sizer) can determine the size distribution, typically in the range 3 nm to 1 μm. The size fractionation is made by a DMA (Differential Mobility Analyser) based on the mobility measurement of the charged particles in an electrical field. Figure 13.10.

13.3 Deposition

Deposition of air pollutants on surfaces of soil, water, vegetation, materials etc. takes place by dry or wet deposition, i.e. either directly as particles or gases or in rain, fog or snow. The deposition is the dominating mechanism of transport of e.g. sulphur or nitrogen compounds from the air to the environment leading to acidification and eutrophication. It is therefore important to have reliable deposition measurement methods.

Wet deposition can be measured by collection in containers with a well-defined upwards opening, e.g. a funnel and a container underneath for the sample. The so-called wet-only sampler is only open when there is precipitation, see Figure 13.11. It is closed during dry weather and opens only when a sensor detects water. The container is often protected against sunlight and possible cooled in order to reduce the risk for chemical transformations in the collected sample. For measurement of metals is it necessary to add acid in order to keep the metals dissolved in the rain water.

The collected samples are analysed by appropriate chemical analytical methods.

Dry deposition is more complicated to measure because the deposition depends strongly on the properties of the surfaces on which the deposition takes place. The deposition velocity depends on the roughness, humidity, biological activity, chemical composition etc. of the surfaces. The deposition velocity depends also on the meteorological conditions and the air concentration of the pollutants.

A very common method for measurement of deposition is the so-called gradient method, where the air concentration of the pollutant is measured at

Figure 13.11 Wet-only sampler, which only opens when it is raining.

two heights. The air concentration will be a little lower near the surface due to the deposition. The dry deposition can be calculated by the difference in the concentration at the two heights and the meteorological condition.

Another commonly applied method is the eddy-correlation method. The flux downwards, F, can be determined by:

$$F = wc + w'c' \qquad (13.6)$$

where w and c are the average vertical component of the wind speed and the average concentration, respectively. w' and c' are the corresponding fluctuations around the averages. The average vertical wind speed is zero for averaging times of half hours or longer; therefore is the first part of the equation zero. The method requires very high measurement frequency, e.g. 10 times per second or faster.

13.5 Strategies and programmes

The pollution concentration at any point is a sum of impacts or contributions originating from different sources on different scales. The monitoring should be stratified according to site types such as rural background, suburban, urban background, industrial- and traffic-influenced. Depending upon the location of the point, the total concentration is a sum of

- The remote/natural background concentration
- The rural/regional background contribution
- The urban background contribution
- The local contribution from nearby sources, such as streets, point sources (industry, power plants), etc.

This is schematically illustrated in Figure 13.12. Important parameters are the concentration increments from rural to urban and from urban to hot spots, which can be related to the different source types and the possible abatement strategies.

Sampling sites must be located at sites considered to be representative for local environments, exposure situations or source activities, such as urban background, residential areas in various parts of the city, industrial or traffic sites, and also extra-urban background sites. An example from the Danish monitoring network is shown in Figure 13.13.

This approach is generally cost-efficient, and especially when dispersion modelling is used as part of the assessment program. It provides measured concentrations, which indicate the exposure in residential areas or hot-spot locations, and provides data for comparison with, and possible adjustment

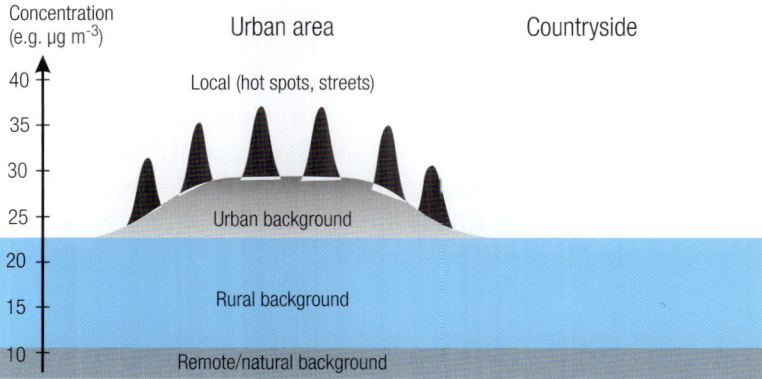

Figure 13.12 Distribution of air pollutants in and around an urban area.

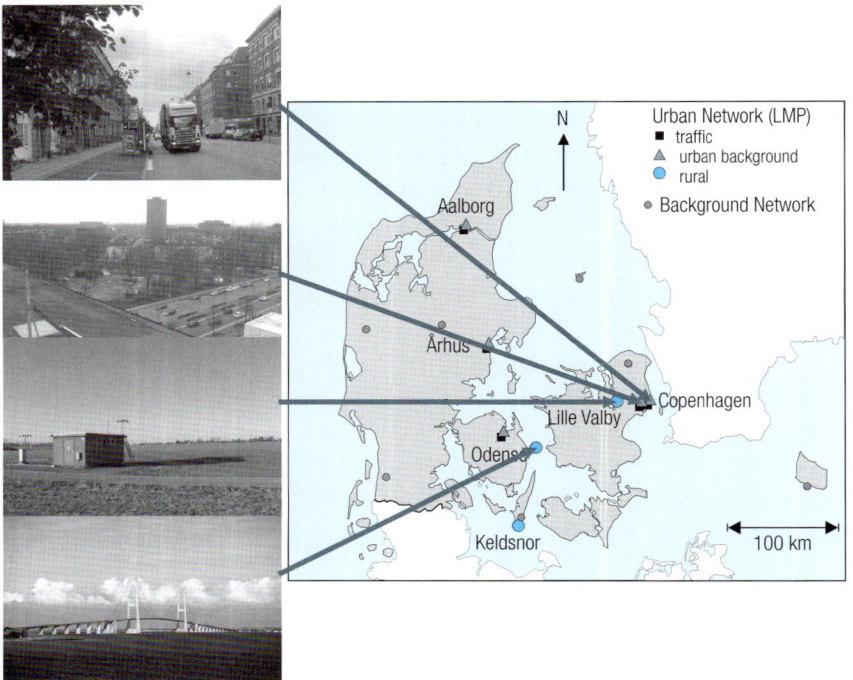

Figure 13.13 Illustration of different types of monitoring stations in Denmark: traffic, urban background, rural background and remote background (Top of the pylon of the bridge over the Great Belt - 254 m above sea level).

of the dispersion model. With proper selection and location of measurement sites, it is possible to use the comparison between measured and model calculated concentrations to actively check emission data for specific sources.

The network design and siting of stations is important, due to cost as well as accuracy/representativity considerations. There are no generally applicable quantitative guidelines, but some harmonisation is in progress in Europe related to the EU-directives on air quality, the EMEP programme and global networks, e.g. under WHO.

Efficient air quality monitoring programmes often include measurements combined with air quality modelling. This is the case in the EMEP programme as well as the EU related programmes.

The EMEP programme

The EMEP network was established in collaboration with UN-ECE and has been a success e.g. resulting in strong reductions in sulphur emissions across

Europe. The regional monitoring activities have been an essential activity to document and underpin the need for abatement measures and have thus contributed to significant improvements in air quality for a number of chemical compounds. Currently, protocols target at emissions of acidifying, eutrophying substances, photochemical oxidants, heavy metals and persistent organic pollutants. Under the revision of the Gothenburg protocol, primary particles have been included. The driving force to including primary particles in addition to the precursors of secondary particles, which were covered in the earlier protocols, is the adverse health effects shown to be related to particle mass.

The main objectives of EMEP are to:
- provide observational and modelling data on pollutant concentrations, deposition, emissions and transboundary fluxes on the regional scale and identify their trends in time;
- identify the sources of the pollution concentrations and depositions and to assess the effects of changes in emissions;
- improve our understanding of chemical and physical processes relevant to assessing the effects of air pollutants on ecosystems and human health in order to support the development of cost-effective abatement strategies; and
- explore the environmental concentrations of new chemical substances that might require the attention of the Convention in the future.

EMEP monitoring programme in the different countries shall be designed to fulfil its purpose. Up to now, EMEP has focused on air pollutants in rural and background areas. The EMEP network is continuously being revised in order to serve the developing needs of the Convention.

The EMEP monitoring programme is organized to allow for monitoring stations having measurements with different levels of scope and complexity. Three levels are proposed, each targeting EMEP objectives in different ways.

The main objective of monitoring at *level-1* sites is to provide long-term basic chemical and physical measurements of the traditional EMEP parameters. Level-1 activities would be the first priority when extending the network to areas not covered by measurements up to now. By undertaking a more demanding monitoring programme, the level-1 stations should gradually be upgraded to level-2 sites.

Level-2 sites will provide additional parameters essential for process understanding and further chemical speciation of relevant components, and represent thus an essential supplement to the level-1 sites. The aim is to establish a total of 20-30 level-2 sites over Europe by 2009. A level-1 site extending its programme to include the level-2 activities for any of the specific topics will be identified as a "supersite" for this topic. Level-1 and level-2 sites will typically be operated by institutions nominated by the respective Parties for implementing their monitoring obligations.

Level-3 activities are research-oriented. The main objective of level-3 sites is to develop the scientific understanding of the relevant physico-chemical processes in relation to transboundary pollution and its control. Level-3 activities may also include campaign data. Level-3 sites are a voluntary component of the new monitoring network.

Figure 13.14 shows a Finnish monitoring station (Utö), which is part of the EMEP network.

Figure 13.14 An EMEP station in Finland, Utö.

EU-directive related programmes

The main focus of the EU air quality monitoring programmes is on protection of human health. The assessment of population exposure with air pollution and determination of the sources of the pollution forms the basis for decisions on cost effective measures for reduction of adverse health effects. An effective monitoring system (a combination of measurements and air quality and exposure models, see below) is crucial for such decisions.

Long- and short-term exposure. Adverse health effects have been documented after short-term exposure to peak concentrations, as well as long-term exposure to relatively low concentrations of air pollutants. The long-term exposure has probably a larger impact on public health than short-term exposure to peak concentrations. Some studies have documented that persons living close to busy roads experience more short-term and long-term effects of air pollution than persons living further away. In urban areas, up to 10% of the population may be living at such "hot spots". The public health burden of such exposures is therefore significant. Many people are exposed to high levels of pollution during transport as pedestrians or by bicycles, cars, trams, buses etc. in busy streets.

Effects of long-term exposure to air pollutants on mortality and morbidity are discussed in (WHO, 2003). It has been estimated that long-term exposure to moderate levels of e.g. fine PM can be associated with a reduction in life expectancy of several months.

Effects of short-term exposure to e.g. PM have been documented in numerous time series studies on mortality and morbidity endpoints (WHO, 2003). Consequently, both short-term and long-term effects of exposure to PM are of concern.

"Hot spots". The question of "hot spots" relates to the relevance of spatial differences in exposure, i.e. the importance of location and proximity to emission sources. Air pollution levels can be significantly elevated near sources, e.g. near busy roads, and especially particle components such as elemental carbon and ultrafine particles which are considerably elevated near traffic sources (WHO, 2004). The vast majority of epidemiological studies characterize exposure as measurements that describe urban background concentrations rather than concentrations at locations influenced by sources in the immediate vicinity. Thus, the effect estimates may not sufficiently include effects due to local hot spots.

The EU Member States have to establish measurements in zones and agglomerations in accordance with the directives (see Chapter 22). The number of measurement points is given in the directives and depends on the population of the zone or agglomeration and the pollutant.

The criteria for determining the location of sampling points for the measurement include macro scale siting, as well as micro scale siting. In relation to protection of human health the macro scale siting shall secure that the measurements provide data on the areas within zones and agglomerations where the highest concentrations occur to which the population is likely to be directly or indirectly exposed for a period, which is significant in relation to the averaging period of the limit values. They should also provide data about the air quality in other areas within the zones and agglomerations, which are representative of the exposure of the general population.

Sampling points targeted at the protection of ecosystems or vegetation should be sited more than 20 km from agglomerations or more than 5 km from other built-up areas, industrial installations or motorways. As a guide-line, a sampling point should be sited to be representative of air quality in a surrounding area of at least 1000 km². A Member State may provide for a sampling point to be sited at a lesser distance or to be representative of air quality in a less extended area, taking account of geographical conditions.

Micro scale siting

The following guidelines should be met as far as practicable:
- The flow around the inlet sampling probe should be unrestricted without any obstructions affecting
- The airflow in the vicinity of the sampler (normally some metres away from buildings, balconies, trees, and other obstacles and at least 0,5 m from the nearest building in the case of sampling points representing air quality at the building line)
- In general, the inlet sampling point should be between 1,5 m (the breath-ing zone) and 4 m above the ground. Higher positions (up to 8 m) may be necessary in some circumstances. Higher sitings may also be appropriate if the station is representating a large area
- The inlet probe should not be positioned in the immediate vicinity of sources in order to avoid the direct intake of emissions unmixed with ambient air
- The sampler's exhaust outlet should be positioned so that recirculation of exhaust air to the sampler inlet is avoided

Location of traffic-oriented samplers:
- For all pollutants, such sampling points should be at least 25 m from the edge of major junctions and at least 4 m from the centre of the nearest traffic lane; for nitrogen dioxide, inlets should be no more than 5 m from the kerbside
- For particulate matter and lead, inlets should be sited so as to be repre-sentative of air quality near to the building line

The following factors may also be taken into account:
- Interfering sources
- Security
- Access
- Availability of electrical power and telephone communications
- Visibility of the site in relation to its surroundings
- Safety of public and operators
- The desirability of co-locating sampling points for different pollutants
- Planning requirements

The minimum number of stations is determined by the size (number of inhabitants) of the urban area and of the pollution level. The number of stations can be up to 10 for the largest urban areas if the pollution level is above the so-called upper assessment level (see Chapter 22 about legislation). The number can be reduced significantly, if a lower level of pollution has been documented.

For zones and agglomerations within which information from fixed measurement stations is supplemented by information from other sources, such as emission inventories, indicative measurement methods and air quality modelling, the number of fixed measuring stations to be installed can be reduced (see Chapter 22 about legislation).

For zones and agglomerations within which measurement is not required, modelling or objective-estimation techniques may be used.

Minimum number of sampling points for fixed measurement to assess compliance with limit values for the protection of human health and alert thresholds in zones and agglomerations where fixed measurement is the only source of information.

13.5 Quality assurance and quality control

Air quality measurements have no value unless their quality is documented. The quality includes the quality of the measurement methods as well as the fulfilment of the requirements about how, where and when the measurements should be carried out. This can be done by using accredited laboratories, which have to document their measurement methods for an independent authority. Very detailed description of the accreditation system is given in ISO/IEC 17025:2005, which specifies the general requirements for the competence to carry out tests and/or calibrations, including sampling. It covers testing and calibration performed using standard methods, non-standard methods, and laboratory developed methods. ISO/IEC 17025:2005 is for use by laboratories in developing their management system for quality, administrative and technical operations. Details will not be given here, but only a short summary with the most important issues.

The system for quality comprises Quality Assurance/Quality Control (QA/QC) with all the activities that assure that a measurement meets defined standards of quality. The quality terms relevant for QA/QC procedures and criteria can be defined as follows (ISO 8402:1994):

- Quality is the totality of characteristics of an entity that bear on its ability to satisfy stated or implied needs
- Quality Assurance involves the management of the entire process, including all the planned and systematic activities, which are needed to assure and demonstrate the predefined quality of data, to provide adequate confidence that an entity will fulfil the requirements for quality
- Quality Control comprises the operational techniques and activities that are undertaken to fulfil the requirements for quality

The Quality Assurance activities cover all the pre-measurement phases, ranging from definition of data quality objectives to equipment and site selection and personnel training, whereas the Quality Control operational functions cover directly measurement-related activities such as routine checks, calibration and data handling.

The principal objectives of QA/QC are to:

- Identify errors
- Quantify errors
- Reduce errors

These errors can be of two types: procedural and technical.

Procedural errors result from unclear and ineffective management, inadequately training staff, improper planning, lack of adequate QA, lack of data tracking and other problems related to how the work gets done.

Technical errors are directly related to the methods and technologies used. They may result from mathematical errors, use of incorrect data, use of incorrect methodology and assumptions, etc. In general, avoiding or minimising procedural errors will reduce the likelihood of technical errors.

The data quality objectives may be defined in terms of the following parameters, which are indicators of data quality, which e.g. are defined in the EU directives (see Chapter 22):

- Precision.
- Accuracy and/or trueness
- Representativeness
- Capture.time coverage

Quality assurance includes procedures for site selection and description of air quality network design. This can include tools for the selection of the stations siting, which could be indicative air quality measurements, emission sources inventories, and application of air pollution modelling.

In general, the quality assurance is taken care of by a central laboratory (accredited), which sets requirements to the staff, adequate education and training of personnel, the methods used, the organisation, the equipment, the calibration facilities and standards etc.

Quality control includes preparation of protocols and implementation of procedures such as site operation and equipment maintenance. This includes routine (and non-routine) site visits, calibration, data validation procedures, reporting and documentation.

13.6 Literature

CEN (2004a). *Ambient air quality – Standard method for the measurement of the concentration of nitrogen dioxide and nitrogen monoxide by chemiluminescence.* prEN 14211.

CEN (2004b). *Ambient air quality – Standard method for the measurement of the concentration of sulphur dioxide by ultraviolet fluorescence.* prEN 14212.

CEN (2004c). *Ambient air quality – Standard method for the measurement of the concentration of ozone by ultraviolet photometry.* prEN 14625.

CEN (2004d). *Ambient air quality – Standard method for the measurement of the concentration of carbon monoxide by nondispersive infrared spectroscopy.* prEN 14626.

Council Directive 96/62/EC of 27 September 1996 on ambient air quality assessment and management.

Council Directive 1999/30/EC of 22 April 1999 relating to limit values for sulphur dioxide, nitrogen dioxide and oxides of nitrogen, particulate matter and lead in ambient air.

Council Directive 96/62/EC of 27 September 1996 on ambient air quality assessment and management.

Council Directive 1999/30/EC of 22 April 1999 relating to limit values for sulphur dioxide, nitrogen dioxide and oxides of nitrogen, particulate matter and lead in ambient air.

ISO/IEC 17025. (2005) *General requirements for the competence of testing and calibration laboratories.*

ISO 8402. (1994) *Quality Management and Quality Assurance – Vocabulary.* http://www.defra.gov.uk/environment/airquality/

14 Air pollution modelling

Ole Hertel and Jørgen Brandt

Measurements carried out under the framework of monitoring programmes provide crucial information about trends in pollution loads as well as exceedances of critical loads and levels at selected sites. Just as crucial are the observations obtained from field studies aimed at improving our understanding of the physical and chemical processes governing the fate of the pollutants in the environment. Application of air pollution models can significantly supplement and extend the information obtained from the various measurements. In monitoring programmes measurements are usually constrained to a limited number of sites due to the limited sources available. In this context models can be applied for estimating pollutant levels at sites that are not covered by measurements in the monitoring programmes. Furthermore, the models may be used as tools in the interpretation of measurements.

In general, air quality models have become integrated tools in environmental monitoring, management and assessment of air pollution (Hertel et al., 2007). An example may be found in the European Councils Air Quality Framework Directive (96/62/EF of 27th September 1996) in which it is stated that assessment of air quality is to be obtained from monitoring or from modelling. This directive represents a major change in policy with respect to applying models in the mapping of the air quality levels in an overall assessment. In general, air quality models may be used for a variety of different purposes in environmental management and assessment. These uses include:

- Providing data for sites and areas not covered by measurements
- Interpretation of measurements
- Source apportionment

- Inverse modelling for determining emissions
- Scenario studies
- Short term prognoses
- Climate change studies

Models are very useful in interpretation of measurements and for the understanding of the governing processes. Determination of the contribution to pollutant loads or levels from different sources is another typical use of models. This is usually done by performing calculations with and without emissions from specific sources or source regions. Subsequently the obtained results from the two sets of model runs are compared.

Another kind of model application is inverse modelling, which aims at determining emissions or emission factors for specific sources or source regions. In inverse modelling, the emission data used in the model calculations are adjusted with respect to obtaining the best possible performance of model results compared with observations. The fundamental assumption is here that the model describes the governing processes sufficiently well – otherwise the obtained emission data or emission factors will not reflect the reality.

Studies of the contribution from various types of sources are important goals for using model tools. However, another and often more important use of models is in exploring what will happen to pollution loads and levels in the future. The impact of a real emission reduction or an addition of new sources etc. can of course only be studied through observations by performing measurements before and after these changes have come into action. Application of models is therefore the only option available for carrying out scenario studies to answer such questions before they are implemented.

Models are also the only option for providing short term prognoses of air pollution at regional or urban scale – e.g. for the next two or three days. Furthermore, model calculations using meteorological input data from climate models may be used for investigating the impact of climate change on air pollution loads and levels. In the following we will show various examples of model applications and discuss the strengths and weaknesses in the different approaches.

14.1 Principles of modelling

Models in the form of generalisations or simple relationships are often used in our daily life without giving it many thoughts. In this context a model is a representation of the real world as it is now or as it would be under certain conditions. A model is only valid within a certain set of conditions that have to be considered with specific address to the studied problem. Otherwise the model may be used in describing situations for which we cannot be certain of its validity, and we may draw totally wrong conclusions.

The persistence method is one of the very simple models, which may be applied for describing a variety of situations; an example: *the pollution level today will be like it was yesterday.* This very simple model actually works for a lot of cases (in weather prediction the persistence method provides the correct results for 60% of the time), but we are often more interested in knowing when more extreme conditions occur, which of course are situations where the persistence method cannot be applied. A variety of different types of models have been developed over the years, and all these model types may be categorised in various ways.

Models may be categorised by the scale of the pollution problem that it describes. Generally the atmospheric processes are divided into:

- Micro (few meters)
- Local (km)
- Meso (hundreds of km)
- Regional (thousands of km)
- Global scale phenomena.

A basic understanding of the general features of the processes on the various scales is necessary, when mathematical models for air pollution simulations are formulated. Assumptions that are valid on one scale may be highly violated, when another scale is considered.

Categorisation of model types by the scale of pollutant problem for which it is applied is thus a solution that in some cases may be very useful. Another categorisation that will be explored a little further in the following is to divide the models into:

- Physical models
- Empirical models
- Statistical models
- Deterministic (physical-chemical) models

Examples of a physical model are a wind tunnel or a water tank e.g. used for observing flow around obstacles like buildings or smog chambers for studying chemical transformations. Such physical models are used as small representations of the real world aimed for understanding selected processes and/

or for determining various parameters and coefficients. Often the final goal is to improve or provide input for deterministic air pollution models. As for any other kind of model, the results from physical models have to be interpreted with care since they can never be full representations of the real world. An example of errors in this connection is that it may be difficult to obtain full turbulent flows in wind tunnels. Still, experiments with physical models are often very useful for understanding and quantifying the governing processes in the atmosphere.

Empirical models

Empirical models are experimentally determined relationships between parameters that may e.g. be used in the analysis of other experimental data. An example of such a relationship is the dispersion that makes a plume expand with a rate which makes the width of the plume equal to approximately 1/10th of the transport distance, i.e. one km downwind from a chimney the plume will be about 100 m wide. Another example is that the chemical conversion of NO_2 takes place with a typical rate of about 5% per hour. Such simple relationships may be useful in interpretation of measurements as well as model calculations, and they may be highly useful to provide a first impression whether obtained results are realistic or not.

Receptor models

Receptor models are statistical models used e.g. for determining the contribution from various types of sources to observed pollutant levels, usually based on knowledge about temporal variation in source profiles. Receptor models have e.g. been used to characterise the contribution to ultra fine particles from diesel and gasoline driven vehicles (Wåhlin et al., 2006). These models are highly useful for the interpretation of measurements.

Deterministic models

In the physical-chemical (or deterministic) models the governing processes are expressed in mathematical equations. The calculations take place in time-steps that may vary from seconds to minutes, hours or days depending on the specific problem for which the model has been developed. The various processes are discretized and solved numerically using a so-called splitting technique. This means that the processes are solved sequentially and eventually using different numerical algorithms and time steps. Careful analyses are necessary in order to reduce the numerical errors and errors introduced from using splitting. This will be explained in more detail later in the chapter.

In most applications there is not a distinct separation between the different types of models. An empirical model may as an example also be categorised as a statistical model, which is not necessarily the case the other way around. Deterministic physical-chemical models always have elements of statistical descriptions of sub-grid scale processes included (often termed parameterisations). Some deterministic models (e.g. Gaussian plume models) also have statistical descriptions of some of the included fundamental processes like the atmospheric dispersion (as a normal or Gaussian distribution), while the dispersion parameters can be denoted as empirical and be expressed as functions of physical parameters.

Motions in the atmosphere

The fundamental equation of motion in the atmosphere is basically non-linear. It contains information concerning atmospheric transport over scales ranging from the largest eddies on global scale down to the very smallest eddies contributing to molecular diffusion. Currently there are no known mathematical techniques for performing an exact integration of this equation. Moreover, feasible observational systems are incapable of resolving or defining all scales of atmospheric motion and transport. Thus, modellers are forced to distinguish between eddies that are in a sense resolvable, either by observation systems or by some form of finite difference representation of the atmosphere, and eddies that are not fully resolved neither observationally nor computationally. Such unresolved eddies can be defined as turbulence and must be parameterized in the model system. Both turbulence and molecular diffusion facilitate transport of tracers, but turbulence and diffusion are in the real world different physical processes and care should be taken to distinguish the words diffusion and dispersion. Here diffusion is used as the term for molecular diffusion and dispersion is used for the parameterization of sub-grid scale transport. The most common parameterization of dispersion in atmospheric models is the K-theory of first order, where K can be parameterized in different ways. Other variables, such as the mixing height, and processes, like wet and dry deposition, also have to be parameterized in the models.

Atmospheric transport of air pollutants is, in principle, a well defined and deterministic process. If information about the state of the atmosphere is given in all details (e.g. infinite accurate information about wind velocity, at all times and with infinite precision, etc.) and infinitely fast computers are available, then the advection equation could in principle be solved exact, and the transport of air pollutants with the wind could thereby be described to any desired degree of detail. This is, however, not the case! The challenge of modelling air pollution is essentially to solve the advection equation when only finite information about the state of the atmosphere is given and only finite computer resources are available.

Roughly speaking, the range of atmospheric motion covers 13 orders of magnitude (10^{-6} m to 10^7 m), and this range is covering from molecular diffusion to large-scale weather systems and with a non-linear continuous interaction between all these scales of motion.

Discretization

Even with the currently available super computers, it is not possible to operate a model with a full description of this range of motions for a relevant model domain. Therefore a discretization of the equations is necessary. With discretization of the problem, a "window" must be determined which covers the most interesting scales of motion of the phenomena, which are to be studied. When simulating atmospheric transport the lower limit for resolving fields is the grid resolution. This lower limit in a typical long-range transport model is of the order of 1 to 100 km. This means that in the present models, in the best case, the scales of atmospheric motion between 10^{-6} m and 1 km (which corresponds to 9 orders of magnitude and includes e.g. turbulence) are not resolved and must be parameterized. Furthermore, in practice discretization of the equations and input data is applied and this introduces some uncertainties and errors in the results.

Uncertainties and tests

Many different issues must be carefully studied in order to diminish these uncertainties and to develop a sufficiently accurate transport and dispersion model which can be used to make reliable simulations of both short- and long-range atmospheric transport and dispersion of air pollutants. Some of these issues are accurate:

• Numerical treatment especially concerning the advection-dispersion equation
• Parameterizations of sub-grid scale phenomena
• Description of the mean meteorological input fields
• Appropriate initial and boundary conditions
• Emission data

An example of a test of the numerical methods against known solutions is given at the end of this chapter. It is more difficult to test the applied parameterizations and the obtained meteorological input data used in the models. One possibility is to test the input data sets and different methods for parameterization of vertical dispersion, mixing heights and depositions by comparing the different model results against measurements. The choice of initial and boundary conditions is less important for short-lived species. For long-lived species the model domain can be chosen, so that the boundaries are far away from the location of the source and the main areas of interest, and the influence of the boundary conditions is small (or none, if the concentrations at the boundaries are zero). The initial conditions are especially important for short-term simulations. The accuracy of the emission data (temporal variations of the source strengths and distribution in different heights) is, however, of major importance for reliable simulations and for validation of model results.

When developing a model there are several steps that need to be handled. With each step dependent of the problem the model is supposed to address:

- The purpose of the model and the necessary precision in the results
- Selection of domain size – is this a local scale, regional or maybe a global scale problem? Should the model e.g. extend into the stratosphere or is the free troposphere sufficient?
- Provision of mathematical equations describing the governing physical and chemical processes that are to be handled
- Discretisizing of the equations by dividing these into discrete elements in space and time. In practise, this also includes selecting a numerical method for solving the equations
- Decision of the spatial resolution in the results, which is resolving the relevant physical and chemical processes
- Definition of the necessary boundary conditions and input data
- Definition of the output formats (spatio-temporal averages, special areas of interest, etc.)

When these steps are handled it is possible to construct a source code in the preferred programming language and subsequently to run the model on a computer. The next part of the work is testing of the code for logical, systematic or programming errors in order to ensure the validity of the model. This is one of the most important and often also most time consuming tasks in the development of the air quality model.

Discretisizing the advection equation

Returning to the discretization of the model; there are many ways of discretisizing the advection equation for solving on a computer. One of the most simple is illustrated here, based on the advection equation in one dimension (in this case the x-direction):

$$\frac{\partial C}{\partial t} = -u \frac{\partial C}{\partial x} \qquad (14.1)$$

This equation states that the change of concentrations C over time t equals the wind speed u times the gradient in C over space x. This equation is continuous. In order to solve the equation on a computer, it has to be discretisized. When doing so, the time and space are divided into small intervals. If one sets $C = \Delta C$, $t = \Delta t$, $x = \Delta x$ and $u = <u>$, where $<u>$ is the mean value of u in every grid cell, and Δ indicates a difference in value of the variable over either time or space, as e.g.:

```
<u₁> <u₂> ...
-|------|------|------|------|------|------|------|-
x₁   x₂   x₃ ....
```

where $\Delta x_1 = x_2 - x_1$, $\Delta t_1 = t_2 - t_1$, $\Delta C_1 = C_2 - C_1$, etc. In this way we obtain the following equation:

$$\frac{\Delta_{(time)} C}{\Delta t} = -<u> \frac{\Delta_{(space)} C}{\Delta x} \qquad (14.2)$$

This means that for every grid cell, we are now able to calculate the change in concentration over time resulting from the transport with the wind:

$$\Delta_{(time)} C = -\Delta t <u> \frac{\Delta_{(space)} C}{\Delta x} \qquad (14.3)$$

This method is very simple and for most practical applications it is too inaccurate. However, it illustrates well the principles of discretization of a mathematical equation.

The following sub-sections will focus on physical-chemical atmospheric models, especially the details concerning the application of these tools in the handling of specific problems. The description of the governing processes and parameterisations will be treated later.

14.2 Local scale models

It is common in air pollution management and research to define local scale models as those describing mainly dispersion processes within a distance of 10 to 20 km from the source. A plume model describing pollutant dispersion of releases from point sources is a typical example of a local scale model, but also models describing dispersion of street pollution and pollutant levels in urban background belong to this category.

Plume models

It is common praxis in air pollution management to apply more or less complex plume models for description of air pollution from larger point sources. These plume models describe the concentration distribution downwind from a continuously emitting source of air pollution. The plume model is usually designed to compute hourly mean concentrations, and often this type of model provides a very reasonable estimate of the pollutant concentration for distances up to about 20 km from the source.

In most cases an assumption of normal (Gaussian) distribution of the plume around its centre line has been applied both in vertical and horizontal direction. The parameters in the applied analytical formulas can be functions of distance to the source, source emission data, and meteorological parameters, geometry of the source and the surface characteristics in the source area as well as in the down-wind direction from the source. The most common plume model is based on the Gaussian distribution that is expressed as:

$$c(x,y,z) = \frac{Q}{2\pi u \sigma_y \sigma_z} e^{-0.5\left(\frac{y}{\sigma_y}\right)^2} e^{-0.5\left(\frac{z}{\sigma_z}\right)^2} \qquad (14.4)$$

where Q is the mass emission rate, u is the wind speed; σ_y and σ_z are the dispersion parameters (the standard deviation in the Gaussian distribution in respectively horizontal and vertical directions) that are functions of the down-wind distance in the x direction. The source is in this case placed in origo in a coordinate system $(x, y, z) = (0, 0, 0)$ with x defined in the direction of the wind.

In Equation 14.4 it is assumed that dispersion of the plume may take place unhindered in all directions from the source. Introducing a reflection from the ground and moving the zero point of the coordinate system from the surface to an effective release height leads to the following expression:

$$c(x,y,z) = \frac{Q}{2\pi u \sigma_y \sigma_z} \left(e^{-0.5\left(\frac{z-H}{\sigma_z}\right)^2} + e^{-0.5\left(\frac{z+H}{\sigma_z}\right)^2} \right) e^{-0.5\left(\frac{y}{\sigma_y}\right)^2} \tag{14.5}$$

where H is the effective height of the release defined as the height of the chimney plus the buoyancy lift of the hot plume. A simple illustration is shown in Figure 14.1.

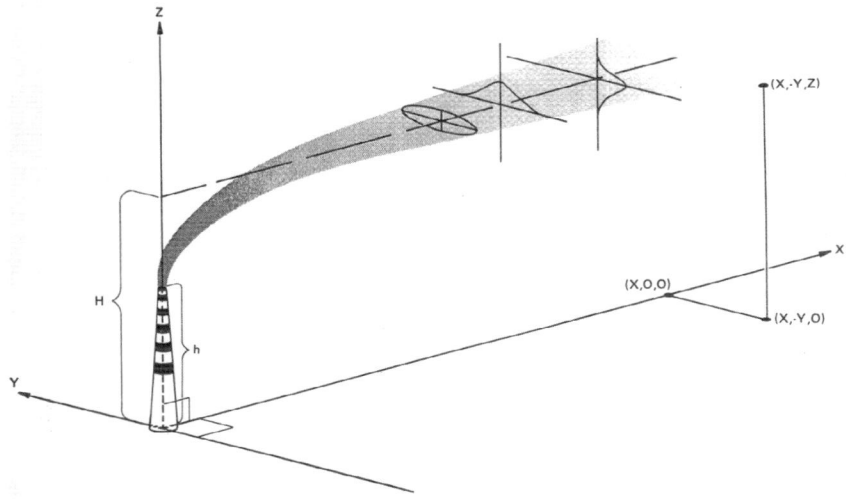

Figure 14.1 Simple illustration of a Gaussian plume model where a normal distribution of the pollutant concentration inside the plume is assumed in vertical as well as in the horizontal direction. A coordinate system is included. h and H are the heights of the chimney and the effective release height (due to buoyancy lifting H may be considerably larger than h).

When the emission and the wind speed are known, and the dispersion parameters are determined at a given distance from the source, the concentration can be computed in a plane perpendicular to the plume. In the literature many formulas have been given for the dispersion parameters σ_y and σ_z as function of the meteorological conditions.

Previously it was most common to apply the so-called PGT dispersion curves (Pasquill-Gifford-Turner curves) (Gifford, 1961). However, since the late 1980s when computer facilities became generally available, plume models have been based on the basic micro-meteorological parameters that describe the turbulence and stability conditions in the boundary layer by using continuous and more precise data. Examples of such models are the American HPDM (Hybrid Plume Dispersion Model) (Hanna and Chang,

1993), the British UK-ADMS (Atmospheric Dispersion Model System) (Carruthers et al., 1994) and the Danish OML (Operational Meteorological Air Quality Model) (Olesen et al., 1992).

Models of this kind are used as routine tools in Air Quality Management for environmental impact assessment of industrial plants, power plants, etc., for determining the height of a chimney on these plants needed for complying with current limit values on pollutant concentrations in the surroundings.

Box models

Box models can be very useful tools e.g. in understanding chemical processes. In the box model it is assumed that the pollutants are fully mixed in a studied air parcel. The chemistry can be treated as in a well-mixed laboratory system and the time evolution followed as a result of the chemical interaction alone. In this case there is no interaction between the air parcel and the surroundings. This approach is useful e.g. in connection with the development and testing of chemical mechanisms and may be directly compared with results from smog chamber studies.

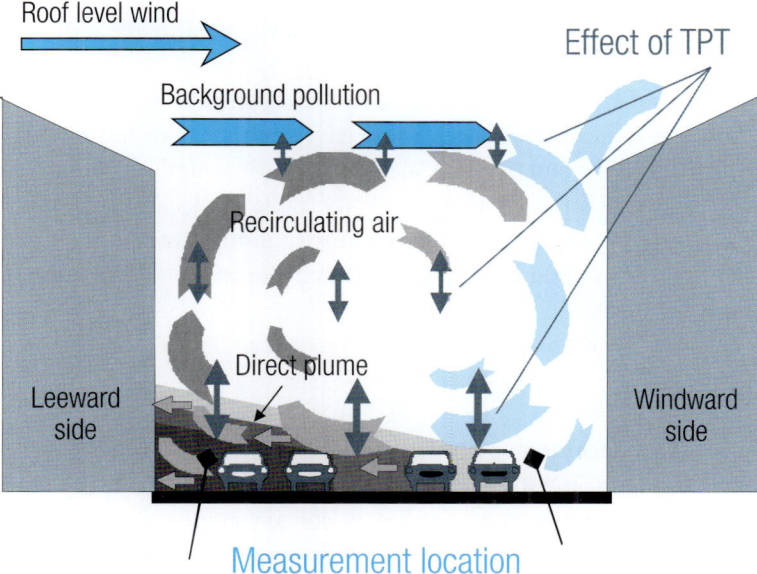

Figure 14.2 Illustration of flow and dispersion in a street canyon. When wind speeds are above about 2 m s⁻¹, the formation of a vortex leads to recirculation of pollutants inside the canyon. Pollutant concentrations on the leeward side are often 5 to 10 times higher than on the windward side of the street. Shown in the figure is also the effect of Traffic Produced Turbulence as well as the usual location of urban street pollution monitoring sites. (Courtesy of Matthias Ketzel, NERI, Denmark).

In other applications a mixing with a fixed concentration field in the surroundings may be assumed or e.g. pollutant emissions or dry deposition
scavenging can be taken into account in the box model. Box models can thus
be applied in process studies, but they can also be used as sub-models in
parameterised models. An example from street pollution modelling is illustrative.

The one-dimensional model is a kind of next step from the box model,
and represents an air column in the vertical. This type of model is designed
to simulate the vertical distribution of atmospheric constituents, disregarding the horizontal variation.

Pollutant dispersion modelling in street canyons requires modelling of
the complex wind flows. The pollution emitted from traffic in the street is
advected by the wind vortex towards the leeward side of the street (Figure
14.2). For the windward side of the street, the impact of emissions from traffic in the street is only from the air that has re-circulated inside the canyon.
This has been parameterised in different ways in the air pollution models for
urban streets.

Applied pollutant models for street canyons aim at describing the main
features of the basic flow and dispersion processes illustrated in Figure 14.2.
The goal behind this kind of model is usually to compute pollutant levels in
2 m height at the pavements where limit values have been defined and air
quality monitoring is usually performed.

There are various types of models that have been applied for describing the
pollutant dispersion in urban streets. Some have been more or less empiri-
cally derived like e.g. the Dutch Car (Calculation of Air pollution from Road
traffic), which is based on the plume type TNO traffic model and analysis of
measurements from a substantial network of urban monitoring stations. See
Berkowicz (1998) for references about the Dutch Car model and the other
street pollution models.

The OSPM model

One of the most modern street pollution models is the so-called Operational
Street Pollution Model (OSPM). It is based on a combination of a plume
model for the direct contribution to pollutant concentrations from traffic in
the street and a box model for the re-circulation of pollutants:

$$C_d = \sqrt{\frac{2}{\pi}} \frac{Q}{u_b} \int_0^{t_{max}} \frac{dx}{\sigma_z(x)} \tag{14.6}$$

where Q is the emission rate (g m^{-1} s^{-1}), W is the width of the street canyon
along the wind path, u_b is the wind speed at street level and $\sigma_z(x)$ is the verti-

cal dispersion parameter at the downwind distance x. The vertical dispersion parameter is calculated taking into account mechanical turbulence generated by wind as well as by the traffic in the street:

$$\sigma_z(x) = h_0 + \sqrt{\sigma_w \frac{x}{u_b}} \tag{14.7}$$

$$\sigma_w = \sqrt{(\alpha u_b)^2 + \sigma_{w0}^2} \tag{14.8}$$

where $(\alpha u_b)^2$ is the mechanical and σ_{w0} is the traffic induced turbulence.

The integration of Equation 14.6 is performed along the wind path and at street level, and therefore depends on wind direction, the extension of the re-circulation zone, and the length of the street. In the latest version of the model, the integration may be limited by the street length, when street intersections are present at shorter distance from the receptor, or the street becomes broader or even open.

The canyon ventilation velocity determines how quickly the pollution is removed from the street canyon. At low wind speeds this ventilation velocity is governed by the traffic induced turbulence (see Equation 14.8), and this is accounted for in the parameterisation of OSPM.

The wind direction in urban areas is seldom constant over e.g. one hour. This applies especially for low wind speed conditions, where large fluctua-

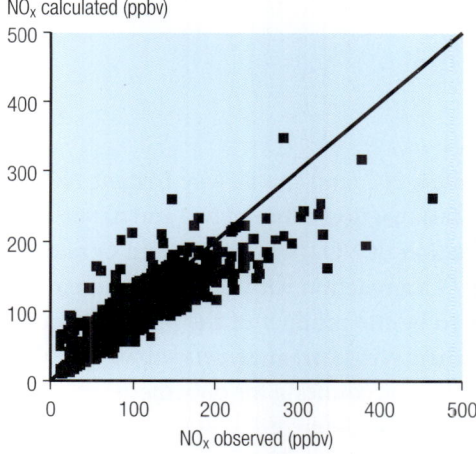

Figure 14.3 Comparison between observed and calculated hourly mean concentrations of NO_x for the monitoring station at Jagtvej, Copenhagen in 2003. Jagtvej is a classic street canyon with 5 storey buildings and about 20 m between building facades. The model calculations are performed with the OSPM. The street has a diurnal traffic of about 20,000 vehicles with only a small fraction of heavy duty vehicles. Only data for working days during daytime (0800-1600) are included. Correlation coefficient (R^2) = 0.8.

tions in wind direction are common. This effect is accounted for in the OSPM by introducing a wind averaging procedure over a wind direction interval which is defined as a function of the wind speed. The interval increases for decreasing windspeed. Furthermore, the model has been designed to handle streets with openings in the building facades, with varying building heights, or even without buildings on one or both sides of the street. An illustration of the performance of OSPM is given in Figure 14.3.

Nitrogen oxide chemistry in urban streets

The residence time inside an urban street canyon is short – in the order of seconds to a few minutes – and only very fast chemical transformations have the time to take place inside the street. One of the few cases in which the chemistry plays a role in urban streets concerns the nitrogen oxides.

Nitrogen oxides (NO_x) are mainly emitted as nitrogen monoxide (NO) and to a lesser extent as nitrogen dioxide (NO_2). Previously, the fraction of NO_x directly emitted as NO_2 was only about 5 to 10% in countries with a little fraction of cars with diesel engines. Due to the use of catalytic converters and an increasing number of cars with diesel engines, this value may in some areas be up to even 40%.

$$NO + O_3 \rightarrow NO_2 + O_2$$
$$NO_2 + h\nu \rightarrow NO + O$$
$$O + O_2 \rightarrow O_3$$
$$\overline{}$$
$$NO + O_3 \rightarrow NO_2 + O_2$$
$$NO_2 + h\nu \rightarrow NO + O_3$$

$$(14.9)$$

The distribution between the harmless NO and the airway irritant NO_2 is therefore largely determined by the fast reaction between NO and ozone (O_3) and the similarly fast photo dissociation of NO_2 back to NO and O radical. The O radical is quickly reforming O_3 in reaction with O_2 and is therefore a good approximation to assume O_3 to be the product of the photo dissociation of NO_2. An analysis of measurements from urban streets shows that NO_2 concentrations may thus be predicted by accounting for only these two reactions. The following three equations may be defined:

$$\frac{d[NO]}{dt} = -k[NO][O_3] + J[NO_2] + \frac{[NO]_v}{\tau_{Ex}} + \frac{[NO]_b - [NO]}{\tau_{Ex}} \qquad (14.10)$$

$$\frac{d[NO_2]}{dt} = k[NO][O_3] - J[NO_2] + \frac{[NO_2]_v}{\tau_{Ex}} + \frac{[NO_2]_b - [NO_2]}{\tau_{Ex}} \quad (14.11)$$

$$\frac{d[O_3]}{dt} = -k[NO][O_3] + J[NO_2] + \frac{[O_3]_b - [O_3]}{\tau_{Ex}} \quad (14.12)$$

where τ_{Ex} is the exchange rate (s^{-1}) between the street canyon and the background air, k and J are the chemical reaction rate coefficients for the reaction between NO and O_3 and the photo dissociation of NO_2, respectively. The index v indicates contributions from direct emission from the car fleet in the street, and the index b the urban background concentration.

In order to solve these three equations, it is useful to introduce the following mass balance expressions:

$$[NO_2] = [NO_2]_v + [NO_2]_b + [O_3]_b - [O_3] \quad (14.13)$$

$$[NO_x] = [NO_x]_v + [NO_x]_b \quad (14.14)$$

$$[NO_x] = [NO] + [NO_2] \quad (14.15)$$

The first of these equations expresses the NO_2 concentration as the sum of the direct emission of NO_2, the contribution from background and the chemically produced NO_2. It is assumed that the only sink for O_3 is from consumption in the reaction with NO. The second equation expresses the NO_x concentration as the emission plus contribution from background. The third equation is the definition of NO_x as the sum of NO and NO_2.

When these 6 equations are combined, an expression for the NO_2 concentration may be derived as:

$$[NO_2] = 0.5 \left(b - \left(b^2 - 4 \left([NO_x][NO_2]_a + [NO_2]_n D \right) \right)^{\frac{1}{2}} \right) \quad (14.16)$$

b is here defined as $b = [NO_x] + R + [NO_2]_a + D$, where $D = \dfrac{1}{\tau_{Ex} \times k}$ is the exchange coefficient and $R = \dfrac{J}{k}$ is the photo chemical equilibrium coefficient. $[NO_2]_n$ and $[NO_2]_a$ are defined as:

$$[NO_2]_n = [NO_2]_b + [NO_2]_v \quad (14.17)$$

$$[NO_2]_a = [NO_2]_n + [O_3]_b \qquad\qquad (14.18)$$

These two help variables provide information about the importance of the chemical conversion from NO to NO_2. The first help variable defines the NO_2 concentration that would appear in case that no chemical formation takes place. The second help variable defines a maximum concentration that would appear in case all O_3 entering the street canyon is converted to NO_2.

The above expression for NO_2 is included in a chemical sub-model implemented in the OSPM in the later 1980s (Palmgren et al., 1996), and later this module has also been included in other air pollution models for local scale problems. The concentration of NO_2 in the urban areas (street as well as urban background) is thus strongly governed by the long range transported O_3. Model calculations are in good agreement with observations, even with this simplification of the NO_x chemistry (Figure 14.4).

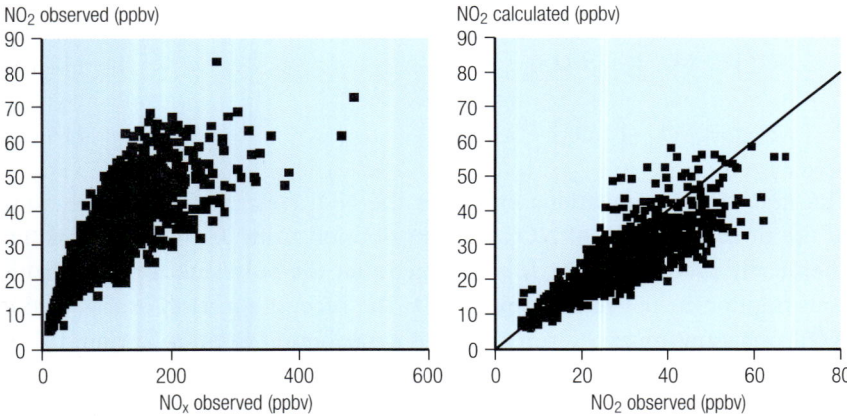

Figure 14.4 The chemistry of NO_x in urban streets. Left plot: the relationship between NO_2 and NO_x. For NO_x concentrations below about 20 ppbv, all NO_x is in the form of NO_2, since the air contains sufficient O_3 for converting all NO to NO_2. For higher NO_x concentrations, only the direct emission of NO_2 contributes to further increase in the NO_2 concentrations. Right plot: Comparison between observed and calculated hourly mean concentrations of NO_2. All data are from Jagtvej in Copenhagen, 2003, and calculations performed with the OSPM. Only working days during daytime (0800-1600) are included. Correlation coefficient $(R^2) = 0.7$.

Urban scale models

The air pollution level in the urban background is, as already stated, the result of both regional pollution and contributions from the city itself. In many air quality and human exposure assessment studies, urban background concentrations may be obtained from measurements performed at one or more urban background monitoring stations in the city. The measured levels in the urban background may then be used to determine the contribution from traffic in the single street (e.g. by subtracting the measured urban background levels from the measured levels in the urban streets), or as input for the calculations of the street pollution levels (see the sections above). Naturally, this method cannot be applied for assessment of the impact of emission regulations, for air quality forecasting, or when calculations are prepared for estimating pollution concentrations for time periods before the monitoring was established. One approach for calculating urban background concentrations in such situations is to apply a simplified Gaussian dispersion model, assuming that regional contributions are known from either measurements or other model calculations. An example of such a model is the Urban Background Model (Berkowicz, 2000), where the concentration is computed using a simplified plume model. The concentration is obtained from the integral:

$$c = \int_0^r \frac{Q}{u\sigma_z(x)} dx + \text{Rural background} \tag{14.19}$$

where Q is the emission density, u is the wind speed, r is the extension of the urban area, and σ_z is the dispersion parameter, which may be calculated as:

$$\sigma_z(x) = h_0 + \sigma_w x \tag{14.20}$$

where h_0 is an initial vertical dispersion determined by the general building height in the urban area, σ_w is the vertical dispersion composed of two contributions: the mechanical (governed by the wind speed) and the convective (or thermal) turbulence given by the free convective velocity scale (for details see Berkowicz, 2000).

14.3 Long-range models

Concentrations of primary pollutants can be high close to their sources. This is especially true for traffic pollution in urban streets and in the areas where plumes from stack releases hit the ground as discussed in the previous sections. However, secondary pollutants may be formed in the atmosphere during transport, which in some cases may last up to several days. Even primary pollutants like nitrogen oxides may be transported up to 24 hours before being either transformed into other pollutants or being deposited to the ground. Air pollution is thus a trans-boundary problem, and a variety of models have therefore been developed for describing atmospheric long range transport of pollutants and all the processes that take place during this transport. Some of these models simulate the fate of individual air parcels along trajectories (so-called Lagrangian models), while other models describe the concentrations and involved processes in a fixed grid (so-called Eulerian models). The following sections explain in more detail these two types of physical-chemical pollutant transport models. Table 14.1 summarises some of the main advantages and disadvantages of these two model types.

Model	Advantages	Disadvantages
Lagrangian	Fast for carrying out multiple model runs that concern a limited number of receptor points.	Short-falls in the description of transport and dispersion on the long range
	Generally easy to apply for most purposes.	The uncertainty in atmospheric transport increases with the distance along the trajectory.
	Generally good description of transport and dispersion on the shorter scale.	Some forward trajectory models can only handle simplified chemistry.
	Can handle point sources and sharp gradients without numerical difficulties.	Computationally demanding for a large number of receptor points.
Eulerian	Generally good description of transport on a larger scale, especially concerning high-resolution three-dimensional models.	Generally computation demanding – especially for three-dimensional models with high resolution, e.g. including nesting techniques.
	Automatically provides results on a grid covering a large domain.	Difficulties in handling large point sources and sharp concentration gradients.

Table 14.1 Advantages and disadvantages of the two most common types of long-range transport-chemistry models: the Lagrangian and the Eulerian models.

In general, Eulerian models have advantages on the large scale, covering a large domain giving results on a grid, but they generally have problems describing the dispersion close to the sources with sharp gradients e.g. from strong point sources. On the other hand, Lagrangian models in general are able to describe the dispersion close to the sources with the ability of handling sharp gradients, but they have problems with the precision in the trajectory calculations on the larger scale.

In the Eulerian models the time integration is carried out by computing the concentration fields for a set of grid points fixed in space, whereas in the Lagrangian models the time integration is performed following the concentration in an air parcel along a trajectory. The difference between these approaches lies in the way the position within a field is described. The Eulerian and Lagrangian approaches are in principle mathematically equivalent. The two different approaches can, however, give different results when solved numerically, due to both numerical and physical reasons.

Eulerian models are in general adequate to describe long-range transport of air pollutants. The traditional Eulerian models have, however, problems in handling sharp gradients caused by a single and strong source, which results in undesired oscillations known as the Gibbs phenomenon. Using a finer grid resolution in the model does not solve this problem, since distributing the emissions on a larger number of smaller grid-cells will just result in even sharper concentration gradients. The problem may be solved by smoothing the emissions or by using various kinds of filtering (as e.g. a Forester filter, which is used in some Eulerian models), but smoothing or filtering are artificial, non-physical solutions. A solution based on physical arguments, like dispersion, is preferable. Dispersion alone is, however, not sufficient to minimize the un-wanted oscillations in Eulerian models since this would require an unrealistic high dispersion coefficient. Furthermore the K-approach, usually used in Eulerian models, is unsatisfactory in the area close to the source.

Lagrangian models do not have problems with sharp concentration gradients and they are able to describe dispersion close to the source reasonably well. Lagrangian models are, however, usually formulated under the assumptions of simplified turbulent diffusion, without convergent or divergent flows and without wind shear. The Lagrangian models therefore have problems with uncertainties in the trajectory calculations on large scales. Although it is in principle possible to carry out calculations in a Lagrangian framework by following a set of marked fluid parcels, on large scales this is not a practical alternative. Shear and stretching deformations tend to concentrate the particles in a few regions, which give difficulties in maintaining a uniform resolution over the domain. Shear and stretching additionally make it very difficult to calculate precise trajectories on large scales. This is especially true near frontal zones where the wind field is divergent or convergent, and in regions of high- and low-pressure systems.

This is a fundamental nature of transport in the atmosphere: a small change in the initial conditions (e.g. starting position) results in major differences over time. Even though the equations describing the transport in Lagrangian and Eulerian models are mathematically linear they exhibit chaotic behaviour. The distance D(t) (which can be considered as a measure for the uncertainty in the trajectory calculations) as a function of time between two trajectories with slightly different initial starting positions, D(0), is exponentially increasing with time t (taken as an ensemble average of many trajectories)

$$D(t) = D(0) \ e^{\lambda t} \tag{14.21}$$

where λ is the so-called Lyaponov exponent. If $\lambda=0$ then the motion is laminar. If the system, on the other hand, has at least one positive Lyaponov exponent then the whole system will exhibit chaotic behaviour. Examples of this chaotic behaviour are shown in Figure 14.5 (p. 314) where different trajectory calculations have been made with almost the same initial positions. After a few days the trajectories are separated by more than 1000 km in these calculations. On shorter scales, an exponential increase can, in fact, be approximated by a linear increase. Therefore Lagrangian models can with good results be used for calculations on short scale.

The basic difference between the two kinds of models is the discretization of the transport equation and of the meteorological fields. The two kinds of models will converge to the same (true) solution as the grid resolution goes to zero. The difference between the two kinds of models is the way they handle the discretization problem. Eulerian models have artificial dispersion (smoothing) of the concentrations, because the concentrations are distributed on the grid-cells and thus representing a grid-cell average. This is in contrast to the Lagrangian models where the transport is described along "one-dimensional" trajectories, even though the trajectories are located in a three-dimensional space. The smoothing has the effect that Eulerian models contain a greater part of the 'true' solution, compared to the Lagrangian approach, especially on large scales. In order to obtain the same effect and to diminish the uncertainties in trajectory calculations, heavy computations on large scale (many puffs/particles) and huge trajectory calculations with some kind of stochastic element included, as e.g. random walk, are needed.

Coupling of a Lagrangian model with an Eulerian model, where the Lagrangian model is used to calculate the initial transport and dispersion of the plume in an area close to the sources, and the Eulerian model is used to calculate transport and dispersion on long range is therefore desirable in many cases.

Lagrangian models

In the Lagrangian framework an air parcel is followed along a transport route. The transport route – the so-called trajectory – is in the most simplified trajectory models using straight line trajectories generated from local wind field data. However, the most commonly applied models are based on 2D trajectories, but there are also several examples of Lagrangian models using 3D trajectories calculated from detailed 3D wind fields.

In order to calculate trajectories, wind speed and wind direction must be known in a given domain. These data may be obtained from measurements, but more often they are provided by a meteorological model. The advection or movement of an air parcel is calculated from the differential trajectory equation and may then be expressed as:

$$\frac{dP}{dt} = V\big[P(t)\big] \tag{14.22}$$

where V is defined as the either 2D (V(x, y)) or 3D (V(x, y, z)) wind velocity vector; depending on the type of trajectory model applied.

A common way to handle the above equation is to expand P(t) in a Taylor series about $t = t_0$ which is evaluated at $t_1 = t_0 + \Delta t$; and about $t = t_1$ which is evaluated at $t = t_0$. The resulting first guess position is then:

$$P'(t+\Delta t) = P(t) + V(P,t) \times \Delta t \tag{14.23}$$

and the final position of the air parcel is then:

$$P(t+\Delta t) = P(t) + 0.5 \times \big[V(P,t) + V(P',t+\Delta t)\big] \tag{14.24}$$

This equation is accurate to the second order, and it has been determined that higher order methods do not yield greater precision.

Computed trajectories have been compared to various types of experimental studies – measurements of various tracer gases which may be detected at very low concentrations as well as GPS data obtained from releases of balloons (Stohl, 1998). These comparisons indicate that for the often applied 96 h trajectories, the errors can be large, and thereby the computed trajectory may end a considerable distance from the real path of the air parcel. The trajectories are particularly uncertain when frontal systems are passing the domain.

The Lagrangian models may be operating in two ways: along forward trajectories to describe the pollutant transport away from selected sources, or along backward trajectories to describe the pollutant concentrations or depositions to a selected array of receptor points. The Dutch TREND model (Asman and van Jaarsveld, 1992) is an example of a Lagrangian type model with first order chemistry and using forward trajectories from places where releases are taking place. The Lagrangian EMEP model is an example of one of the early models based on backward trajectories from selected receptor points distributed over the European area. Examples of back-trajectories are shown in Figure 14.5. The three trajectories are drawn for the same receptor points, but for different wind conditions. In the first example air masses are received from both north and south despite the receptor points are placed very close to each other. The explanation is that a front system is passing the area. This is a situation where the trajectories are rather uncertain. In the next two examples air masses are received from south, but in the last of these examples the received air is somewhat colder as the air takes a path to the north before arriving at the receptor points.

Figure 14.5 Three examples of computed 2D back-trajectories for nine receptor points in the North Sea north of Great Britain. Each plot contains 9 trajectories generated with a distance of 150 km between the starting points. The three plots are all for the same receptor points, but the first plot shows trajectories computed for the 21st of June 2000, the second is for the 23rd of July 2000 and the third is for 25th of July 2000. (Courtesy of Ambelas Skjøth, NERI, Denmark).

In many Lagrangian models the atmospheric dispersion taking place along the trajectories are represented by puff or particle dispersion models. In the first type of models the dispersion is parameterised by a number of puffs, each representing a mass and a spread and the pollution is diluted as it is advected along a trajectory. The particle dispersion models are also known as Lagrangian Monte Carlo or stochastic models. The turbulent dispersion is simulated by perturbating a large number of particles as they are advected. The flow is usually computed with a meteorological model and represented by the mean flow and the turbulent fluctuations.

The pollutant release in such models can be simulated to take place sequentially as in a plume or simultaneously as in a puff. Especially in the past, it has been common to apply Lagrangian transport-chemistry models based on 2D trajectories for describing mesoscale to long range transport of pollutants. These models include more or less detailed chemical trans-formation schemes together with processes like emission, and dry and wet deposition. One of the un-physical features of this type of model is that air parcels do not have any exchange of mass with the surroundings except for emission and deposition, although various parameterisations may to some extend compensate for this. The change in concentration in a Lagrangian model may be expressed as:

$$\frac{dc}{dt} = P_i - L_i c_i - \Lambda_i \frac{p}{h} c_i - v_d c_i \tag{14.25}$$

Where i is an index for the considered specie (i.e. a chemical compound or particulate material), c is the concentration of the specie, P and L are the chemical production and loss terms, respectively, Λ is the wet scavenging coefficient, p is the precipitation amount, h is the height of the considered air mass, and v_d is the dry deposition coefficient. The common parameterisations of dry and wet scavenging processes are described in Chapter 10.

The first two terms on the right hand side of the equation express the chemical production (including emission), and loss terms, the third and fourth terms express loss as a result of wet and dry deposition, respectively. This equation was applied in the Lagrangian version of the EMEP model (Iversen et al., 1990) applied in the 1970s and 1980s.

The chemical production (P) and loss (L) terms are functions of the concentration of chemical species in the system that take part in reactions that either lead to formation of removal of the considered species. This may be expressed as:

$$P_i = \sum_j k_j c_j + \sum_{jk} k_{jk} c_j c_k + \sum_{jkl} k_{jkl} c_j c_k c_l \tag{14.26}$$

$$L_i = \sum_j k_j + \sum_k k_k c_k + \sum_{jk} k_{jk} c_j c_k \tag{14.27}$$

where the three terms are expressions for the first, second and third order reactions leading to the formation or loss of the chemical specie i.

A weakness in the concept of the Lagrangian transport-chemistry models is the inability of accounting for dispersion of the air parcel. There are, however, various ways to indirectly account for the dispersion in the handling of the input of emissions. A first approximation is to assume that a plume in average has a width of one tenth of the travelled distance (Hanna et al., 1982). An indirect handling of dispersion is therefore to perform an averaging of the emissions over a square with a side length which is equal to one tenth of the travelled distance (Figure 14.6). Thereby the model handle emissions

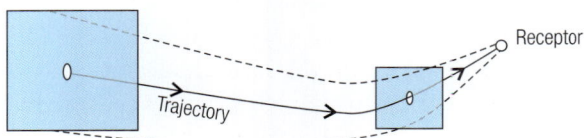

Figure 14.6 Illustration of the simplified indirect parameterisation of horizontal dispersion. The full line curve is the trajectory of the one-dimensional model column. The patterned squares represent the areas over which emissions are averaged. The sides of the square are defined as 1/10th of the travelled distance along the trajectory to the receptor point.

from the area that represents the horizontal dispersion and these emissions will contribute to the pollution at the receptor point. This is of course a rather crude simplification since the dispersion of the plume depends not only on the travel distance but also on the actual meteorological conditions. However, for many practical applications this method has been shown to work well.

Another difficulty of the Lagrangian model is the handling of vertical dispersion. The ACDEP model is in many ways similar to the EMEP Lagrangian model, although in this case a column with 10 vertical levels is advected along the trajectory (Figure 14.7).

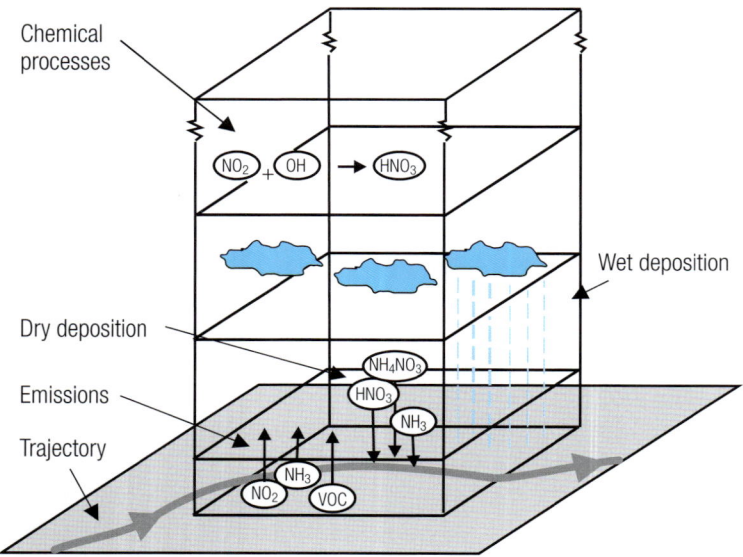

Figure 14.7 Simple illustration of a Lagrangian transport-chemistry model where a vertical column is advected along the trajectory. During the transport, the air column receives emissions from the source areas. Gases and particles are dispersed in the vertical, chemical transformation takes place and gases and particles are removed by dry and wet deposition. Only few vertical grid boxes are shown, but the model may contain 10 or even 20 boxes.

Eulerian models

In the Eulerian models, the conservation of mass equation is solved usually for a grid in a 2D or 3D space (Figure 14.8 illustrates the grid cells in a full 3D Eulerian model).

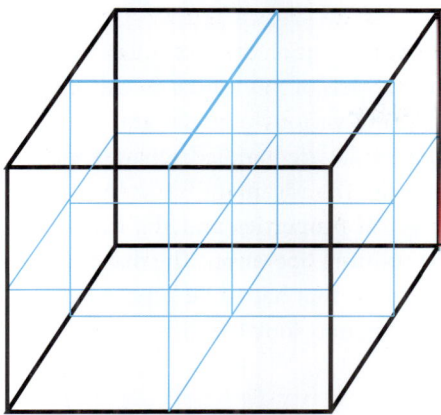

Figure 14.8 Simplified 3D grid as it may look in a full 3 dimensional Eulerian model. In this case the grid consists of 2 x 2 x 2 grid cells, however, normally a grid will consist of many hundreds or even thousands of grid cells. The grid is fixed and the model describes the transport in and out of each cell, together with emissions, chemical transformations and dry and wet deposition.

When the atmospheric transport is split into terms for advection and turbulent transport, the conservation of mass equation may be expressed as:

$$\frac{\partial c_i}{\partial t} = -\left(u\frac{\partial c_i}{\partial x} + v\frac{\partial c_i}{\partial y} + w\frac{\partial c_i}{\partial z} \right)$$
$$+ K_x\frac{\partial^2 c_i}{\partial x^2} + K_y\frac{\partial^2 c_i}{\partial y^2} + \frac{\partial}{\partial z}\left(K_z\frac{\partial c_i}{\partial z} \right) \qquad (14.28)$$
$$+ E_i(x,y,z) - \Lambda_i c_i - v_{d_i} c_i$$
$$+ P_i - L_i c_i$$

where i is the index of the chemical species in question, c is the concentration of the chemical species, u, v, w are the wind components in the x, y and z directions, respectively, K_x, K_y and K_z are the dispersion coefficients. Typically, the horizontal coefficients are defined as $K_x = K_y$ = constant and K_z is defined as a function of height z. E_i is the emission of the species, v_d and Λ are the dry and wet deposition scavenging coefficients, and P and L are the chemical production and loss, respectively.

For one specific location (a grid cell), the left hand side of the equation provides the change in concentration of chemical species, i as function of time. The first term on the right hand side of the equation describes the change in concentration due to transport by the wind (advection in x, y and z direction), the next three terms describe the dispersion, then follows emission and wet and dry deposition. Finally, there is the chemical production and loss of the species.

Numerical handling

For the transport-chemistry models the conservation of mass equation needs to be handled numerically. In almost all models of this kind a so-called splitting is introduced. Splitting means that the various physical and chemical processes that are parameterised in the model are handled separately using different time steps and numerical methods. The advantage is that the various parameterisations have different numerical properties and the calculation time may be optimised by applying the splitting operation. The disadvantage is that splitting introduces a numerical error that has to be kept as small as possible. In the Eulerian frame, the splitting may look like this:

$$\frac{\partial c_i}{\partial t} = -u\frac{\partial c_i}{\partial x} - v\frac{\partial c_i}{\partial y} - w\frac{\partial c_i}{\partial z} \tag{14.29}$$

$$\frac{\partial c_i}{\partial t} = K_x\frac{\partial^2 c_i}{\partial x^2} + K_y\frac{\partial^2 c_i}{\partial y^2} \tag{14.30}$$

$$\frac{\partial c_i}{\partial t} = \frac{\partial}{\partial z}\left(K_z\frac{\partial c_i}{\partial z}\right) \tag{14.31}$$

$$\frac{dc_i}{dt} = P_i - L_i c_i - \Lambda_i c_i \tag{14.32}$$

The first of these equations describes the advection, the second describes horizontal diffusion, the third describes vertical diffusion and the fourth describes chemical transformation and wet deposition.

In the splitting procedure, the calculations are performed in a sequence where the calculated concentrations from the advection part are used as input for the horizontal diffusion, which is then used as input for the vertical diffusion and so forth.

The different processes can be solved using different time steps. A typical time step for the horizontal advection depends on the so-called Courant-Friedrich-Lewi (CFL) stability criteria, which in the simple case states that the time step should be less than the grid resolution divided by the maximum wind speed U in the domain ($\Delta t < \Delta x/U$). If the grid resolution is e.g. $\Delta x = 50$ km and the time step is set to $\Delta t = 900$ seconds, the model is stable as long as the wind speed anywhere in the model is below ~55 m s^{-1}. This criteria also implies that increasing the grid resolution e.g. by a factor of two in the two horizontal directions, the computing time will increase with a factor of eight, since there will be twice as many grid cells in each direction and the time step is decreased by a factor two!

A stiff operator describes a process for which the time scales span over several orders of magnitude. Numerical studies have shown that the stiffness of a system has a stabilising effect on the splitting, which is important to take into account. These numerical studies have also shown that the error associated with splitting is complex and often will oscillate with a diurnal cycle.

Chemical reactions take place on time scales that range from fractions of a second to several month or years. This means that the set of equations describing the chemical transformations constitute a stiff system of ordinary differential equations, which in the splitting procedure therefore should be handled as the last part in the sequence. A number of fast methods for solving the stiff system has therefore been introduced over the years; the Quasi Steady-State, the Eulerian Backward Iterative, the 2-step method and various other solvers. For grid cells where emission takes place, this may be included as a production term in the chemistry part. Similarly, the lower boundary condition for the vertical diffusion is set equal to the dry deposition.

When diffusion is parameterised using first order K-theory (like in the equations 14.30 and 14.31), there is no need for very accurate solvers to handle this part of the model. Stability of the scheme is important and implicit methods like the θ-method are often applied.

An accurate algorithm for the advection is more crucial, and should be performed having a small numerical diffusion. A variety of schemes exist, and different types of models apply to different schemes. All advection schemes create some numerical diffusion, although some perform better than other. Unfortunately, schemes with high accuracy may as previously mentioned in some cases result in unwanted and unphysical oscillations (known as Gibbs phenomena), which may even result in negative concentrations when sharp gradients are present in the advected field. These problems can be reduced or even eliminated using filtering techniques, although this may also introduce some additional numerical diffusion.

Nesting techniques

The increasing computer power has made it possible to obtain still higher resolution in the transport-chemistry models. However, operating with very high spatial resolution for a large domain is very computer resource demanding. In this case nesting – operating with a sub-domain of higher resolution and a mother domain with relatively coarser resolution – is the common approach. The simplest form of nesting is one-way nesting where a coarse resolution model is operated for the entire domain, and the results from this model are used as boundary condition for a high resolution model operating on the sub-domain. In the two-way nesting the Eulerian model has a zooming capability that allows higher resolution in the sub-domain for the model.

In other words, there is two-way communication between the domains with high and coarse resolution.

Another way of nesting is to couple two different model approaches, e.g. an Eulerian approach for a coarse resolution domain and a Gaussian plume model for a high resolution domain gaining advantages of both types of models. The nesting of one model type into another is often also referred to by the terms "integration" or "coupling".

Figure 14.9 shows an example where the use of one-way nesting of a local scale plume model with a long range Eulerian chemistry-transport model is improving the results considerably compared with the calculations using a coarse resolution Eulerian transport-chemistry model on its own.

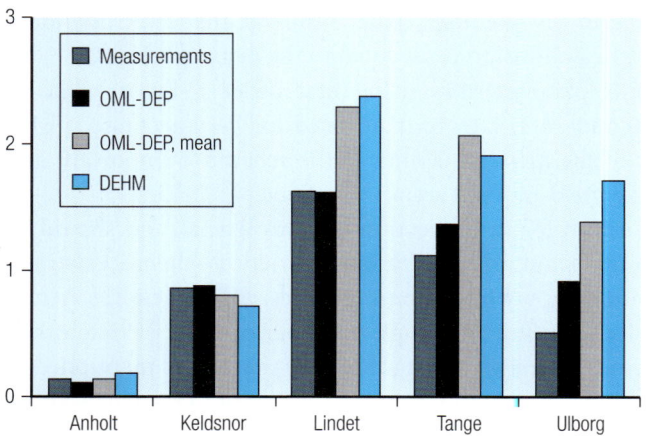

Figure 14.9 Measured and calculated NH_3-N concentrations (μg m^{-3}) at five Danish rural stations in 2005. The Eulerian model, DEHM, calculations represent the mean value for a 16.67 km x 16.67 km cell where the station is placed. The Gaussian plume model, OML-DEP, represents the 400 m x 400 m grid cell where the station is placed. The "OML-DEP, mean", represent the average value of the OML-DEP calculations for the 400 m x 400 m cells in the 16.67 km x 16.67 km grid cell from the DEHM model. The measurements of NH_3-N on Anholt and Tange are performed with filter pack samplers, whereas the measurements at Keldsnor, Lindet and Ulborg are performed using the denuder method. (Courtesy of Ellermann et al.).

This example actually involves both one-way and two-way nesting. The DEHM (Danish Eulerian Hemispheric Model) is an Eulerian 3D model covering the entire northern hemisphere with a 150 km x 150 km resolution for the coarse domain. For the European area a 50 km x 50 km grid resolution is applied and over Denmark an even higher resolution of 16.67 km x 16.67 km was used. In the example the results from the DEHM model on the 16.67 km x 16.67 km grid are used as input to a Gaussian plume model OML-DEP which describes the dispersion of NH_3 emissions from single local farms in the sub-domain.

It is seen that coupling the local scale model to the regional scale model (the "OML-DEP" results) gives a very good performance compared with measurements. On the other hand the validity of the model results are also emphasised by the fact that taking an average of grid cells of the local scale model, covering the same area as one grid cell in the regional scale models provide very similar results.

Figure 14.10 shows an example of model results obtained from the two-way nested Eulerian model DEHM. Annual mean concentrations of particles

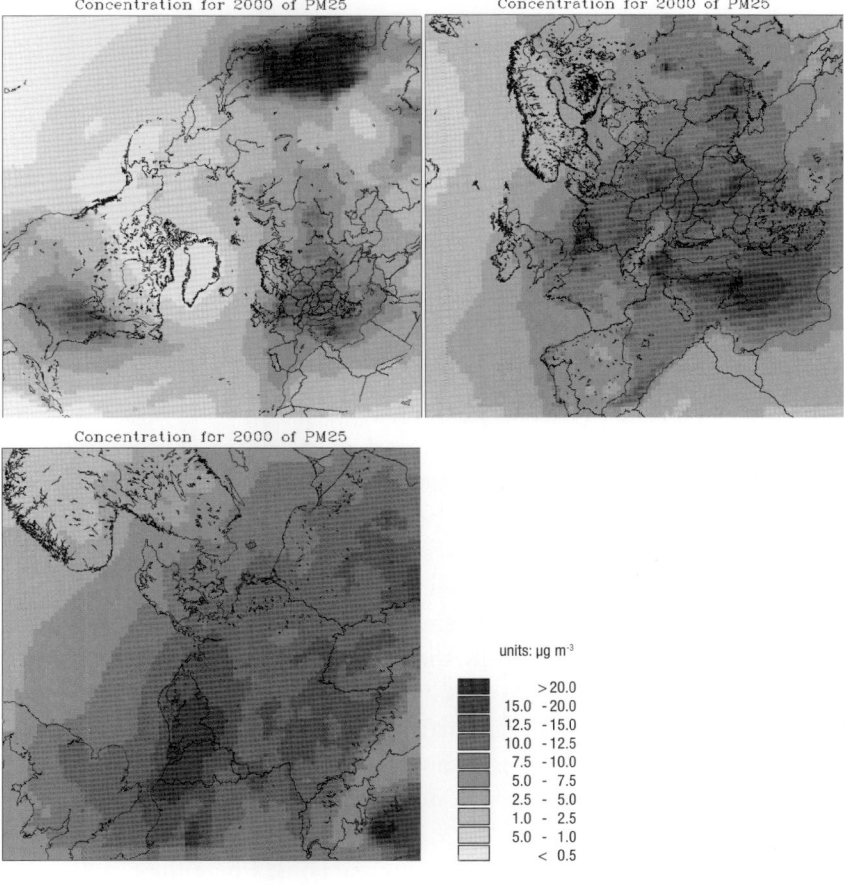

Figure 14.10 Annual mean concentrations of PM$_{2.5}$ for the year 2000 calculated with two-way nested model – the Danish Eulerian Hemispheric Model, DEHM. The three different plots show results from three different domains with increasing spatial resolution: 150 km x 150 km for Northern Hemisphere, 50 km x 50 km for the European area, and 16.67 km x 16.67 km for the area around Denmark. (Courtesy of Jesper H. Christensen, NERI, Denmark).

with an aerodynamic diameter less that 2.5 µm ($PM_{2.5}$) are shown for the three different model domains with increasing spatial resolution. The difference in the degree of detail is clearly seen in the plots. Given that the model works well, the additional information resulting from using higher resolution input like meteorological data, emission data and land cover data, together with a more accurate solution from the numerical methods, will lead to better results also when compared with measurements.

14.4 Testing the models

It takes time to decide between the various model parameterisations, to program these in a programming language as routines (developing a model code), organise the input and output routines and make the entire air pollution model system work on a computer. One may think that the job is then done, and yet the most important and time consuming tasks are still remaining. The model has to be tested, verified and validated. This part of the job may include:

- Verification of the programming code, including systematic and logical tests. Every subroutine in the code should be tested for expected results e.g. using typical or idealized input data
- Verification of the model, where typical values are given as input to the model and the results are investigated for consistency
- Numerical testing of individual processes for accuracy against known solutions. This could also imply testing for convergence of model results when the temporal or spatial resolution is increased
- Sensitivity studies where e.g. different parameters in the model are varied within a known distribution and the effect on the results are investigated
- Testing the model results visually by making 2D and 3D visualisations of the model results as well as animations are very useful tools for verifying that the model is actually performing as intended
- Validation of the model against measurements looking at variations in both time and space
- If possible to perform model inter-comparisons, where the results from the model are compared to results from similar models

Comparisons of model results with measurements are essential in order to validate the model performance when simulating real situations. Comparison with measurements should, however, be connected with estimation of different statistical parameters in order to determine the best performing combination of numerical methods, meteorology and parameterizations.

Also comparisons with other models and other types of models are important in order to verify and to validate the applied methodology. This has e.g. been the essential idea behind a project around the European Tracer Experiment, where more than 30 air pollution models from many different countries have been inter-compared and validated against the measurements. In this way it is possible to get an impression of which models and/ or methods that show the best performance and indicate the direction for future research.

Finally, the importance of visualizations and animations of model output should be emphasized. A typical simulation of the long range transport and transformation of air pollutants gives a data output in the order of Gigabytes. Without two- and three-dimensional visualizations and animations of e.g. the concentrations and the different meteorological parameters, it may be very difficult to get an overview of the performance of the models, and to obtain an understanding of the model characteristics and dynamics. Three-dimensional animations also make it much easier to find possible errors or identify a poorly performing methodology applied in the model.

Numerical tests where the model results are compared to known solutions are important to verify numerical stability, avoid large numerical noise or as a basis for correcting the numerical implementation. An example of the so-called Molenkamp-Crowley rotation test is shown in Figure 14.11. The

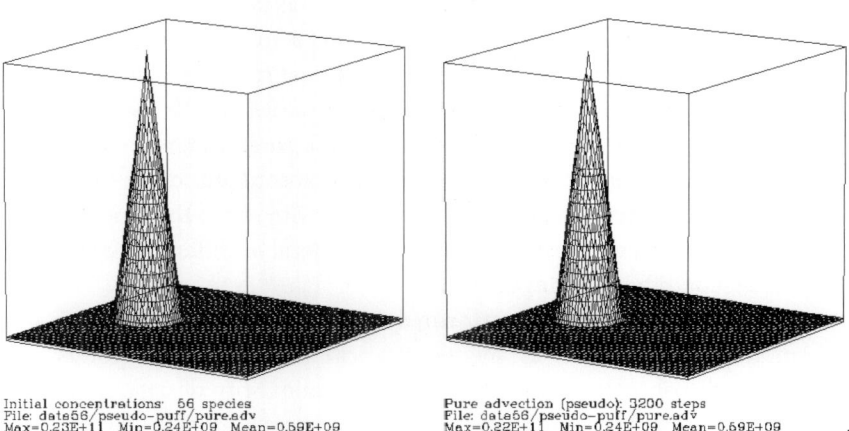

Initial concentrations 56 species
File: data56/pseudo-puff/pure.adv
Max=0.23E+11 Min=0.24E+09 Mean=0.59E+09

Pure advection (pseudo): 3200 steps
File: data56/pseudo-puff/pure.adv
Max=0.22E+11 Min=0.24E+09 Mean=0.59E+09

Figure 14.11. An example of the Molenkamp-Crowley rotation test performed in a Eulerian model. Left figure shows the initial concentrations which have been defined in a way so they form a cone. The right figure shows the same cone after a full rotation around the centre of the grid. The two cones look very similar, showing that the advection algorithm – in this case the pseudo-spectral method – has a good performance.

rotation test can be useful in finding a well balanced advection scheme in which the relation between precision and computing time is optimized.

Comparison with measurements

Validation of models by performing a comparison of calculations against measurements may be analysed both in time and space. When all data are considered both temporally and spatially, the analysis is termed global. Temporal analysis is employed for comparison of time series of measurements with corresponding time series generated by the model. Direct comparison of time series is a powerful way to examine the qualitative behaviour of the model like e.g. temporal variations, extremes or biases. Global analysis is e.g. the case where global scatter diagrams are employed and different periods from many different measurement stations can be compared. It is, however, important in all cases to use statistical parameters and statistical tests for comparison in order to determine the best performing setup of the model.

One point should be emphasized when using statistical tests: The model calculations are not themselves the only source of variability. The observations are also subject to certain levels of error that in many cases are difficult to quantify. The uncertainty e.g. in the measurements during the first release of the European Tracer EXperiment (ETEX) carried out at NERI, has been estimated to be approximately 25-30% with a reproducibility of approximately 15%. The uncertainty in a specific model result can therefore hardly be estimated to be smaller than the uncertainty in the measurements, when the only way to determine the model uncertainty is to compare model results with measurements. It should also be stressed that measurements generally represent the value at a specific point and the model results usually represent a spatial average in a grid cell. Thus, local factors, which can have great influence on the measurements, are often not accounted for in the model simulations. The implication of this is that even when both the model results and the measurements are exact, they will in general be different just because they represent different things.

When statistical tests are performed, it is necessary to make the usual assumptions in statistics, as the stochastic independence of points, the identically distribution of variables, and the stationarity of data. Stationarity means that it should not be possible in a specific data set by physical arguments to determine where a point is. It is therefore important that the data do not contain periodic variations (like diurnal- or annual cycles) when tested statistically.

14.5 Future directions

Present air pollution models have already reached a level of high complexity. A state-of-the-art model usually consists of many thousand lines of model codes, and includes many different physical and chemical processes in order to describe the phenomena as correctly as possible. Present models can be stated to be very comprehensive with respect to:

- The descriptions of the various processes as e.g. emissions, advection, dispersion, chemical transformations, deposition, etc.
- The level of integration with other models as e.g. weather forecast models
- The level of application, where present models constitute an important contribution to decision making of e.g. emission reductions, exceedances of critical levels or loads
- Assessment of impacts on human health and marine and terrestrial ecosystems

The applicability of models is now mature in the sense that decision makers often use results from such models. However, the uncertainties in the model results are still relatively high for many applications. Large efforts are therefore put into the further development of the models. Typical new developments are within the areas of:

- Increasing the resolution in the air pollution models for obtaining more detail and higher accuracy in the results
- Increasing the number of chemical species and chemical reactions in the model in order to diminish the effects from incompleteness of the models. Present models typical include 50-100 chemical species, but several thousands are known
- Increasing the complexity with respect to processes. An example is the description of particles, where processes such as nucleation, coagulation, evaporation, condensation are important as well as handling the particles in different particle size classes (size distribution)
- Deriving better input data for the models – as e.g. better emission data, better meteorological data and better land cover data
- Increasing the level of integration with other kinds of models, as e.g. models for impacts on human health, impact on the ecosystems as well as the economical valuation of the impacts

14.6 Literature

Asman, W.A.H. and Van Jaarsveld, J. A. (1992): *A Variable-Resolution Transport Model Applied for Nhx in Europe.* Atmospheric Environment Part A-General Topics, *26*, 445-464.

Berkowicz, R. (1998): *Street Scale Models.* Chapter 12 pp. 223-251. In: J. Fenger, O. Hertel and F. Palmgren, Urban Air Pollution – European aspects. Kluwer Academic Publishers, Dordrecht, The Netherlands

Berkowicz, R. (2000): *A simple model for urban background pollution.* Environmental Monitoring and Assessment, *65*, 259-267.

Carruthers, D.J. et al. (1994): *Uk-Adms – A New Approach to Modelling Dispersion in the Earths Atmospheric Boundary-Layer.* Journal of Wind Engineering and Industrial Aerodynamics, *52,* 139-153.

Gifford, F.A. (1961): *Use of routine meteorological observations for estimating atmospheric dispersion.* Nuclear Safety 2, 47-51.

Hanna, S.R. et al. (1982): *Handbook on Atmospheric Diffusion, Atmospheric Turbulence and Diffusion Laboratory, NOAA..* DOE/TIC-11223. USA Techical Information Centre, US Department of Energy.

Hanna, S.R. and Chang, J.C. (1993): *Hybrid Plume Dispersion Model (Hpdm) Improvements and Testing at 3 Field Sites.* Atmospheric Environment Part A-General Topics, *27*, 1491-1508.

Hertel, O. et al. (2007): *Integrated air-quality monitoring – combined use of measurements and models in monitoring programmes.* Environmental Chemistry, *4*, 65-74.

Iversen, T. et al. (1990): *Calculated budgets for airborne sulphur and nitrogen in Europe.* The Norwegian Meteorological Institute. EMEP/MSC-W Report 2/90.

Olesen, H.R. et al. (1992): *An Improved Dispersion Model for Regulatory Use – The OML Model.* Proceedings of the NATO CCMS on Air Pollution Modeling and its Application IX., Springer, Secaucus, NJ,,USA

Palmgren, F. et al. (1996): *Effects of reduction of NOx on the NO_2 levels in urban streets.* Science of the Total Environment, *190*, 409-415.

Stohl, A. (1998): *Computation, accuracy and applications of trajectories – A review and bibliography.* Atmospheric Environment, *32*, 947-966.

Wåhlin, P. et al. (2006): *Characterisation of traffic-generated particulate matter in Copenhagen.* Atmospheric Environment, *40*, 2151-2159.

V

Impacts

Air pollution impacts various geographical scales ranging from pure local over regional to global. Generally the time scale varies along with the spatial scale, with most urban impacts being more or less instantaneous, regional impacts having a time horizon of years to decades and global impacts may be centuries.

Cities are by nature concentrations of humans, materials and activities. They therefore exhibit both the highest levels of pollution and the largest targets of impacts. Today nitrogen oxides, hydrocarbons and small particles from traffic attract the attention. Most important are health impacts (Chapter 15), whereas materials damage (Chapter 16), mainly related to sulphur dioxide, is of declining importance. Also impacts on vegetation are declining. In the industrial world the London smog is a thing of the past. Still, however, some reductions of visibility (Chapter 17) can be observed in European Cities, but mostly as an aesthetic loss.

Impacts on vegetation are generally on a regional scale (Chapter 18). The impacts are mainly on natural and agricultural ecosystems in the form of acidification and eutrophication. Also impacts of ozone have been observed in the form of reduced agricultural productivity.

Compounds with a lifetime of years can be dispersed over the entire globe. The depletion of the ozone layer (Chapter 19) and the increasing greenhouse effect (Chapter 20) are the two most important global environmental problems. They are normally treated independently in international negotiations on mitigation. In scientific respects they are, however, connected. The problem with the depletion of the ozone layer is in principle solved by phasing out the use of CFCs and similar compounds, although it may take decades before the natural situation is restored. The human impact on the natural greenhouse effect by emissions of greenhouse gases and the ensuing global climate changes is a more complicated issue. The expected climatic impacts influence differently in different areas of the Earth, and are not always negative. Likewise the attempts to prevent the climate changes by emission reductions influence all sectors of society, and they are in many cases in direct conflict with wishes and plans for economical and technological development.

15 Health Impacts

Anders Carlsen

Health effect in the form of odour, irritation of eyes, coughing and dyspnoea provoked by air pollution has been described for centuries. Symptoms are more severe in connection with physical activity, lung disease or prolonged repeated exposure, where e.g. exposure to ozone may eventually result in chronic effects on the lungs. There is great individual sensitivity, children and elderly being generally the most sensitive.

Smog episodes are now rarely seen in the developed parts of the world, but are frequently seen in developing countries. Still, even in the developed countries a connection between the severity of air pollution and morbidity and mortality can be demonstrated. This is reflected in indicators like increased use of medicine, absence because of disease and contacts to physicians especially for people with heart or lung diseases as well as in registries of morbidity and mortality.

One of the main sources for air pollution is traffic, but also burning of wood, coal and waste account for a considerable morbidity and mortality. In developed countries the pollutants of current interest are ozone, particles and nitrogen dioxide, while sulphur dioxide has less interest due to decreased concentration as a result of extensive to abatement measures.

In many developed countries a rise in the prevalence of so-called hyper reactive lung diseases has been observed to involve now about 10 – 20% of the population. The reason for the increase is not fully known, but a factor could be changes in surface properties of particulate allergens like pollen affected by air pollutants like acids, promoting their allergenic potencies. Other possibilities are harmful effects on the immune system or the cell types involved in hyper-reactive reactions in the respiratory tract. Changes in indoor climate may also play a major role.

Investigations on the impact of air pollution on human health are difficult to conduct and to interpret because many coinciding factors influence health to a varying degree. Generally it can, however, be concluded that:

- For some diseases, where air pollution is assumed to be part of the reason for their evolvement, the incidences seem to be rising, being most pronounced in city areas. This applies to hyper reactive diseases like asthma, chronic bronchitis, hay fever, and certain cancers and reduced reproductive fertility. Even moderate air pollution affects the disease for these persons
- The health effects are most pronounced for sensitive groups like children and the elderly and particularly for those with an already existing lung disease
- There seems to be a synergetic effect when more than one pollutant is present. As an example some investigations suggest a synergetic effect by exposure to SO_2, NO_2 and particles
- The health effects during air pollution episodes increase with increasing duration of exposure to the air pollution, suggesting that cumulative exposure has stronger effects on mortality than estimated from associations between day-to-day variations in air pollution and mortality
- Generally exercise and consequently mouth breathing exacerbates the effects in the lungs as the nose acts like a protective filter

The effects on children are of special concern, but little is known about their presumably greater sensitivity. It is estimated by WHO (2004) that the total effect of air pollution on the health of children is very large and that air pollution accounts for 6-8% of the mortality of children in Europe. Air pollution primarily exerts its effect via provoking or promoting asthma, which is a widespread disease of children of increasing prevalence and concern.

15.1 Exposure routes

Before a pollutant can exert its harmful effects it must be transported to the "target" for the effect. It can be a direct effect on skin, eyes, upper airways or lungs or an indirect effect after being coughed and subsequently swallowed and absorbed via the alimentary tract.

Air pollution may be deposed on soil or plants that are swallowed or eaten e.g. by playing children. Air pollution components may be accumulated in plants or animals and some times be concentrated in food chains ending in food items. Examples are certain persistent pollutant like PCBs (Poly Chlorinated Bi-phenols), mercury and pesticides. These partly gaseous pollutants are e.g. brought to the peoples in the Arctic, where some of the highest blood concentrations of these chemicals have been found, although the chemicals are produced and emitted in the industrial countries far from the Arctic.

When a substance is taken up by the organism it can be distributed via the blood to different organs and be harmful at the "target" sites. Some persistent pollutants are deposed in organs (often fat) from where they may exert harmful effect over long periods. Dioxins have e.g. a halftime for elimination from the human body of 7 – 10 years.

Some pollutants are metabolised in the body to more toxic substances or to harmless end products, which are excreted via urine or faeces. There are great individual genetic differences in these mechanisms.

The harmful effects from air pollution can occur at different locations and anatomic levels in the body depending of the physical and chemical properties of the components (Table 15.1).

Exposure routes	Component properties	Major effect
Local exposure of skin and eyes	Acidity (pH), reactive capacity	Irritation, inflammation or etching with breakage of chemical bonds between molecules.
Local exposure of nose epithelium	Reactive capacity, particle size and properties	Smell, irritation, inflammation, cancer
Local exposure of respiratory tract epithelium and linings	Reactive capacity, particles	Irritation, asthma, bronchitis, inflammation, cancer,
Uptake from respiratory system	Toxicity of compounds	Systemic toxicity involving different organs
Uptake from alimentary system after coughing and swallowing	Toxicity of compounds	Systemic toxicity involving different organs

Table 15.1 Examples of impacts of air pollutants.

Systemic toxicity can occur at different levels as
- Alteration of biological processes in the cells like receptors, chemical equilibrium, binding of substances and enzymatic activity
- Altered transport over cell membranes
- Damage of cells e.g. by influencing immune responses
- Cellular necroses in organs
- Functional alterations of organs e.g. affecting the function of the cardiovascular system, liver or central nervous system
- Malaise like nausea altering wellbeing and behaviour

15.2 Anatomy and organ/cellular responses

Skin

The human skin has a total area of about 2 m^2 and apart from functioning as an effective protective barrier for the body against most harmful substances, it controls heat balance and minimise loss of water. The outer layer, stratum cornea, consists of dead keratinised skin cells with low water content. This layer is renewed in two weeks and functions as a barrier against especially hydrophilic substances.

However, some substances easily penetrate the skin and can give systemic toxic responses. Examples are chlorinated phenols and some pesticides. Reactive substances or detergents and organic solvents may affect the skin to react by irritation or inflammation whereby the skin looses keratin and becomes less protective against substances and micro-organisms.

As an indirect consequence of air pollution the skin is being more exposed to harmful UV radiation, when the ozone layer is depleted (Chapter 19).

Eyes

The eyes may react by irritation and blurred vision, when chemicals are taken up though the epithelium. However, the eyes are relatively well protected by eyelids and lachrymal fluid, which dilutes and clear away harmful substances.

Nose

The nose acts as a filter, holding back particles and absorbing highly water-soluble gases. It also has an important function in humidifying and warming the inhaled air. Nasal epithelia can metabolise or activate some pollutants to more toxic substances affecting the epithelium, the olfactory epithelium – the organ for smell sensing – being particularly vulnerable.

At the same time the nose is vulnerable to irritating substances, which can produce swelling of the mucous membranes and secretion disturbing the respiration. Hereby the protective action of the nose may be bypassed. When held back in the nose for longer periods some pollutants like wood dust may be carcinogenic to the cells in the mucous membranes of the nose. Furthermore some substances are easily taken up through the membranes in the nose to the dense capillary blood vessel system. It has also been suggested that ultrafine particles may be transported to the brain through the nerve that signals the smell sensations.

The nose is also the site of the olfactory nerve endings with receptors for smell and leading via nerve fibres directly to the brain. The sensation of smell may be either reinforced or suppressed by longer lasting exposure and there may be large individual differences in the perception of smell.

The sensation of smell is one of the earliest, most important and sensitive senses for survival. In some occasions smell may be a valuable indicator of acceptable/unacceptable risk like the smell from oil pollution and of styrene from repair of plastics in e.g. sewer pipes.

Humans can smell a very large number of chemical substances, certain substances even at concentrations in the air far below what any technical device can detect. Smell may provoke a number of reactions ranging from discomfort, nausea, and loathing of food to wellbeing and high spirits. Smell therefore must not be considered a pure annoyance without health relevance, as it may constitute a threat to health and wellbeing.

In practice regulation on emission of smell can be based on limit values set by using a group of individuals (a smell or olfactory panel) and calculating the concentration that 50% of the panel can sense.

Trachea and lungs

Trachea and the upper parts of the airways in the lungs are kept open as stiff tubes by crescent-shaped rings of cartilage. The airways bifurcate repeatedly ending in about 500,000 respiratory bronchioles and further on to about 250 million alveoli (Figure 15.1).

This results in a total area for diffusion between inhaled air and the lung tissue of about 140 m^2. As the airways branches out the sectional area of each individual branch decreases, but the total sectional area of all the branches increases from about 3 cm^2 for trachea to about 400 cm^2 for the respiratory bronchioles and to about 9500 cm^2 for the most peripheral.

The wall of the airways consists of epithelium containing ciliated cells, mucous cells and serous cells. The mucous and serous cells together produce a fluid in which pollutants are trapped and transported upwards toward the pharynx by the coordinated movements of the cilia, which are controlled by the central nervous system. From pharynx the pollutants may be swallowed or expectorated.

Particles over PM_{10} in size are removed in this way from the upper airways. Smaller particles may reach the more peripheral airways, where they may deposit by sedimentation or impaction. In recent years ultra fine particles able to reach the alveoli, have been in focus as an important factor for promoting chronic lung diseases.

Gas exchange occurs in the alveoli representing 80 – 90% of the parenchy-

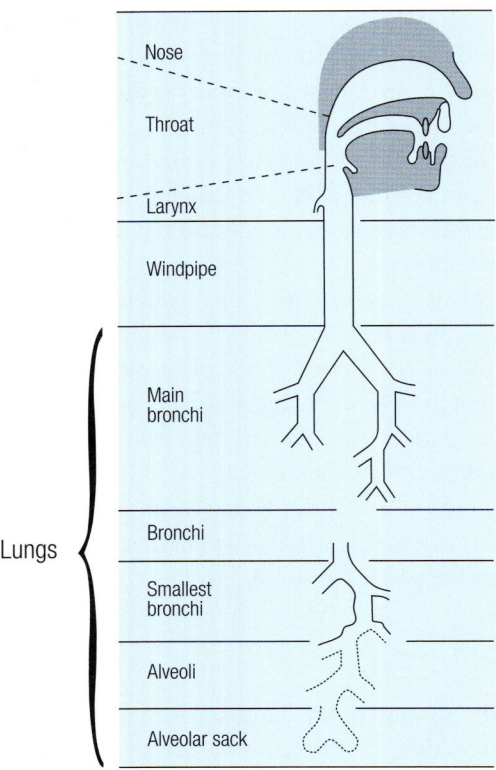

Figure 15.1 Simplified structure of the human respiratory system.

mal lung volume. In the alveoli only a thin cell layer consisting of epithelial, interstitial and endothelial components, the latter forming the wall of the blood vessels separating inhaled air and blood. The diffusion of gases, most important of course of oxygen, over this barrier may be compromised by thickening of the alveolar walls and accumulation of liquid in the alveolar space as is seen as a result of chronic toxicity.

The most important adverse effect of many toxic inhalants is due to seque-lae (complications) of the imposed oxidative stress resulting in diseases as chronic bronchitis, fibrosis, emphysema and cancer. The oxidative stress is often mediated by free radicals carried on particles, ozone, NO_2 and tobacco smoke. Also when particles are phagocytised (taken up by cells) potent oxidants are released.

Numerous airborne micro organisms and low- and high-molecular weight antigenic materials may stimulate the immune system to react both by specific and non specific responses producing bronco-constriction and chronic pul-

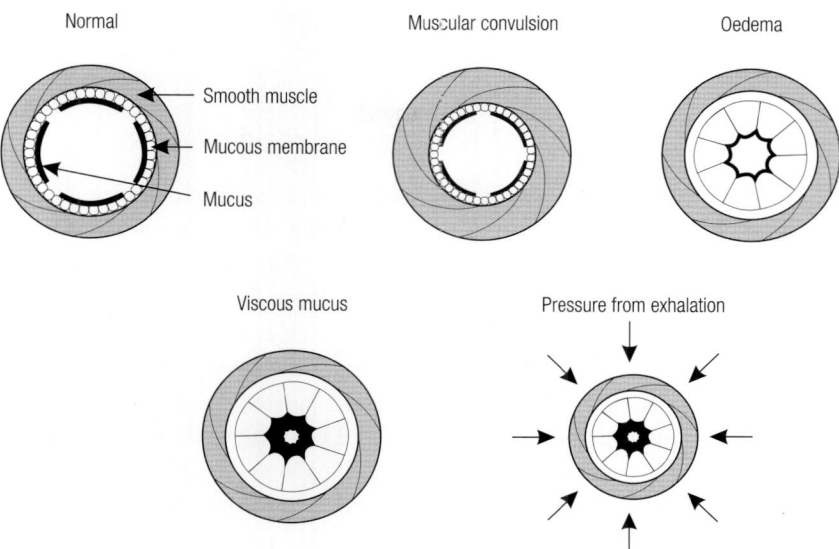

Figure 15.2 Cross section of the walls of the respiratory system with four factors contributing to narrowing of the airways by irritation.

monary diseases as asthma, hay fever and farmers lung. When first acquired, asthma symptoms may be provoked by irritation from a variety of substances, by infections or physical exercise even without exposure to specific antigens.

The reduced sectional area of the middle sized bronchi in asthma is due to a combination of bronco-constriction, increased secretion of mucous fluid of higher viscosity and swelling of the epithelium because of oedema and inflammation in combination with increased intra thoracic pressure during expiration to overcome the increased resistance. This results in the well known wheezing, coughing, extended expiration and shortness of breath. The effects are especially severe for already weakened persons leading to increased morbidity and mortality during pollution episodes.

Cardiovascular system

Urban air pollution, especially particles, nitrogen oxides and carbon monoxide is known to be linked to coronary heart disease and associated morbidity and mortality. Also here already weakened persons are at special risk. The mechanism appears to be related to inflammation and oxidative stress mediated by free radicals. This may activate factors that influence blood clotting and vascular function. However the mechanisms are not yet fully understood.

15.3 Health impacts from important sources of pollution

Health impacts from stoves and ovens in residental housing

Residential heating and cooking with wood-stoves is quite common also in developed countries. The lighting up the fire and the period when the fire dies out are especially critical, as the most toxic substances are created and emitted during these phases, but also the conditions of the burning between these phases are important. As domestic burning is often not optimal, air pollution from burning of wood and coal constitute a relatively large contribution to health-affecting air pollution. The substances emitted include dioxins, PAH, carbon monoxide and particles, and the health effects are linked to these components (see below). Especially in developing countries the fuel in ovens inside domestic houses also constitutes a considerable indoor pollution health problem. The global number of death due to indoor air pollution from solid fuel use is estimated to be more than 1.6 million annually and the global number of deaths due to acute respiratory infections in children under five years of age as a result of this exposure has been estimated to be nearly 800,000 per year (WHO Health Report 2002).

Health effects of urban atmospheric pollution

Clear correlations have been found between more intense air pollution in urban areas and cardiovascular diseases, lower lung function, hospitalization for pneumonia, acute bronchitis and asthma in children.

The important components are particles, nitrogen, sulphur oxides, ozone and carbon monoxide. These compounds act as irritants but may also after persistent high exposure lead to chronic damage like fibrosis and reduced immunological defence against infections.

Especially air pollution from traffic is a serious health problem. It has been estimated that it accounts for 5-10 times as many deaths due to respiratory and cardiovascular diseases as traffic accidents. Air pollution from traffic is considered the dominant air pollutant risk factor for cancer in cities.

From a health point of view the most important components of pollution from traffic are particles, nitrogen oxides, ozone, PAH, benzene, 1,3-butadiene, ethane, propene, formaldehyde, acetaldehyde and acroleine, all often found in concentrations well above safety level standards. In some developing countries lead (Pb) is still used as additive in gasoline and thus contributes as an important component. Especially particles have recently been in focus as a major factor in the health effects from traffic (see later).

In population studies the indicator of exposure is often estimated from the distance of the living house to a street. Numerous investigations have found residence near high-traffic areas to be linked to higher mortality from heart and lung disease, childhood asthma hospitalization, respiratory illness in small children, including wheezing, ear-, nose- and throat infections and asthma and rhinitis as well as premature and low-weight newborns.

Because of the carcinogenic components of traffic emissions there is a considerable higher cancer risk associated with living close to traffic.

Health effects from industrial pollution

In many developed countries regulation of pollution from industry has reduced exposure to below levels where effects are expected to be found. However there may still be harmful effects. Some important examples are dry-cleaning establishments emitting perchloroethylene to the nearby residents and waste incinerators and district heating stations emitting dioxins and PAH (see later).

Emission of biological material like proteins may constitute a health problem of still unknown magnitude as the emission may provoke establishment and aggravation of allergy even at very small emissions.

Health effects from air pollution from agriculture

Pesticides can enter the atmosphere by spray drift and volatilisation during application, by post-application volatilisation and by drift of soil particles onto which they are adsorbed. Pesticides have been found in rain water. Likewise aerosols from spreading of manure and fertilizers can be transported by the wind. The health risks caused by agricultural activities are associated with inhalation of pesticides, organic dust and microorganisms as well as smell and harassment because of swarms of flies.

Pesticides are, in their nature, toxic and some pesticides are carcinogenic or may affect the hormone systems. Most of the most troublesome are banned in developed countries, but are still be used in developing countries.

Health effects from industrial accidents

The Monsanto, Seveso and Bhopal accidents in 1949, 1976 and 1984 are well-known examples of sudden emission of large quantities of very toxic chemicals to the air, e.g. dioxins and isocyanates. In these instances the acute effects of the chemicals become relevant in contrast to the normal situation where industrial plants continually emit small amounts of chemicals, and where it is the repeated exposure of small doses that are relevant to manage.

In the Bhopal (India) accident very large quantities of methyl-isocyanate was released and spread to the nearby town leading to severe lung- and eye damage among the inhabitants. About 5000 died and 60,000 survived with persistent lung- and eye-damages.

In accidents at a Monsanto plant (West Virginia, USA) and in Seveso (Italy) dioxins were released together with other chemicals from the production of herbicides. Dioxins are very potent carcinogenic substances. The acute effect is a skin disease named chloracne. After the Seveso accident the nearby population were investigated for many years. About 20 years after the accident is has been possible to document increased cancer incidence in the exposed group in accordance with the expected period necessary to develop cognizable cancer.

Another example of industrial accidental release is the Chernobyl accident in 1986, where a cloud of radioactive substances, including caesium-137 and iodine-131, were released into the air and transported over a distance of thousands of km. About 30 persons died immediately, and afterwards more than 700 children developed cancer in the thyroid gland.

To cope with and manage chemical accidents three concentration intervals has been defined:

* From zero concentration to a concentration below which only slight reversible effects are expected
* An interval above which irreversible effects may be expected and/or orientation and ability to take flight is severely impaired
* The upper concentration interval, where most people will die

Even for very well-known substances the setting of these intervals has, however, been difficult because of lack of reliable human data and great individual differences in sensitivity.

When these three intervals are used in combination with weather forecast, dispersion models, geographic information systems and data on the amount of chemical released, it is possible to assess risk zones and arrange for rescue activities as effective as possible. These instruments are also used in the planning phase of siting of industrial plants and in planning of the preparedness for accidents.

Health effects from volcanic eruptions

Besides the direct catastrophic effects volcanic eruptions create far-reaching, "natural" air pollution. The main components of this type of air pollution are

SO_2, aerosols composed of primarily sulphuric acid and other sulphate compounds, dust particles and small amounts of metals like selenium, mercury, arsenic, iridium. The visible haze created by interaction between SO_2, other volcanic gases, atmospheric oxygen, moisture, dust and sunlight during minutes to days is called "vog" and provokes irritation of eyes, nose, throat and skin and may also produce respiratory distress depending on concentration and circumstances. The typical symptoms are breathing difficulties, watery eyes, sore throat and flu-like symptoms.

15.4 Health effects from selected pollution components

Sulphur dioxide

The emission of SO_2 has generally been reduced significantly in developed countries and is no longer a dominant air pollution health factor. In developing countries SO_2 is still a major pollutant with major health implications. Sulphur dioxide is predominantly an upper airway irritant being able to stimulate bronchial constriction and mucus secretion. Asthmatics are more sensitive than healthy persons. When deposited most of the SO_2 dissolves in the airway and lung fluids as sulphites and sulphuric acid which are irritating to the mucous membranes and furthermore promotes the irritation responses to ozone and inhibits the upwards mucus ciliary transport of pollutants like soot. Sulphites are taken up in the blood and distributed in the body.

Nitrogen oxides

Nitrogen oxides are gases relatively insoluble in water and therefore penetrate deep into the lungs and elicit toxic responses in the alveoli. They are converted in the lungs to nitrate, nitrite, nitric acid and nitrous acid, which are all irritants to the mucus membranes. In laboratory studies nitrogen dioxide has demonstrated capacity to activate oxidant pathways. After long-term exposure the lung structure, the cellular metabolism in the lungs and the resistance against bacterial infections may be affected. The nitrogen dioxide also enhances airway responses to inhaled allergens in asthmatics. Especially for persons with chronic bronchitis and asthma exposure may result in reduced lung function.

Ozone

Ozone is a potent reactive and oxidizing gas generated by photochemical reactions involving nitrogen dioxide and organic pollutants. The reverse reaction occurs when ozone reacts with nitrogen oxide emitted from diesel engines forming nitrogen dioxide. Ozone has been shown to affect lung development and induce higher risk of sensitization by allergens resulting in asthma-like symptoms in newborn monkeys. Ozone may also affect cardio-vascular and pulmonary morbidity and mortality. Large individual geneti-cally based differences may exist in the responsiveness to inhaled ozone.

Particles

Particles are increasingly being viewed as very important health risk compo-nents in the emission from a variety of sources both for acute illness and for chronic diseases.

Particles are often carriers of a great number of hydrophobic highly toxic substances like PAH, dioxins and heavy metals depending of the source of emission. Particles also have an intrinsic toxic effect because they act as irri-tants or harm the cells in the linings of the respiratory system in the different parts of the lungs depending on their form and size. When transported to the distant parts of the lungs the substances can be taken up by the blood and transported to other parts of the body.

In recent years studies on particles and health have focused on the form of the particles, especially if they are thread formed like asbestos fibres, and their size, which is currently divided into three groups:

PM_{10}, $PM_{2.5}$ and $PM_{0.1}$, i.e. particles smaller than 10, 2.5 and 0.1 μm respectively. Particles greater than PM_{10} are deposited in the upper airways, whereas $PM_{0.1}$ can reach the most peripheral parts of the lungs and be taken up by the blood.

For technical reasons in most investigations only the concentration of $PM_{2.5}$ and PM_{10} have been measured. The ultra fine particles, which compose the dominating number among the particles and are supposed to be the most important in relation to health, are only measured in very few investigations (Chapter 13).

The interest has, however, increasingly been concentrated on the smallest, ultra fine particles $PM_{0.1}$ being able to penetrate to the smallest and deepest parts of the lung. Their surface area in relation to volume is larger than for greater particles, which mean that the chemical constituents on the surface can easily react with molecules on the cell surfaces and after uptake inside the cells.

Investigations of effects of air pollution in US cities and in Europe has documented a statistical strong correlation between concentration levels of particles and mortality from all-causes, cardiovascular, lung diseases and lung cancer. Similarly an association between exposure to particles and hospital admissions has been demonstrated. Other investigations have pointed to an increase in the prevalence of bronchitis and chronical coughing associated with increase in particle concentration.

Other studies have confirmed that long-term exposure to particulate pollution is linked to lung cancer and cardiopulmonary mortality as well as progression of atherosclerosis. Especially exposure to particulate pollution from diesel engines has been shown to increase cancer risk.

WHO has estimated that pollution with particles today globally accounts for 1% of hearth and lung diseases and 3% of cancer of lower airways and lungs. This results in 0.6 mill. (1.2%) premature deaths and the loss of 7.4 mill (0.5%) disease adjusted life years (DALYs). An investigation in the European Union indicates 288,000 too early deaths per year in EU.

WHO has estimated that a short term increase in exposures to PM_{10} of 10 $\mu g\ m^{-3}$ will produce an increase in mortality of around 0.5% in both developed and developing countries. WHO has also estimated that an increase in the average $PM_{2.5}$ level of 10 $\mu g\ m^{-3}$ in developed countries will result in a reduction in the expected average lifespan of about one year.

New and more refined investigations have documented effects at still lower exposures. No lower threshold has been identified for particle concentration in relation to all-cause and cardio-respiratory mortality, which means that no safe limit exists below which there are no health effects. Thus the WHO guidelines (Table 15.2) represent realistic goals rather than limits that offer full protection.

$PM_{2.5}$	10 $\mu g\ m^{-3}$	annual mean
	25 $\mu g\ m^{-3}$	24-hour mean
PM_{10}	20 $\mu g\ m^{-3}$	annual mean
	50 $\mu g\ m^{-3}$	24-hour mean

Table 15.2 WHO guidelines for particles in air.

Polyaromatic hydrocarbons, PAH

This is a group of organic compounds with three or more condensed aromatic rings that are formed in incomplete combustion of biomass or coal. They have low solubility in water, but high solubility in fat and are in air pollution found adherent to particles, mostly soot particles. When adhered to the surfaces they can react in the atmosphere with e.g. sunlight, ozone, nitrogen oxides and SO_2 to form other toxic substances. PAH can be absorbed by airway epithelium. The far most important effect from inhalation is lung cancer and the risk is dependent of the composition of PAH. One particular PAH, benz(a)pyrene, or BaP for short, has been used as indicator and the toxicity been expressed as BaP equivalents. However the toxicity of PAHs is not additive, but synergistic, meaning that a mixture of different PAHs is more toxic than the sum of the single PAHs. This means that the use of BaP equivalents as toxicity indicator is restricted to mixtures of approximately the same composition. The risk evaluation of PAHs is still under debate, as it is discussed whether a lower threshold for the carcinogenic effects of PAH exists or not and this is crucial for the setting of limit values.

Benzene

Blood levels of benzene have been found to be higher in children living in high-traffic-density areas than in children who lived in low-traffic-density areas. Aplastic anemia and leukaemia are known to be associated with excessive exposure to benzene.

1,3-butadiene

This gas is taken up by inhalation and converted to the genotoxic carcinogen 1,2-epoxybutadien in the organism.

Aldehydes (formaldehyde, acetaldehyde and acroleine)

Being easily dissolved in water most of these substances are taken up by the epithelia in the upper airways. They are highly irritating to eyes and upper airways, mutagenic and suspected carcinogens.

Carbon monoxide

The effect of CO is the consequence of the strong binding of CO to haemoglobin on the binding sites normally occupied by oxygen, reducing the transport of oxygen to the cells in the organism. When more than 2-5% of the haemoglobin is occupied by CO neurological symptoms of increasing severity become evident beginning with concentration difficulty and headache and death at about 40 - 50%. Persons with cardiac diseases are especially vulnerable and may present angina already at 2-4% and may die from a heart attack at about 15%. Smokers normally have 5 – 10% CO-Haemoglobin.

Dioxins

Polychlorinated dibenzo-*p*-dioxins (PCDD´s) and dibenzofurans (PCDF´s), in short dioxins are compounds with two benzene rings connected by one or two oxygen atoms and with 1 – 8 chlorine atoms replacing hydrogen atoms in the benzene rings. They are created in all combustion processes, where chlorine and biological material is involved especially in the temperature range 200 - 800 °C. Incineration of municipal solid waste is today a major source of emission to the air. Most of the dioxin in the air is bound to particles and aerosols. Dioxins are extremely persistent and highly lipophilic substances that accumulate in food chains with bio concentration factors from 200 – 70,000 meaning that more that 90% of human exposure occurs through the diet. The half-life of dioxins in the human body is very high ranging from 5 – 50 years depending of the type of dioxin, body composition, age and sex. Dioxins are transported through the placenta barrier to the foetus and via breast milk to the newborn.

Dioxins may give rise to a variety of health effects. Chloracne occurs shortly after exposure by skin contact, ingestion or inhalation. Other health effects that have been observed are increased risk of cardiovascular disease, diabetes, liver disease, and cirrhosis, immunological alterations, chronic bronchitis, respiratory infections, alteration of thyroid hormone production, and affection of the eye. Transient nausea, vomiting and abdominal pain and neurological symptoms have also been observed.

In investigations of workers that had been exposed to dioxins, increased risk has been found for a number of cancers like soft tissue sarcoma, lung cancer, non-Hodgkin sarcoma, and digestive tract cancers. Dioxins are extremely toxic even in very small doses. The daily intake that increases cancer mortality by 1% has been calculated by different scientists to be in the range of 5 – 40 pg day^{-1} per kg bodyweight. In animals also developmental and reproductive effects have been observed as well as tumours and cancers. The carcinogenic effects of dioxins are additive meaning that a Toxic Equivalent (TEQ) system can be used to evaluate the toxicity of a mixture of different dioxins.

Lead

Although lead has been phased out as a gasoline additive in most developed countries, this is not the case in some developing countries. Lead affects many body organs including the nervous, haematopoietic, gastrointestinal, vascular and reproductive systems as well as the heart, liver and kidneys. Of great concern are the neurological effects in children, who are particular susceptible, including effects on development, intelligence quotient (IQ) and behaviour. There does not seem to be a threshold below which there are no effects on the central nervous system, and the most toxic exposures occur at chronic low levels.

Mercury

Mercury is toxic to brain and the nervous system when in the form of metallic or "organic mercury". All mercury compounds are toxic to the kidneys and may promote allergy. Methyl mercury has been shown to accumulate in food chains. On the Faroe Islands is has been demonstrated that this can affect the IQs of children of mothers eating whales. In certain parts of the world, where Hg emissions are rising because of increasing burning of coal, this element may in the future be an important air pollution component spreading over long distances.

Combination effects

Air pollution consists almost always of a mixture of several substances. The combined exposure may result in effects different from that expected from the single substances. Very little precise knowledge exists on how to estimate combination effects, which may be additive, antagonistic or synergistic. The possible mechanisms by which substances may affect the effects of other substances are numerous. One substance may enhance the uptake, metabolism or secretion of the other or compete for binding to a receptor transmitting the effect to intracellular organelles. It is known that toxicities of dioxins are additive and that the PAHs act synergistic but these examples are exceptions.

It should be remembered that in many of the investigations on air pollution it is the health effects from the combined exposures to many substances that has been evaluated.

15.5 Limit values

The aim of setting limit values for ambient air is to create a tool for restricting emissions of pollution to protect human health and nature. The setting of limit values is not an exact science due to a number of circumstances:
- Humans exert considerable differences in sensitivity to chemical substances
- Data on effects on humans are rather few

- Data most often exist only from animal experiments of differing quality and always with a standard error of 5 - 10% because of the number of animals (most often 50 animals in each group)
- Data consist only of visible harmful effects on the animals during their (short) life span or post mortal
- There may be important differences between animals and sensitive human groups which are not known

When setting the limit values different objectives have been brought forward:

- The public health point of view putting most emphasis on prevention often chaired by physicians and consumer organisations: As we don't know for sure the effects of the pollutants and the pollutants furthermore exist together with thousands of other pollutants with possible synergistic effects great care must be taken to prevent harmful exposures. This means that limit values must be far lower than the values causing the smallest effect shown in animal experiments. This is particularly important for persistent substances as the exposure and effects will last for a long time after the use and emissions have ceased
- The "document list" point of view often chaired by industry representatives and to some extent by toxicologists: The public health policy is not scientific, as it does not give direct reproducible answers and makes to a great extent use of rough estimates. It imposes expenses to the industry with no strong indisputable justification and is altogether too expensive for the society

These types of arguments are well known presently represented in the discussion on existence and causes for the climate change. In practice often the limit values have been set as a compromise between the two objectives, most weight though been given to the public health objective. For some substances existing and general background values make a purely preventive limit value meaningless. However it must be expected and is for some substances well known, that existing limit values also for other substances per se does not secure against harmful effects. This is partly due to modification caused by documental arguments and partly because of emerging new evidence since setting the limit values.

In the EU this has been expressed in the Treaty on European Union, Article 174, where it is stated that the level of protection of health in the field of the environment shall be high ensuring that everyone can live safely in the society, even persons being more vulnerable to polluting substances due to an e.g. inherited sensitivity.

The precautionary principle shall further be respected i.e. if there is suspicion that a harmful effect exists, this must be taken into consideration even when there is still no proof of the effect. Therefore different types of safety margins are generally used in assessing environmental protection, like in the form of un-safety factors. Likewise doubt about the safety shall be taken into consideration.

The goal is that the EU regulation shall be preventive and secure a sustainable development. The environment must not gradually be degraded by pollution and exhaustion.

Setting of limit values requires comprehensive knowledge and experience in the fields of human health, toxicology and administration. Therefore they should be set based on the current knowledge after discussion between different experts.

An internationally agreed practise and tradition for setting limit values with the use of un-safety factors has evolved, originally starting in working groups within the WHO in the 1970s.

Data collection and hazard assessment

The first phase of the risk evaluation is collecting data on the pollutant in general and assessing the general hazards that the substance might exhibit. The data includes knowledge on physico-chemical properties and effects on humans, animals and cell cultures. Often toxicological data can only be found from experiments with animals like mouse and rat and to a lesser extent rabbits, dogs and guinea pigs supplemented by simple tests on bacteria.

Based on these data the hazards are described qualitatively like carcinogenic, irritating, toxic or corrosive. If possible the hazard is quantified by relating sizes of effects to doses attempting to establish a dose-response curve.

Population investigations may be designed in various ways. In the cohort design population groups exposed to different degrees of pollution are followed and compared for health effects after correction for confounders like social differences. In other types of population investigations – the case control studies – the degree of exposure for pollution is estimated for groups of people with certain diseases like asthma and compared with the exposure of the general population.

In both types of design the aim is to assess a connection between the amount of exposure and the effects, if possible by constructing a dose-effect curve, but most often by estimating odds-ratios for the probability for disease between exposed and unexposed population groups.

Population investigations are not very sensitive, therefore rather great effect levels are required before they can be documented. Furthermore there are great practical difficulties connected to the estimation of exposure, effects and confounders.

On the other hand, if a statistical association has been documented in a population investigation this gives very strong evidence for a true connection.

For a few chemicals like lead, mercury, cadmium, and nitrate the evaluation is made on the basis of human (normally epidemiological) data, but most often only qualitative or no knowledge exists about the direct effect on humans. Therefore the assessments of toxic effects are for most chemicals based on experience from animals, especially lifetime animal experiments supplemented by investigations on cell cultures. Only for a few chemicals the experiments have been conducted by the respiratory route after exposure in a gas chamber. By far toxicities of most chemicals through the respiratory route have been estimated from toxicity observed after intake with food. This fact of cause increases the error of the data used as input to the risk evaluation.

When the assessment is based on long term animal experiments one traditionally operates with two models.

The NOEL-model

For some substances there appear to exist a dose, the so called NOEL (No Observed Effect Level) below which no harmful effects are observed in the animal experiments, even if the animals are exposed to this dose daily for their entire lifespan.

Under these circumstances based on a WHO tradition three so-called un-safety factors UF-I, UF-II and UF-III to extrapolate from animal data to presumed worst case human values. The un-safety factors aim to compensate for that:

- Humans may generally be more sensitive than laboratory animals because of e.g. differences in uptake, enzyme systems, metabolism and excretion etc. An example is the far longer half-life time for dioxins in human compared to mice and rats (UF-I)
- Some individual humans may be far more sensitive than the average population (UF-II)
- The quality of the data may vary considerably or be less relevant. E.g. most data for limit values in air are based on food exposure (UF-III)

If the differences mentioned above are not well documented one uses as default values UF-I = 10, UF-II = 10 and UF-III from 1 to 10,000 depending on the quality of the animal data. From these a Tolerable Daily Intake (TDI) can be calculated:

$$TDI = \frac{NOEL}{UF\text{-}I \times UF\text{-}II \times UF\text{-}III} \; \mu g \; (kg \; bodyweight)^{-1} \; day^{-1}$$

Extrapolation models when a NOEL cannot be identified

For other substances i.e. some carcinogenic chemicals it is presumed that no lower limit exists for the doses that can promote cancer and in these circumstances a statistical model is used. It is then estimated which dose given daily over a lifespan will give a probability of 1:1,000,000 for an individual for development of cancer. This dose is taken as the Tolerable Daily Intake, TDI for these substances.

Assessing the limit values

Many substances exist not only as a pollutant in air but also in foodstuff, drinking water or soil. If the limit values for each of these media were set independently it might result in a too high total intake. Therefore after concrete evaluation for each chemical certain roughly estimated fractions of the TDI is *allocated* to each of the possible pathways air, food, drinking water etc. When an additive or synergistic effect has been documented for groups of chemicals this is also taken into consideration. An example where synergism is assumed is the group of air pollutants SO_2, NO_2, particles, acidic aerosols and ozone.

The limit value (LV) is calculated from NOEL in animal inhalation experiments as

$$LV = \frac{NOEL}{UF\text{-}I \times UF\text{-}II \times UF\text{-}III} \; \mu g \; m^{-3}$$

if or when an NOEL exists, or from an extrapolation model.

Most often inhalation data are missing and then data from exposure with food or drinking water is used. It is assumed that a standard person weighs 70 kg and inhales 20 m^3 air per day. Then the limit value is calculated as:

$$LV = \frac{TDI \times 70}{20} \; \mu g \; m^{-3}$$

15.6 Literature

Barnett, A.G. et al. (2005): *Air Pollution and Child Respiratory Health.* American Journal of Respiratory and Critical Care Medicine, *171,* 1272-1278.

Brunekreef, B. et al. (1997): *Air pollution from truck traffic and lung function in children living near motor-ways.* Epidemiology, *8,* 298-303.

Brunekreef, B. and Holgate, S.T. (2002): *Air pollution and health.* Lancet, *360,* 1233-1242.

Daniels, M.J. et al. (2000): *Estimating particulate matter-mortality dose-response curves and threshold levels: an analysis of daily time-series for the 20 largest US cities.* American Journal of Epidemiology, *152,* 397-406.

Dockery, D.W. et al. (1993): *An association between air pollution and mortality in six U.S. cities.* New England Journal of Medicine, *329,* 1753-9.

Ezzati, M. and Kammen, D.M. (2001): *Indoor air pollution from biomass combustion and acute respiratory infections in Kenya: an exposure-response stud.,* Lancet, *358,* 619-624.

Hoek, G. et al. (2002): *Association between mortality and indicators of traffic-related air pollution in the Netherlands: a cohort study,* Lancet, *360,* 1203-1209.

Lin, S. et al. (2002): *Childhood Asthma Hospitalization and residential Exposure to State Route Traffic.* Environmental Research, Section A, 88, 73-81.

Maitre, A. et al. (2006): *Impact of urban atmospheric pollution on coronary disease.* European Heart Journal, *27,* 2275-2284.

McMichael, A.J. and Smith, K.R. (1999): *Seeking a global perspective on air pollution and health.* Epidemiology, *10,* 1-4.

Pope, C.A. et al. (2002): *Lung cancer, cardiopulmonary mortality and long-term exposure to fine particulate air pollution.* JAMA, *287,* 1132-41.

Pope, C.A. et al. (1995): *Particulate air pollution as a predictor of mortality in prospective study of U.S. adults.* Am J Respir Crit Care Med., *151,* 669-74.

Pope, C.A. and Dockery, D.W. (2006): *Health Effects of Fine Particulate Air Pollution: Lines that Connect.* Journal of the Air & Waste Management Association, *56,* 709-742.

Smith, K.R. et al. (2000): *Indoor pollution in developing countries and acute lower respiratory infections in children.* Thorax, *55,* 518-532.

Valent, F. et al. (2004): *Burden of disease attributable to selected environmental factors and injuries among Europe's children and adolescents.* Environmental burden of Disease Series No. 8, World Health Organization, Geneva.

WHO (World Health Organization) **(2002):** *World Health Report,* Geneva.

WHO (World Health Organization) **(2004):** *Burden of disease attributable to selected environmental factors and injuries among Europe's children and adolescents.* Environmental burden of disease series no. 8.

WHO (World Health Organization) **(2006):** *Air quality guidelines, global update 2005,* World Health Organization, Regional Office of Europe, Copenhagen.

Wilhelm, M. and Ritz, B. (2002): *Residential Proximity to Traffic and Adverse Birth Outcomes in Los Angeles County, California, 1994-1996.* Environmental Health Perspectives, *111,* 207-216.

16 Materials damage

Jes Fenger

Changes in temperature, frost erosion, wind and precipitation as well as fungal attacks and other natural phenomena will always degrade artefacts down to basic materials. But in a polluted atmosphere the rate may become unacceptable high. As for many other environmental problems it is not a new phenomenon, but as stated in Chapter 2 it has been known for millennia. Material degradation is on the other hand a disappearing problem in the industrialised world. Partly because the cities have become less polluted, especially for the most important pollutant SO_2, partly because modern materials are more resistant against attacks and are easier to repair or substitute (Figure 16.1). Finally, many artefacts are constructed with an intended

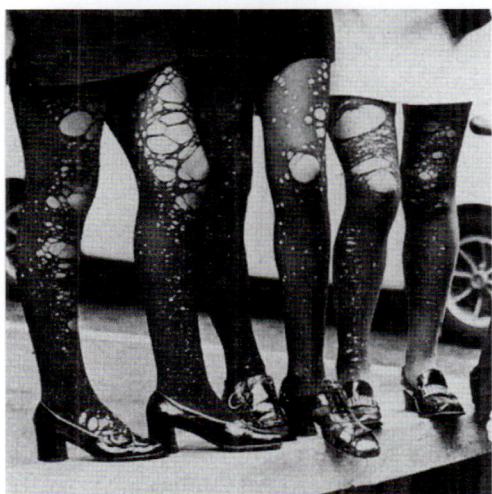

Figure 16.1 Nylon tights damaged by acidity in air. A sight from 1970 not seen anymore. Partly because sulphur dioxide is a disappearing pollutant, partly because more resistant materials have been developed.

lifetime shorter than determined by air pollution. For these reasons materials damage is not taken directly into account in the establishment of standards and limit values related to air pollution.

An important exception, however, is the cultural heritage – e.g. statues or building ornaments or objects in museums that differ from common materials in various ways that make them more vulnerable. The material composition can be more primitive and not adapted to a modern polluted atmosphere, and the original processing or a later influence from wind and weather can have produced cracks in the surface that facilitate further attack. The value of the artefacts can also be in the form of a thin processing or decoration on the surface layer and not in the mechanical strength of the basic material. Unfortunately, even the best restoration cannot prevent the loss of value and originality.

16.1 Pollutants

The crucial compounds are the major air pollutants especially sulphur dioxide, nitrogen dioxide, particles and ozone. Further hydrocarbons as such can have impacts. Attack by sulphur dioxide is easily detected by the formation of sulphates, whereas attack by nitrogen dioxide is less obvious, possibly because nitrates are more soluble and thus more easily washed away. There is, however, evidence of a synergistic effect with sulphur dioxide (Arroyave, Morcillo, 1995).

Particles act mainly by soiling, which has two types of indirect impact: They keep surfaces dirty and thus easier humid – especially the ammonia salts are important, and they call for cleaning, which in itself can be harmful.

Ozone (and other oxidising compounds) mainly attacks organic material. Other hazardous air pollutants are generally present in concentrations too low to have any significant impact.

16.2 Targets

Stones

There are numerous examples of degradation of stone decorations on important historic buildings: Taj Mahal in India, Acropolis in Athens, castles in Germany and many less known buildings in all larger cities.

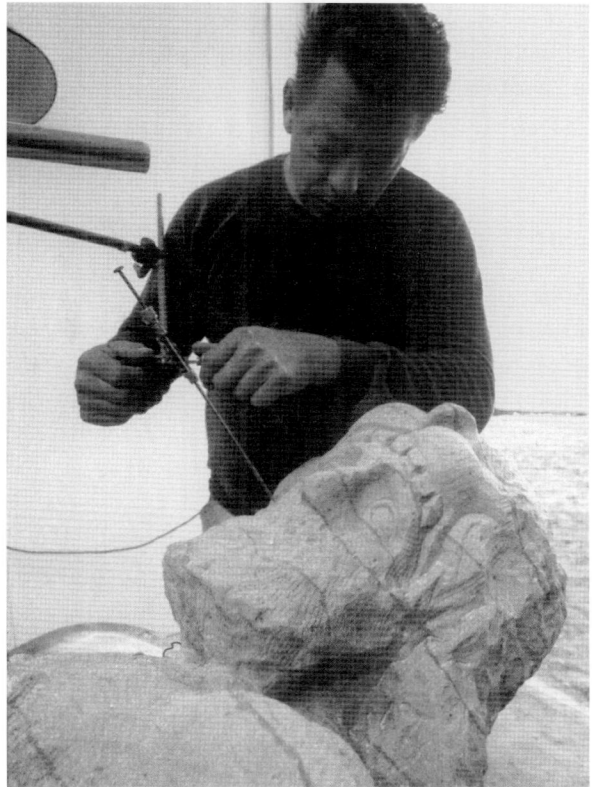

Figure 16.2 Re-creation of decorations on the "Marble Bridge" at Christiansborg in Copenhagen.

Earlier a combination of sulphur dioxide and particles was the most important agent. Particles soil the surface and keep it humid and sulphur dioxide is oxidised to sulphite and later to sulphuric acid that attacks the surface:

$$SO_2 + H_2O \rightarrow H^+ + HSO_3^- \rightarrow 2H^+ + SO_4^{2-} \tag{16.1}$$

Decorations on buildings are the most sensitive as they are often made of soft sandstone. This is more or less porous, and crystallisation of salts with varying numbers of water molecules deeper in the material cause the most serious destruction. Materials degradation can continue for a long time after the exposure to pollution has stopped.

In the last couple of centuries many decorations have been completely destroyed. They are now in many cases being re-created in a more resistant material (Figure 16.2). Sometimes the original is kept in a museum and substituted by a copy outdoors.

Metals

Atmospheric corrosion of metals (Figure 16.3) is electrochemical and can be separated in an anode and a cathode process:

$$Me \rightarrow Me^{n+} + ne^- \tag{16.2}$$

$$O_2 + 2H_2O + 4e^- \rightarrow 4OH^- \tag{16.3}$$

This requires that the surface of the metal is humid. Completely clean surfaces will only be humid at relative humidity of 100%, but under normal conditions, where a surface is more or less soiled, the relative humidity where corrosion becomes significant is 50-90%. The attack will normally start with e.g. sulphur dioxide being absorbed in the water film, where it is oxidised to sulphate:

$$SO_2 + O_2 + 2e^- \rightarrow SO_4^{-2} \tag{16.4}$$

The electrons come from (16.2), and the net result for iron is thus:

$$Fe + SO_2 + O_2 \rightarrow FeSO_4 \tag{16.5}$$

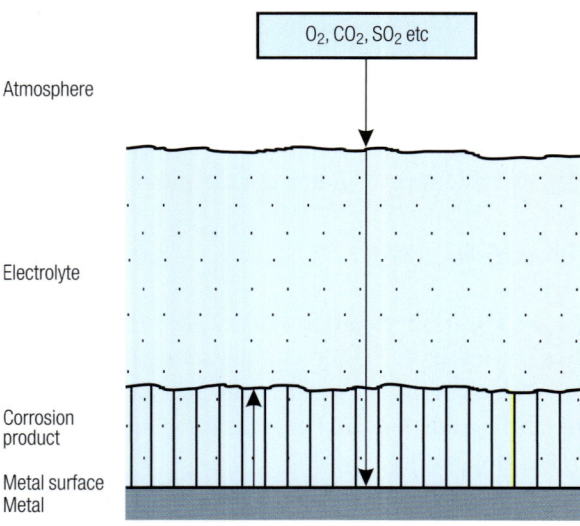

Figure 16.3 Corrosion on a metal surface takes place in a thin electrolyte film, where gases can be dissolved.

Ferrous sulphate can be oxidised to ferric sulphate and transformed to rust (Fe_2O_3). Sulphate is thus released and can attack more metallic iron. A single molecule of sulphur dioxide can therefore transform many iron atoms, before it is washed away or is removed by other means.

For other metals, e.g. copper, more complicated compounds are formed. In a relatively clean atmosphere copper roofs will in some years be covered with a layer of green basic copper carbonate that protects against further attack. The bright green patina that is observed on urban copper roofs, on the other hand, consists mainly of basic copper sulphates and copper oxides and is less protective. For statues of bronze the impact is twofold: The surface as such can be attacked and the original details are lost, but the attack can also be uneven (and unattractive) due to uneven exposure to acid rain (Figure 16.4).

Figure 16.4 Statue by Utzon Frank discoloured by acid rain.

Textiles

Textiles are a complicated group of materials, both with respect to chemical constitution and treatment and to application. The evaluation of impacts is therefore very uncertain. Sulphur dioxide attacks the chemical bonds in both cellulose (cotton wool and other plant fibres) and polyamides (nylon) with resulting loss in strength. Some processes only appear to be important under sunlight and humidity; essentially outdoors. Nitrogen oxides hardly influence the strength, but can result in changes in colour. As a strong oxidant ozone can attack many fibrous materials. Earlier damage to everyday clothes is unimportant today, as the pollution levels are lower, and clothes are normally discarded long before air pollution impacts become important.

Organic materials

Degradation of other organic materials is normally attributed to ozone. Most important are reactions with polymers containing double bonds. Here chain scissoring can give a reduction in molecular lengths and a loss of tensile strength or a cross-linking can give loss of elasticity. Some synthetic compounds (polybutadien and polyisoprene) are very sensitive whereas other with saturated structures or with content of chlorine are more resistant.

Notably rubber in e.g. car tires has been attacked and degradation of rubber has been used as a measure for ozone. The impact, however, may be difficult to distinguish from direct sunlight damage. The importance of the impact is reduced with the appearance of more resistant materials. Rubber is thus more resistant with addition of aromatic amines or phenols.

Paint

Painting of surfaces serves two purposes: Protection of the material and aesthetic improvement. Therefore not only degradation, but also discolouring will be perceived as a loss. Cement and chalk based paints are most sensitive to attack by sulphur dioxide forming gypsum (calcium sulphate). This weakens both the structure and the adherence to the underlying layer. Also the mechanical strength of oil- and alkyd paints can be affected because the hardening- or drying process can be prolonged by sulphur dioxide in the air. A series of colour- and filling compounds can be affected, but the problem is often solved by use of other (more expensive) substitutes.

16.3 Indoor pollution

Indoor pollution can arise both from indoor sources such as evaporation of solvents or use of open fire and from outdoor sources, where the pollution seeps in through windows, cracks etc. Typically the indoor levels of pollution from outdoor sources are of the order of half that of the outdoor levels. In modern airtight built houses it may be somewhat less. Indoors the pollution is absorbed on carpets, furniture, various types of decoration etc., potentially degrading the materials. Nowadays these materials are often changed for other reason than degradation by air pollution.

These effects can, however, be more important in museums, where books, leather-furniture and other items of organic material have suffered by degradation of the protein structure. Textiles do not significantly change appearance after degradation with air pollution, but are weakened. This becomes apparent e.g. if historical costumes are cleaned after being soiled by particles.

As many museums have now filtered ambient air the impacts are declining.

16.4 Economic evaluation

It is difficult and controversial to evaluate environmental impacts in monetary terms, but for materials damage it is at least possible to measure the damage on utilities and relate it to pollution levels and abundance of materials. On one hand, however, clothes can be discarded long before they are worn out, or buildings can be torn down due to traffic regulations – this leads to an overestimation. On the other hand, impacts on e.g. electronic equipment can as direct damage be modest, but it may lead to serious loss of function time or even disasters – thus underestimating the economic loss. A special problem is the damage on cultural values, where a calculation in terms of restoration costs does not take the loss of originality into account.

Nevertheless, early estimates have shown that a reduction of material damage was a significant argument for reducing pollution levels. Thus it was calculated (Cowell, Apsimon, 1996) that the reduction in costs due to damage to buildings in Europe after implementation of the second sulphur protocol was about 9,500 million US dollars per year. A similar calculation made to day would at least for West European Countries give a much less significant value.

A more detailed treatment has been given by Tidblad and Kucera (1998).

16.5 Literature

Cowell, D., Apsimon, H. (1996): *Estimating the cost of damage to buildings by acidifying atmospheric pollution in Europe.* Atmos. Environ., *30*, 2959-2968.

Arroyave, C., Morcillo, M. (1995): *The effect of nitrogen oxides in atmospheric corrosion of metals.* Corrosion Science, *37*, 293-305.

Tidblad, J., Kucera, V. (1998): *Materials Damage.* pp.343-361 in Fenger, J. et al. (eds.) Urban air Pollution – European Aspects. Kluwer Academic Publishers. Dordrecht.

17 Reduction of visibility

Jes Fenger

It is aesthetically pleasing, when the air is clear, so we can see far away, and it can also be practical e.g. for fast-going traffic. A reduction in visibility is therefore generally inconvenient and can in extreme cases be catastrophic.

When we see an object in our surroundings the object emits or scatters light that hits sensitive cells in the retina of our eyes and releases nerve impulses there. These impulses are processed in the brain to pictures. The eyes can to a large extent adapt to the luminosity. How we perceive an object of a given size is therefore not so much the luminosity as such, but more its contrast to the background. This contrast is normally defined as the ratio between light intensity from the object and the background. It is for various reasons reduced when light passes through the air, and the reduction increases with the degree of pollution.

Our visual perception therefore depends upon what happens to light in the air. Light can be absorbed and will therefore disappear with respect to visual impact. It can be scattered but still be registered, but as coming from a different direction than that from which it was emitted. Both processes can be caused by particles and gases. Since the processes that reduce the visibility depend upon the wavelength of the light, they will also change the apparent colour of the objects.

The human eye can normally perceive contrasts down to 0.02 – 0.04. The largest distance where an object can be perceived is called the *sight*.

17.1 Impacts of gases

Light absorption by gases in the troposphere is in practice only due to nitrogen dioxide, which preferably absorbs the shorter, blue wavelengths and thus makes the red radiation relatively more pronounced. Atmospheres containing larger amount of nitrogen dioxide therefore appear reddish-brown.

Scattering by gaseous molecules is dominant in atmospheres that are relatively free of aerosols and light absorbing gases. Since gas molecules have a typical size of about 0.0005 μm, and visible light has the much longer wavelength of about 0.5 μm, all molecules are so-called Rayleigh scatters, where the scattered light is sent out in all directions. The scattering cross section of a molecule is inversely proportional to the fourth power of the wavelength. Therefore gas molecules scatter short-wave (blue) light better than long-wave (red) light. This explains why the Sun appears red in the morning and afternoon, when the sunlight has to pass trough a thick air mass, and appears white at noon, when the light path in the atmosphere is short. It is the blue light that is scattered from the direct beam, and scattered again by other gas molecules that makes the sky look blue.

The scattering reduces the contrast of an observed object, but in pure air, where the scattering is only due to oxygen and nitrogen the visibility can be of the order of 300 km. When the air is polluted by nitrogen dioxide the absorption can reduce the visibility to below 100 km at a concentration of about 100-500 μg m^{-3}. This may be of significance in some US national parks, while in Europe, where the visibility is normally reduced by topology, and the nitrogen dioxide concentrations are much lower, this impact can largely be ignored.

17.2 Impacts of aerosols

Suspended particles (aerosols) are the most important factor in visibility reduction with a relative contribution of about 80%. Black soot can absorb light, but generally the particle absorption is of minor importance, and the particle scattering dominates. The effect depends upon the material and the size of the particles, particles with a diameter of 0.1 – 1 μm being the most efficient.

A dust concentration of 60 μg m^{-3} may give a visibility in dry air of about 40 km. 100 μg m^{-3} (a fairly high value) may reduce it to 20 km. Thus particle reduction of visibility is not a problem as such in modern cities, but as the particles act as condensation nuclei for water vapour, the humidity is also important. In practice the air is normally rather humid (in Denmark the average relative humidity is e.g. 83%) and the visibility can be much more reduced. Fig. 17.1 shows the impact in an example from Vienna on a day with stagnant air. It appears that the visibility is reduced by a factor of 4 in the morning and the evening. This may partly be due to increased traffic in the rush hours, but clearly humidity plays a role.

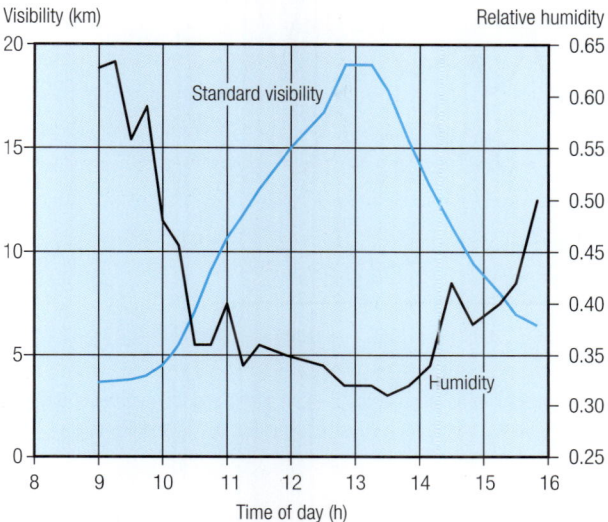

Figure 17.1 Variation of visibility and humidity in Vienna on a day with stagnant air. Both show a variation by a factor of 4. (Horwarth, 1998)

17.3 Visibility in cities

The total contribution to the optical attenuation at a wavelength of 550 nm in a typical European urban atmosphere can be distributed as:

Air 6%, particles 88%, NO_2 5%, SO_2 ~ 0% and O_3 1%. Particles are thus absolutely dominating, although their impact may vary with many factors, and a range in visibility of 2 – 60 km is easily possible.

The influence on visibility from the improvement in the urban air quality by the end of last century is clearly demonstrated in Figure 17.2 from Tokyo. The particulate pollution was reduced by a factor of nearly 10 and the visibility increased correspondingly.

In megacities in developing countries particulate pollution is still a major problem, both for visibility and health.

Although the London "Pea Soup" type of visibility reduction to a few meters is thus a thing of the past in the industrialised countries, particle

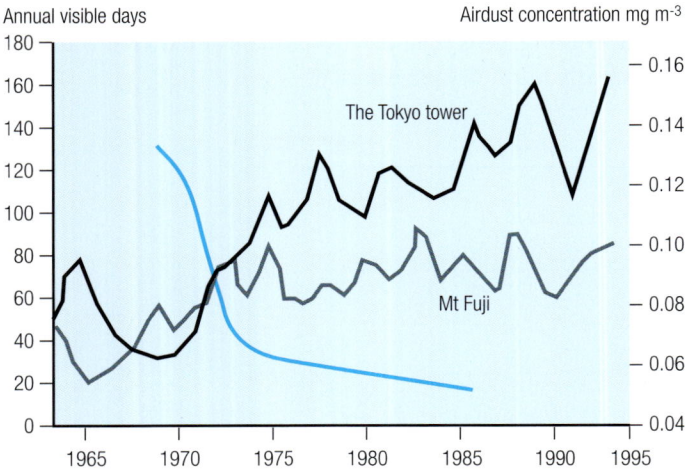

Annual visible days   Airdust concentration mg m^{-3}

Figure 17.2 The concentration of particulate pollution in Tokyo (right scale) and the number of days when the Tokyo Tower and Mount Fuji were visible from a suburban site (Kurashige and Miyashita, 1998).

induced visibility reduction can still be a nuisance. Especially in Southern Europe photochemical air pollution may be important. Today, however, normal fog is much more important.

A more detailed treatment has been given by Horwarth (1998).

17.4 Literature

Horwarth, H. (1998): *Reduction of visibility.* pp. 373-384 in Fenger, J. et al. Urban Air Pollution – European Aspect. Kluwer Academic Publishers, Dordrecht.

Kurashige, Y. and Miyashita, A. (1998): How many days can Mt. Fuji and The Tokyo Tower be seen from a Tokyo suburban area? Journal of the Air and Waste Management Association, *48*, 763-5.

18 Biological effects

Ib Johnsen and Helge Ro-Poulsen

This chapter deals in biological impacts and effects on a regional scale, as they have occurred in the past and occur at present. Focus is on ecosystem effects, with a few comments on human health aspects. Most adverse air pollution effects on ecosystems are due to long-range transport and subsequent deposition of atmospheric pollutants. The nature of air pollution permits this long-range transport and therefore ecosystem effects are nearly always at the regional scale – some even global.

18.1 Impacts and compounds

Direct and indirect effects

The direct effects on vegetation are usually due to uptake of air pollutants via the leaf stomata (Figure 18.1). The indirect effects are based on changes in general living conditions for the organisms in the ecosystems.

Distinction between direct and indirect effects is important in understanding ecosystem reactions to air pollution. In the following, two main examples are given to illustrate this.

Firstly, direct and indirect effects in connection with the term *acidification* shall be discussed. Two primary pollutants are responsible for acidifica-

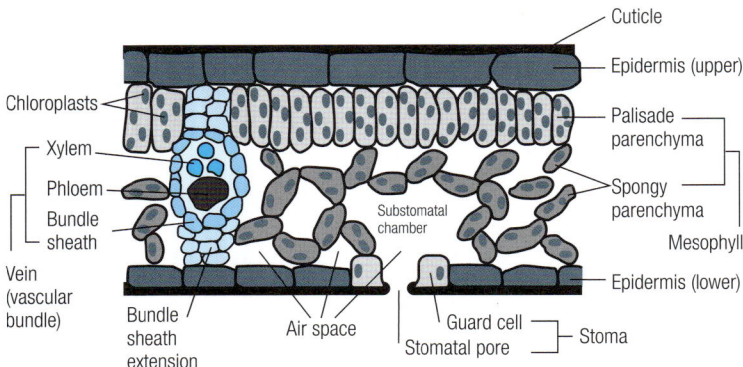

Figure 18.1 Cross section of a typical dicotyledonous leaf. The exchange of gases and very small particles take place through the stomata pores. The epidermis effectively prevents uptake (and loss) of water and minerals.

tion of the environment at present: SO$_2$ and NO. The first may cause severe direct plant damage and has been the main causal agent in elimination of epiphytic lichens (Figure 18.2) from urban and industrial areas, mainly in the past (in Europe into the 1980s). NO, however, has no direct phytotoxic properties. Both gases may react with water before and after oxidation to SO$_3$ and NO$_2$, respectively, forming strong acids in the receptor cells – and thus contributing directly to acidification. At the ecosystem level, however, the

Figure 18.2 Two species of terrestrial lichens: *Cladonia macilenta* and *C. diversa*, with red apothecia.

indirect acidification due to formation of strong acids in the precipitation is of main concern. These strong acids are sulphuric and nitric acid and are reaction products of SO_2 and NO. The supply of strong acids to ecosystems may indirectly result in increased uptake e.g. of heavy metals – causing toxic effects – or leaching of nutrients, resulting in mineral deficiencies.

Secondly, *tropospheric photochemical air pollution* has a direct and indirect dimension. Again, the primary pollutant is NO, being oxidised to NO_2. These nitrogen oxides are directly rather harmless to plants. NO_2, however, is precursor for O_3 formation, which is extremely biotoxic, and this photochemical reaction product constitutes thus a potential indirect effect of nitrogen oxide emissions.

Distinction between direct and indirect effects is also relevant concerning greenhouse gas emissions and climate change. Greenhouse gases may – with the exception of CO_2, see below – not interfere directly with ecosystem performance, but indirectly by changing the climatic conditions for life.

The individual compounds

The lower atmosphere (the troposphere) has been altered significantly by man, and the following compounds are the main contributors to biological effects. In the pre-industrial atmosphere, these compounds played only a minor role in biological processes:

- Sulphur compounds: sulphur dioxide: SO_2/sulphur trioxide: SO_3/sulphuric acid: H_2SO_4 (l)
- Nitrogen compounds: nitrogen monoxide: NO/nitrogen dioxide: NO_2/ nitric acid: HNO_3 (g,l), ammonia: NH_3/ammonium: NH_4^+
- Photochemical oxidants: ozone: O_3/peroxyacetyl nitrate: PAN = $CH_3C(O)O_2NO_2$)
- Metallic compounds such as heavy metals (e.g. Pb, Cd, Hg, V and Ni) often defined as the metals with a specific gravity greater than 5 g cm^{-3}, some 70 elements. Most of them are, however, only rarely considered as pollutants. The heavy metals of environmental concern form insoluble salts with e.g. carbonates, sulphides and phosphates and form very stable complexes with many functional groups (as ligands) present in the living cells. This is the reason for the toxicity of heavy metals. It also means that most heavy metals are not easily available for the ecosystems
- Carbon dioxide: CO_2, steadily increasing (1% per year) due to human emissions (2007: 385 ppmv)

18.2 Acidification

Precipitation has always had pH lower than 7 due to the CO_2 from natural sources in the atmosphere. In modern times, however, pH in rain has dropped as low as about 4,2 due to the human emissions of SO_2 and NO/NO_2, which are oxidised to the highly water soluble strong acids, sulphuric acid (H_2SO_4) and nitric acid (HNO_3). Sulphuric acid is strongly hygroscopic, and only nitric acid occurs as a gas in the atmosphere.

Even the primary gaseous pollutants SO_2 and NO/NO_2 may cause injury to plants and animals, and in fact acidify plant and animal tissue through their solution in and reaction with water. SO_2 is very harmful to plants, but in rural areas less important for animals and humans. Some of the most sensitive plants are lichens, which are small symbiotic double organisms of fungi and (typically) green algae, which are so called photobionts responsible for lichen assimilation of CO_2. The lichens are evergreen, like some higher plants such as coniferous trees. It has been shown that among the higher plants, the evergreen tree species are particularly sensitive, as they are biologically active when exposed to the highest levels of SO_2, i.e. during winter. SO_2, as one of the main precursors for acid rain, is well capable of causing regional plant injury, resulting in e.g. forest die-back. Normally, the direct effect of SO_2 is restricted geographically to urban and industrial areas and their immediate surroundings, as SO_2 is easily soluble in water and react with other air pollutants, like NH_3.

NO and NO_2, unlike SO_2, are not very harmful to plants, but of much higher concern to humans. As especially NO_2, like SO_2 reacts with water forming strong acids, NO/NO_2 contributes to pH decrease in cell tissues after uptake.

Besides acidification, the reason for the harmful effects of these gases is their reducing powers. Especially SO_2 is a strong reducing agent, being oxidised to sulphate. This process interferes with the many important electron transfer processes in the cells.

Both SO_2 (in small amounts – best when transformed to sulphate) and NO/NO_2 may also have some beneficial effects on vegetation, as they supply the plants through stomata with sulphur and nitrogen, which are easily incorporated into biologically active organic compounds; indeed, sulphur and nitrogen are often required by the plants to maintain high productivity levels – especially in crops.

Acid rain

The main regional impact of nitrogen and sulphur oxides is, however, their contribution to acidification of rain. Most particles also contribute to lowering of pH, but the gas ammonia counteracts acidification, being a strong base.

This neutralising effect of ammonia is restricted to the close vicinity of the outlet, since ammonia reacts rapidly with water and/or SO_2. In both cases, ammonium ions are formed, with properties of a weak acid. Furthermore, once deposited into the soil, ammonium ions are normally being oxidised to nitrate (nitrification; *Nitrosomonas* and *Nitrobacter* bacteria responsible under aerobic conditions), according to this process:

$$NH_4^+ + 2O_2 \rightarrow NO_3^- + 2H^+ + H_2O \qquad (18.1)$$

i.e.: one equivalent of ammonium produces two equivalents of hydrogen ions. This process causes acidification of soils receiving ammonium, in particular agricultural soils (through fertilizers, synthetic or organic); farmers have to neutralize this reaction in the soil by regular supply of chalk to their fields.

The ecological problem with acid rain is that water with low pH is a very strong solvent and leaches minerals from plant surfaces and soils. Furthermore, low pH in soil water may release large amounts of toxic cations, and especially the dissolution of aluminium ions has been thought to have inflicted adverse effects on root growth in particular in conifers. The solubility of aluminium increases substantially when soil pH is lowered from around 6 to 4, which happened in many central European soils in the second half of the 20th century. The toxic effect is well known, but no clear evidence for the significance of the 'Aluminium toxicity' hypothesis was found at ecosystem level. There is, however, no doubt that the increasing solubility with lower pH of not only aluminium cations but also iron (II) and manganese (II) ions, have had some negative impact on root growth. Indirectly, the release of these cations may also have interfered with root uptake of phosphate, as they all form insoluble phosphates; this mechanism is very critical in phosphate poor soils, which were widely common in forests throughout Europe.

Depletion of mineral nutrients.

The latter aspect – phosphate deficiency – leads naturally to the main adverse ecosystem effect of acid rain: Depletion of mineral nutrients. By rain water contact with leaves and bark, there is some loss of nutrients in throughfall and stem-flow, but generally this effect is insignificant. By far the most important loss takes place in the soil, due to two mechanisms, both requiring a surplus of water leaching through the soil profile.

The first mechanism is the action of hydrogen ions to release cations, electronically attached to negative sites on humus and/or clay soil colloids, into the soil solution by an ion exchange process, and the solution of mineral salts, like e.g. carbonates, sulphates and phosphates, into this same soil solution. The vertical flow of water through the soil profile results in transportation of these minerals away and eventually out of reach from the ecosystem. In most of Northern Europe the percolation of water through the soil is of the magnitude 100 mm per year out of app. 700 mm precipitation.

The second mechanism is also related to the net downwards flow of the acidified (hydrogen ion enriched) water through the soil. The anions dissolved in the percolating water are always accompanied by the exact equivalent amount of cations. Since the hydrogen ion, due to its small size (ionic radius), is stronger attached to the above mentioned negatively charged colloidal sites, the metal cations there are being replaced by hydrogen ions in the soil solution. Thus the ratio hydrogen ions/metal cations decreases with soil depth, leaving the top soil with falling amounts of minerals needed by the ecosystem.

The result of all this is ecosystem malfunction, mainly seen in reduced productivity and stress resistance (stress due to e.g. parasitic attacks, extreme weather etc.). The soils most susceptible to mineral depletion by acid rain are brown earth soils developed on relatively course basic parent material. Large parts of NW Europe have such soils. On the other hand, ecosystems with adaptation to acid conditions do not suffer from acid rain. Such systems are raised bogs (dominants *Sphagnum* spp.; pH in bog surface water may be as low as 3) and heaths (with *Calluna vulgaris* and *Cladonia* spp.) on so called podsol soils with a well developed mor layer (sour humus).

Long-range transport

Particles suspended in the air are normally carrying a surface film of acid, when the relative humidity is normal to high. Some of the particles are formed by the reaction between ammonia and SO_2 forming small particles of $(NH_4)_2SO_4$. Long-range transport of acids to remote areas is mainly by particulate matter. The effect of acidification may therefore be observed far away from the source.

Fresh water systems

Fresh water systems are affected by acidification and nutrient poor lakes with low pH buffering capacity are affected. Lakes with clear water and low mineral content are characterised with a benthic flora of rosette plants like *Lobelia dortmanna* (Figure 18.3), *Isoëtes* spp. and *Litorella uniflora*.

The vegetation of such lakes changes and becomes gradually dominated by *Sphagnum* spp. and algae. In Denmark, this development is seen in Central and Western Jutland, where these Lobelia-lakes occur. Also more mineral

Figure 18.3 Flowering *Lobelia dortmanna* with leaf rosettes (not visible) below the water surface.

rich lakes dominated by *Potamogeton* spp. are threatened by acidification, and have practically disappeared in Denmark – but probably mostly due to eutrophication (see Section 18.3).

The sea

The sea is so far not affected by acidification being a well-buffered system (the $CO_2/HCO_3^-/CO_3^{2-}$ buffer system) with a pH around 8. Sea ecosystems, such as coral reefs, may however be affected by the increasing CO_2 concentration as the organisms may not be able to aggregate chalk which is the hard constituent of the reefs.

18.3 Eutrophication

By eutrophication is understood excessive supply of nutrients to ecosystems causing them to change, often irreversibly. The nutrients responsible for these changes are the same as used in intensive agriculture to maximize crop production: Nitrogen-compounds, phosphate and potassium (NPK). The effects of eutrophication on natural ecosystems are reduced biodiversity and change in species composition due to increased dominance of few, highly productive species able to utilize the ample supply of nutrients. In the case of fresh water systems and even the open sea disastrous incidents due to oxygen deficiency may occur. The effects of eutrophication are widespread and seen regularly throughout Europe and North America, unfortunately still at an increasing rate. The two main reasons for eutrophication are intensive agricultural production and the high energy consumption based on fossil fuels in the traffic, industrial and common household sectors.

Threats to natural ecosystems

In most of Europe and Eastern North America, eutrophication may be regarded as the most serious threat at present to natural ecosystems. Many high priority nature types are under change as a consequence of the atmospheric deposition of elevated amounts of mainly nitrogen and phosphorous. These nature types include ombrotrophic mires (= raised bogs, built of *Sphagnum* spp.), heaths (Figure 18.4) and dry grasslands, which are all given special attention in the NATURA 2000 network of nature types of highest importance within the European Union.

Figure 18.4 Coastal heath near Bjålum, Western Jutland, sensitive to eutrophication.

Fresh water systems, ground water and coastal sea regions suffer from eutrophication as well. The fauna (fish and invertebrates) in streams and rivers is, however, generally only little affected by eutrophication, since the critical levels of oxygen dissolved in water mainly is determined by the content of easily decomposable organic material, e.g. from sewage treatment plants. The flora along the stream and riverbanks may, however, be affected by eutrophication, and dominance e.g. of the plants *Epilobium hirsutum*, *Urtica dioica* and *Phragmites australis* reveals the increased supply of mainly nitrate in the running water.

Ground water may receive increased amounts of especially nitrate, when the water leaching through the soil profile is becoming enriched above the reducing capacity of the soil. Finally, coastal sea regions suffer regularly from oxygen depletion often causing catastrophic events to the fauna. The oxygen depletion is a result of high production of plankton, when sinking to the bottom and there being decomposed utilizing most of the oxygen.

The atmospheric contribution to the eutrophication is significant. In large parts of NW Europe the average present atmospheric N-deposition is about $20 \, kg \, N \, ha^{-1} \, y^{-1}$. Close to point sources, such as pig farms and other industries, the N-deposition reaches much higher levels, and values over $150 \, kg \, N \, ha^{-1} \, y^{-1}$ has been recorded. This is much higher than the natural background deposition believed to have been about and probably less than $1 \, kg \, N \, ha^{-1} \, y^{-1}$. Mainly natural ecosystems suffer severely, but even agricultural systems (receiving about $180 \, kg \, N \, ha^{-1} \, y^{-1}$ in fertilizer) and coastal waters are believed to be affected by atmospheric N-deposition.

The regional effects of eutrophication are due to long-range transport of especially nitrogen compounds in the atmosphere. More local effects on natural ecosystems vis-à-vis intensive farms, highways and certain industries are also frequent, and the introduction of buffer zones in order to minimize these is an administrative tool under development, as the understanding of the range and impact on the neighbouring ecosystems becomes operational. The buffer zone between cultivated land and natural ecosystems, with respect to protection from ammonia emissions from agricultural production, is at present set at app. 250 m.

Nitrogen versus phosphorus

In recent years the focus has shifted somewhat from nitrogen deposition to the phosphate contribution. The reason is that nitrogen deposition has been so high for many years that many ecosystems are more or less saturated with available nitrogen; nitrogen is no longer the limiting factor for growth. The fundamental Liebig's Law of limiting factors states: Growth is controlled not by the total of resources available, but by the scarcest resource. Phosphate, being next in line as the scarcest resource, is thus promoted to be the trigger of substantial biological effects of the general eutrophication. This is true for terrestrial, fresh water and marine systems.

The atmospheric transportation of phosphate is based on burning of coal and straw as well as suspension of phosphate containing particles into the air by strong winds. Usually dry and light top soil is the source of these particles being blown across land covering several hundreds of kilometres. This phenomenon often occurs on cultivated agricultural land prepared for spring crop production during dry early summer months, April and May in particular. The total deposition of phosphate is estimated to 0.02-0.04 kg P ha^{-1} y^{-1}. Regional impacts of P-deposition are seen in a number of terrestrial and fresh water ecosystems causing the vegetation to shift composition from specialists to opportunists.

It is paramount to deal with phosphate in the environment and new initiatives in that direction have been taken by the EU commission in connection with VFD-directive (2000/60/EC).

18.4 Photochemical air pollution

The term photochemical air pollution covers in this context the formation of highly oxidising substances in the troposphere (the lower 12 km of the atmosphere at 50° northern latitude). The main controversial biological agents formed are ozone and PAN (Peroxy Acyl Nitrates; PAN is also used as a specific abbreviation for peroxy acetyl nitrate). The biological effects are due to the extreme oxidative capacities of these compounds. Ozone is taken up by plants through their stomata (Figure 18.5) and known to destroy cell

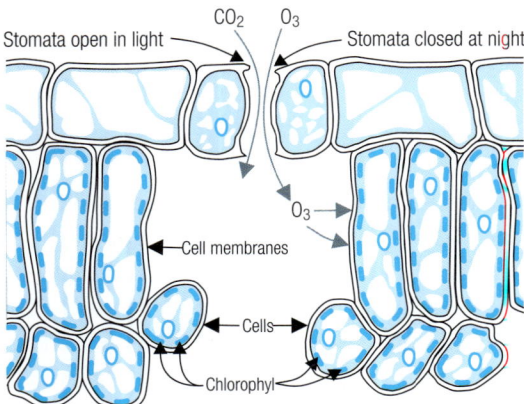

Figure 18.5 Penetration of ozone through the stomata resulting in damage on the chlorophyll carrying cells.

membranes by reaction with C=C double bonds in the cell wall components. The result is destruction of tissue around the stomata cavity (the opening just inside the stomata) seen as whitish flecks on the leaves. The number and size of these is a function of ozone dose (Figure 18.6. See further in Section 18.4).

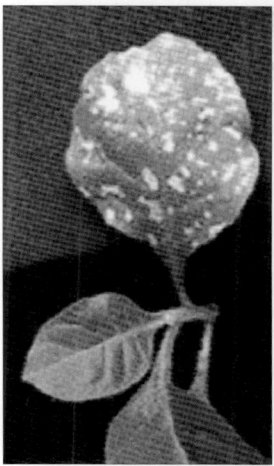

Figure 18.6. Typical damage (white spots) on a leaf of the tobacco variant Bel W3, the most sensitive higher plant towards ozone.

During summertime episodes of high ozone immission occur, and it has been documented by OTC (= Open Top Chamber) studies that there is substantial crop loss consequences from these episodes (Figure 18.7). National Crop Loss Assessment Network (NCLAN) studies in the USA have revealed heavy economic consequences of ozone exposure, in forests (especially *Pinus ponderosa* in the eastern USA) as well as in field crops, dicotyledons (cotton, soybean, peanuts) being most sensitive. In Europe estimates of annual crop losses are in the order of magnitude of several 100 million €.

Furthermore, forests may also be affected. Some conifers in particular have been shown to be sensitive to ozone. This was observed early in the Western USA (*Pinus ponderosa*). Also in Europe natural ecosystems may be affected. Some studies indicate that there are sensitivity differences between selected wild plants belonging to *Fabaceae*: *Lotus corniculatus* and *Medicago lupulina*. This may result in vegetation changes in natural ecosystems.

Figure 18.7 Open Top Chamber. Air pollutants are supplied at the bottom of the chamber at known flux rates to provide the desired average concentration around the experimental plants in the chamber. At the same time the plants receive approximately the natural precipitation and sunlight.

18.5 Metallic pollutants

The toxicity of metals to organisms is due to their important role in bio-chemical processes in all cells. A number of enzymes activated by metal ions and metallo-enzymes are known. The first mentioned group comprises iron, cobalt, chromium, vanadium and selenium. Copper, zinc, iron, cobalt and molybdenum are able with a stronger bond to form metallo-enzymes, metallo-proteins, metallo-porphyrins and metallo-flavins. As pollutants are particularly cadmium, lead, vanadium, nickel and mercury in focus, because of their extremely high toxicity (Table 18.1).

Loss of heavy metals by run-off and drainage from agricultural land varies highly between the metals. Results of a study of the removal of heavy metals by run-off and drainage from a typical cultivated clay soil are shown in Table 18.2.

Heavy metals are bound to clay particles due to their ion exchange capacity and to hydrated metal oxides, such as iron sesquioxides (Fe_2O_3): As, Cr, Mo, P, Se, and V, and manganese sesquioxides (Mn_2O_3): Co, Ba, Ni and lanthanides. Calcium phosphate is furthermore able to bind As, Ba, Cd and Pb in alkaline soil.

Iron	Fe	Essential to chlorophyll-synthesis and several redox enzymes
Beryllium	Be	Non-essential; >0.5 mg l^{-1} toxic to plants; extremely toxic to mammals
Boron	B	Essential; unknown function
Manganese	Mn	Essential to oxygen-dynamics; involved in several enzymes
Zink	Zn	>60-400 mg l^{-1} toxic to plants; mostly beneficial to mammals
Copper	Cu	Essential to respiration and photosynthesis; >0.5-8 mg l^{-1} toxic to plants
Molybdenum	Mo	Essential in nitrate-reductase and nitrogenase
Cobalt	Co	Essential to nitrogen-fixing organisms
Lead	Pb	Non-essential; >3-20 mg l^{-1} toxic to plants; very toxic to man and animals
Cadmium	Cd	Non-essential; >0.2-9 mg l^{-1} toxic to plants; very toxic to man and animals
Mercury	Hg	Non-essential; extremely toxic to all organisms
Vanadium	V	Essential to some organisms; >10-40 mg l^{-1} toxic to plants normally
Nickel	Ni	Essential to all vertebrates and some plants; >0.5-2 mg l^{-1} toxic to plants
Chromium	Cr	Essential to mammals; 0.5-10 mg l^{-1} Cr(VI) toxic to plants
Selenium	Se	Essential to all vertebrates and some plants; >1-2 mg l^{-1} Se(IV) toxic to plants
Arsenic	As	Essential to *Rhodophyceae* (red algae) and mammals; >0.02-7.5 mg l^{-1} toxic to plants. Non-essential, very toxic to humans
Aluminium	Al	Non-essential; >0.1-30 mg l^{-1} toxic to plants

Table 18.1. Metals of biological importance. Beryllium and Boron are not heavy metals. Concentrations refer to aquaculture.

	Removal (mg m^{-2} y^{-1})	Annual removal % of pool
Pb	0.5	1-3
Cu	1.2	2-3
Zn	15.9	30-50
Cd	0.07	15-30

Table 18.2. Difference in mobility for four heavy metals through the soil (M.F. Hovmand, unpublished results 1980).

Fulvic acid (molecular weight about 1000) and humic acid (molecular weight about 150,000) are able to form complexes with a number of heavy metals such as Hg, Cu, Pb and Sn.

The mobility of heavy metals is dependent on a number of factors. The pore water contains soluble organic compounds (acetic acid, citric acid, oxalic acid and other organic acids) partly excreted by the roots. These small organic molecules form chelated, soluble compounds with metal ions such as Al, Fe and Cu. Activity of living organisms in soil may also enhance the mobility of heavy metal ions. Fungus and bacteria may utilise phosphate and thereby release cations. Formation of insoluble metal sulphides under anaerobic conditions from sulphate implies a reduced mobility. The lower oxidation stages of heavy metals are generally more soluble implying increased mobility, than the higher oxidation stages.

The many possibilities for binding heavy metals in soil explain the long residence time. Cadmium, calcium, magnesium and sodium have the most mobile metal ions with a residence time of about 100 years. Mercury has a residence time about 750 years, while copper, lead, nickel, arsenic, selenium and zinc have residence times of more than 2000 years under temperate conditions. Tropic residence times are typically lower for all heavy metals, about 40 years.

The biological effects of the heavy metal pollution occur at two levels: at organisms level and at the higher level, first of all the ecosystem level.

The plant toxicity is very dependent on the presence of other metal ions. For instance Rb and Sr are very toxic to many plants, but the presence of the biochemically more useful K and Ca is able to reduce or eliminate the toxicity. The toxicity of arsenate and selenate can be reduced in the same manner by sulphate and phosphate.

Bioavailability

Formation of metal complexes by reaction with organic ligands reduces the toxicity due to reduced bioavailability. The plant toxicity of heavy metals in soil is consequently also understood as concentration of heavy metal ions. The heavy metals that are most toxic to plants are silver, beryllium, copper, mercury, tin, cobalt, nickel, lead and chromium. With exception of silver and chromium, the divalent form is the most toxic. For silver it is Ag+ and for chromium it is chromate and dichromate that are most toxic. Silver and mercury ions are very toxic to fungus spores, and copper and tin ions are very toxic to green algae; lethal concentrations may be as low as 0.002-0.01 mg l^{-1}.

One of the key processes on ecosystem level is the mineralisation processes, because they determine the cycling of nutrients. Heavy metals can inhibit the mineralisation due to blocking of enzymes. The effect is known not only to the enzymes produced in the organisms but also for extra-cellular enzymes – exo-enzymes – originating from dead cells or excreted from roots and living micro-organisms. As the various processes forming the cycling of nutrients are coupled, the entire mineralisation cycle is disturbed if only one process is reduced. It is therefore possible to determine the change of the mineralisation cycle by measuring the respiration, the transformation of nitrogen and the release of phosphorous. A concentration of copper as low as 3-4 times the background concentration may lead to a reduced soil respiration. A few hundreds of mg copper per kg soil is furthermore able to diminish the nitrogen release rate by one half.

The most sensitive mineralisation process is the phosphorous cycling. Biological material binds phosphorous as esters of phosphoric acid. The phosphate is released by hydrolysis of the ester bond, a process catalysed by phosphatase. This process is inhibited by the presence of heavy metals.

The inhibition is decreasing in the following sequence:
Molybdate (VI) > wolframate (VI) > vanadate (V) > nickel (II) > cadmium > mercury (II) > copper (II) > chromate (VI) > arsenate (V) > lead (II) > chromium (III).

The inhibition of exo-enzymes by heavy metals does not form a clear pattern. It is therefore difficult to generalise. Most experiments give, however, a clear picture of the influence of heavy metals on the mineralisation of soil constituents: the rate of mineralisation may be reduced significantly with a consequent reduction of the productivity of the entire ecosystem.

Atmospheric deposition

The atmospheric deposition of metals typically causes in Northern Europe an average annual increase of the total content of heavy metal in soil between 0 and 0.6%, but varies greatly from location to location. The general trend is that metal deposition in the developed world is diminishing in time due to abatement measures in many countries.

In accordance with the many possibilities for side reactions of heavy metals in soil, including adsorption to the soil particles, the amount of heavy metals ions that is available to plants is only a fraction of the total content. If only the bio-available heavy metal is used as basis, the annual percentage increase in the concentration due to atmospheric deposition is probably higher.

Most lead in soil is not mobile and cannot be transported via the root system to the leaves and stems. This is in contrast to cadmium that is very mobile. About 50% of the cadmium in soil will be found in the plants after the growth season, although the concentration may be very different in different parts of the plants. The cadmium in grains has for instance not increased in proportion to increased atmospheric deposition of cadmium.

The heavy metal pollution of soil has been and is one of the major challenges in environmental management in industrialised countries. Due to the many diffuse sources of heavy metal pollution, the solution of the problem requires a wide spectrum of methods, but first of all application of cleaner technology. In other words, it is necessary to reduce the total emission of heavy metals. Dilution (for instance higher chimneys) is not an applicable solution. Moreover, as pollution, particularly air pollution has no borders, it is necessary continuously to take international initiatives and agree on international standards particularly for the most problematic heavy metals, i.e., cadmium, mercury, nickel, chromium and vanadium. Initiatives generally include these points:

- A national and international environmental strategy is accepted
- Agreed international standards and long term goals
- Monitoring programmes to assess the pollution levels and compare the measured concentrations with standards

18.6 Impacts of CO_2

Carbon dioxide is mainly recognized as a greenhouse gas, but is, however, the basic carbon source in primary production by green plants and elevated levels of CO_2 affect the rate and volume of photosynthesis. The rate of photosynthesis in plants is governed by mainly five general factors: The genetic set-up of the plant physiology; the availability of water; the availability of light; the availability of essential nutrients; the atmospheric concentration of CO_2. The physiology related to photosynthesis is very complex and results from evolutionary processes starting at the very beginning of life on Earth. The present situation reflects a large number of adaptations to the multitude of habitats – and that is nearly everywhere – populated by photosynthetic organisms (= autrotrophs; that means capable of using CO_2 as carbon source). Some plants may utilise increasing CO_2 levels more efficient than others due to their way to capture CO_2 in the initial photosynthetic process. They are called C4-plants and are rare in Europe. Species with C4 photosynthesis possess a competitive advantage to the usual C3 plants under dry, hot and N+C enriched conditions. Two widespread species having C4 photosynthesis are maize and sugar cane. Several species of tall, tropical grasses have C4 photosynthesis as well. Water is a main limiting factor for growth and drought may inhibit photosynthesis completely, with the exception of plants adapted to desert like conditions.

Availability of sufficient light is essential, but outside forests plant growth is generally not limited by lack of light. Low nutrient supplies may be limiting for plant growth. Macronutrients (N,P,K) and micronutrients (mainly some of the heavy metals) are necessary to complete the life cycles of plants (and animals). In most cultivated systems addition of nutrients is needed to compensate for losses by crop removal and to the surroundings. In natural ecosystems, however, sophisticated recycling and containment take place to secure ecosystem functioning.

Finally, atmospheric CO_2 is one of the important limiting factors. If the above four factor complexes are optimized, increasing CO_2 levels shall increase plant productivity (Fig. 18.8). This fact is applied in greenhouse cultivation, where burning of propane enhances the CO_2 immission in the green house resulting in higher plant productivity.

Net primary production

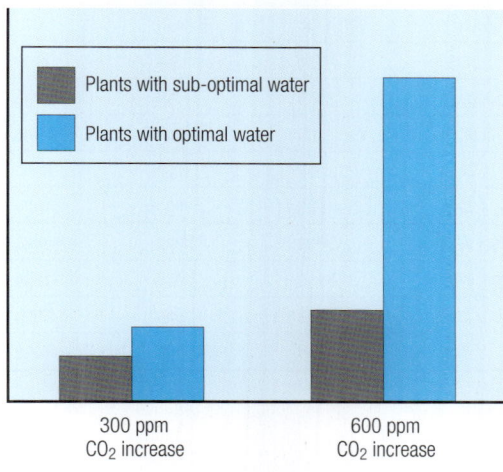

Figure 18.8 Relative plant growth with increasing CO_2 levels.

In conclusion, elevated CO_2 levels may increase plant productivity and thus the formation of organic matter – which in turn may result in accumulation of minerals in organic top soil and succession to ecosystems with generally higher productivity and based on fewer opportunistic species. This trend is very much like the one observed due to eutrophication alone.

18.7 Combined effects on ecosystems

Ecosystems are always subject to a multitude of simultaneously acting environmental factors, including air pollutants. It is therefore important to understand the interaction of the various individual components of the humanly altered atmospheric chemistry. With respect to biological effects, this interaction may be antagonistic, additive or synergistic. It is often very difficult to predict combination effects, and our present knowledge is based on observations at the ecosystem level of changes induced by industrial emissions or experimental exposures (Table 18.3).

	Effect on ecosystems	Effect on plants	Contribution to acidification	Contribution to eutrophication	Contribution to photochemical air pollution
CO_2	High	High	Low	Medium	None
SO_2	High	Very high	High	Medium	None
SO_3	High	High	High	Low	None
H_2SO_4	High	Medium	Very high	Medium	None
NO	Low	Low	None	Low	High
NO_2	Low	Medium	High	High	High
HNO_3	Medium	Medium	Very high	High	None
NH_3	Low	High	Low	High	None
NH_4^+	Medium	Low	High	High	None
O_3	High	Very high	None	None	Very high
PAN	Medium	Very high	None	None	High
Heavy metals	High	Low	None	Low	None
Pesticides	High	High	None	Medium	None

Table 18.3 Ranking of the individual air pollutants according to their biotic effect potential, and their contribution to three important pollution scenarios.

It is evident from Table 18.3 that the nitrogen containing compounds, and especially the nitrogen oxides, are of particular concern. The abatement of air pollution in the troposphere should mainly aim at these air pollutants.

Agriculture

With respect to the pollutants causing acidification, only SO_2 is of concern. This pollutant has practically been removed from the air in most developed countries by effective abatement measures since the 1970s. It is therefore necessary at present to add sulphate to fertilizer. Acidification is a problem in itself, but is dealt with by ample and regular application of lime to the fields – mainly, however, to neutralize the acidifying effect of the nitrification process stimulated by addition of fertilizer. The acid precipitation plays only a minor role in this picture. With respect to nitrogen oxides, they may even be beneficial as nitrogen supply through stomata. But the photo-chemically polluted atmosphere with ozone and PAN is strongly phytotoxic and may easily override the beneficial effect above. It is believed, that significant crop losses occur every year in Europe and USA due to ozone and PAN episodes during summer. Heavy metals may become more mobile in the acidified soils, and thus causing higher levels in crops like cereals. Especially the levels of Cd in agricultural produce in Northern Europe is alarming.

Forests

Forests are perennial, complex systems. Long-lived ecosystems are among the most vulnerable to air pollution. In contrast to agriculture, it is not customary to add lime to the forest floor to avoid acidification. Instead, tree species are selected for production according to their capacity to grow at the given, actual conditions.

In a few cases hot spots with extreme heavy metal deposition may result in inhibition of mineralisation leaving the litter and its mineral content on the soil surface unavailable for the forest ecosystem, thus decreasing primary production.

Normally, a high degree of natural acidification takes place in the forest top soil. Forests are also accumulating ecosystems, and the supply of nitrogen compounds from the atmosphere may lead to nitrogen saturation resulting in deficiency of other elements, causing abnormal growth or even forest die-back. The most worrying development in growth conditions for forests over the recent decades is the combination of high levels of oxidizing compounds like ozone every summer in combination with rapid climatic changes and extreme weather.

During the last half of the 20th century there were many observations of forest death (*"Forest die-back"*) in Central Europe. The cause was initially quite clearly SO_2 emissions, but the rapid decrease at the end of the 20th century in the emissions of this pollutant did not stop the forest die-back; furthermore, similar observations began to occur in the USA. Today, the occurrence of deficiency of minerals necessary for growth in the heavily N- and P-enriched environment in Europe and USA is believed to be responsible for the bad shape of forests – conifers in particular. In Western USA and Southern Europe ozone may be the main cause of forest die-back, but even at higher latitudes in Europe adverse effects on forests due to photochemical air pollution are seen. The impact appears to be disappearing alongside with the reduction in sulphur dioxide emissions.

Non forest terrestrial ecosystems (dry grasslands, heaths and wetlands)

The main threat from air pollution to these ecosystems is eutrophication by N-, P- and K-compounds supplied mainly from the neighbouring intensively farmed agricultural landscape. In Europe the critical loads for nitrogen has been exceeded for several decades already; thus the limiting factor now is phosphate, which is carried via the atmosphere in particulate matter such as top soil dust. Together with present lack of extensive cultivation of these semi-natural ecosystems, air pollution has become the main threat to the survival of these high biodiversity systems despite that most of them are protected internationally through EU directives implemented in laws of the member countries.

Lakes

Most lakes in heavily populated areas (mainly urban areas) are strongly eutrophied due to former outlet of household waste water before establishment of efficient waste water treatments plants. Again, the combination of atmospheric deposition of N- and P- compounds together with acidification causes changes – often irreversible – to growth conditions in *Lobelia* and *Potamogeton* lakes. Furthermore, the massive losses of nitrogen and phosphorous compounds from agriculture result in destruction of these clean water systems – the original lake types of NW Europe.

18.8 Critical values

The critical load and level concept was introduced to identify deposition goals to be achieved in order to protect natural ecosystems from adverse effects. Critical load is a quantitative estimate of an exposure to one or more pollutants below which significant harmful effects on specified sensitive elements of the environment do not occur according to present knowledge. It is now more than 20 years since the critical loads concept was developed. The concept has been a very successful tool for the development of the effect-based cost-effective regional air pollution strategies in Europe. In Table 18.4 is given the critical N-loads for sensitive natural and semi-natural ecosystems.

Critical load means the maximum deposition of a pollutant in e.g. kg ha^{-1}yr^{-1} that is tolerated by the ecosystem in question. If the critical load is exceeded, the ecosystem changes its composition and dynamics, often irreversibly. For the sensitive natural ecosystems it is paramount to reduce deposition below the critical loads, if their existence shall prevail.

	Ecosystem	Critical Load
Wetlands	Ombrotrophic mires (*Sphagnum* spp.)	10 kg N ha^{-1} y^{-1}
	Extreme poor fens	15 kg N ha^{-1} y^{-1}
Lakes	With benthic flora of *Lobelia dortmanna*, *Isoëtes* spp., *Litorella uniflora*	5-10 kg N ha^{-1} y^{-1}
	With benthic flora of *Potamogeton* spp.	5-10 kg N ha^{-1} y^{-1}
Terrestrial	Heaths, high cover of *Cladonia* spp.	12-15 kg N ha^{-1} y^{-1}
	Dry grasslands	15-20 kg N ha^{-1} y^{-1}
	Coniferous forests	20-30 kg N ha^{-1} y^{-1}

Table 18.4 Critical loads for selected sensitive ecosystems. (Mainly from UK NFC 2003).

Critical level, however, means the immission e.g. in µg m^{-3} of a certain pollutant above which injuries to organisms or ecosystems occur. These levels are often applied when defining air pollution standards. Obviously, a time factor needs to be included here, as the systems and organisms react differently to short or long term exposure to high levels of air pollutants. Therefore, the critical level concept is comparable to the approach used in air quality standards, where normally two values are given: The 50% and 90% percentiles meaning that all concentration levels should be lower than these values in respectively 50% or 90% of the actual measurements over time.

18.9 Biological monitoring

The term biological monitoring (sometimes also called biological indication) refers to the use of organisms to estimate environmental quality, i.e. the contents of pollutants in the environment, by observation and analysis of effects on these organisms. Effects include changes in organism performance as well as in tissue levels of pollutants. Measurements of air pollution by physicochemical methods yields no direct information about adverse biological effects.

Living organisms, dead biological tissue with special properties as well as ecosystems may be used as biological indicators and monitors of air pollutants. The purpose of biological monitoring is to observe and measure environmental changes in biological systems, and to interpret these changes in order to predict future developments and prevent injuries and damages to living organisms, including man.

Often biological monitoring is accompanied by – or coordinated with – physico-chemical measurements as well as supplementary ecological and physiological studies, in order to reveal causal relationships and understand mechanisms behind the responses.

Biological monitoring takes place at different levels:

- Measurement of chemical, biochemical and/or physiological changes in organisms
- Measurement of population changes, such as changes in distribution and occurrence of indicator species
- Measurement of ecosystem changes

Some pollution scenarios cause changes observable at several levels. Low concentrations and doses of some pollutants may cause biochemical disturbances, not immediately changing the performance of the whole organism. With increasing levels observable injuries may occur, eventually causing death of the organism. Deaths of individual organisms do only interfere with population dynamics, if the reproductive capacity of the population is reduced to such an extent, that the losses no longer are replaced.

The selection of biological parameters to study in a monitoring program must be careful and in accordance with the monitoring purpose. If the need is to map the air pollution for sulfur dioxide, one needs to use the indicators (in this case lichens) having the largest sensitivity range towards sulfur dioxide. It is mandatory to understand the relation between exposure and effect on the monitor organism in question.

The presence and performance of a plant (and mostly plants are used for biological monitoring of air pollution) is determined by a number of environmental conditions, abiotic as well as biotic. Indicator plants are selected due to 1) their known specific reaction and thus sensitivity to exposure to one or a few air pollutants, 2) their known specific adaptation to growth conditions, which may be altered by air pollution and 3) their known strong ability to accumulate pollutants.

The following categories (with examples) of monitor plants may be identified:

- Species sensitive to even very small concentrations and doses of one pollutant. Example: Tobacco (*Nicotiana tabacum* var Bel W3), being extremely sensitive to ozone
- Groups of plant species with high sensitivity combined with large interspecific differences in tolerance to increased exposure to a certain pollut-

ant. Example: Epiphytic lichens (lichens growing on trees) and SO_2, used to map levels of sulfur dioxide in cities and around industries

- Species that accumulate certain pollutants in their tissue, normally without any impact on their growth and performance. Such species are so to speak biological filters, and they reflect e.g. the deposition level of heavy metals and pesticides; the plants must be analysed in the laboratory after field collection of *in situ* plants or transplants. Examples are lichens and bryophytes, in particular *Sphagnum* spp., which accumulate heavy metals efficiently by ion exchange, *Achillea millefolium* which absorb particulate matter from the air (SPM) efficiently by impaction and *Gladiolus gandavensis* ssp, which accumulate fluoride in the leaf tips
- Species which are frequent in areas with strong air pollution load. E.g. one lichen and one bryophyte specie in Europe are rather tolerant to sulfur dioxide and thus occur in many city centers. They may be used to map deposition levels of e.g. heavy metals in these cities. The species in question are: The lichen *Lecanora conizaeoides* and the moss *Dicranoweisia cirrata*

Monitor or indicator organisms and systems may provide important information about general environmental problems, such as eutrophication, acidification and photochemical air pollution. Other organisms may be used in inventories aiming at describing a specific air pollution problem with suspected adverse biological effects. The advantages of using biological monitoring as opposed to physicochemical measurements are:

- It is a reaction of a living organism or tissue from a living organism that is observed
- Biological monitoring integrates the environmental impact over a shorter or longer time span
- The biological effect of air pollutants is registered directly
- Provides information of pollutant cycling in food webs and bioaccumulation
- Measures the combined impact of more pollutants, thus revealing synergistic or antagonistic effects, otherwise impossible to identify
- Biological monitoring is based on simple and cheap technology, and the methods may often be quickly applied

The disadvantages, however, are:

- Suitable organisms may be absent in the study area. This may be overcome by use of transplantation of monitor organisms from a background area to the study area, usually retaining their original substrate

- Information about the degree of specificity towards individual air pollutants may be limited. Usually an organism reacts to several air pollutants in different ways while being indifferent to others

Measurements of environmental change are easier at the species level rather than at the ecosystem level. The need for simplification and precision regarding monitoring information at the community (ecosystem) level has led to the development of biological indices of air pollution.

Index of Atmospheric Purity

The most well known example is the saprobic system developed for limnic systems (fresh waters), based on macroinvertebrates and their various oxygen demands. In air pollution monitoring the "Index of Atmospheric Purity" (lAP) has been developed. The index is based on the frequency of epiphytic indicator lichens and the species co-occurring with these (LeBlanc and DeSloover, 1970).

Index of Atmospheric Purity is defined as:

$$IAP = \sum_{1}^{n} \frac{1}{10} Q_i f_i$$

where n = the number of species at the site, Q_i = ecological index of toxiphoby of the ith species expressed as the average number of species found with it, f_i = frequency-coverage on a scale of 1 to 5. The divider 10 is reflecting the frequent number of species examined. Larger values = cleaner air.

A high IAP value is obtained, when the epiphytic lichen community is well developed and diverse, as it would normally be in unpolluted areas, and low values reflect some degree of disturbance of the community. The index is not directly transferable from region to region, mainly because of geographic differences in species distribution.

Most biological indices are based on biodiversity calculations including records of species occurrence as well as importance value (frequency, cover etc.). For some ecosystems, it is a sign of instability, that the species diversity increases (ombrotrophic mires; coastal heaths) when the few, but extremely well adapted species characteristic for these sensitive ecosystems are being replaced partly by opportunistic and more widely distributed species. This is e.g. seen as a typical outcome of eutrophication of natural low productivity natural ecosystems.

18.10 Literature

Anonymous (2003): *Status of UK critical loads, Critical load methods, data and maps*, UK National Focal Centre, CEH Monks Wood, in collaboration with a range of UK experts, Version 3 (May 2003).

Anonymous (2000): Directive 2000/60/EC of The European Parliament and of The Council of 23 October 2000 establishing a framework for Community action in the field of water policy

Bell, J.N.B. and Treshow, M. (2002): *Air Pollution and Plant Life.* 2nd ed. Cinchester:John, Wiley & Sons, Inc.

Heagle, A.S. (1989): *Ozone and crop yield.* Annual Review of Phytopathology, *27*, 397-423.

Heck, W.W., et al. (1984): *Assessing Impacts of Ozone on Agricultural Crops: I. Overview.* J. Air Pollut. Control Assoc., *34*, 729-735.

Heck, W.W., et al. (1986): Effects on Vegetation: Native, Crops, Forests. In: Air Pollution. 3rd Ed. Vol. VI. Supplement to Air Pollutants, Their Transformation, Transport and Effects. A.C. Stern, ed., pp. 248-333. Academic Press, New York.

Heggestad, H.E. and Middleton, J.T. (1959): *Ozone in high concentrations as cause of tobacco leaf injury.* Science, *129*, 208-210.

Hogsett, W.E., et al. (1997): *An approach to characterizing tropospheric ozone risk to forests.* Environmental Management, *21,* 105-120.

Idso, S.B. (1995): CO_2 *and the biosphere: The incredible legacy of the industrial revolution* (technical bulletin). University of Minnesota.

Jacobsen, J.S. and Hill, A. C. (eds.) (1970):.*Recognition of air pollution injury to vegetation: a pictorial atlas.* Air Pollution Control Assoc. Pittsburgh, PA.

Johnsen, I., and Søchting, U. (1973): *Influence of air pollution on the epiphytic lichen vegetation and bark properties of deciduous trees in the Copenhagen area,* Oikos *24*, 344-351.

LeBlanc, F. and DeSloover, J. (1970): *Relation between industrialization and the distribution and growth of epiphytic lichens and mosses in Montreal,* Can. J. Botany, *48,* 1485–1496.

Middleton, J.T., et al. (1950): *Injury to herbaceous plants by smog or air pollution.* Plant Disease Reports, *34*, 245-252.

Richards, B.L., et al. (1958): *Air pollution with relation to agronomic crops. V. Oxidant stipple of grapes.* Agronomy Journal, *50*, 559-561.

U.S. Environmental Protection Agency (1996): *Air Quality Criteria for Ozone and Related Photochemical Oxidants,* EPA/600-90/004bF, National Center for Environmental Assessment, Office of Research and Development, Research Triangle Park, NC.

19 Depletion of the ozone layer

Ole John Nielsen

Approximately 90% of the ozone in the Earth's atmosphere resides in the stratosphere where it forms the so-called "ozone layer". It is, as already stated, a bit misleading to state that the ozone *is* in the stratosphere. The ozone actually *forms* the stratosphere by absorbing solar radiation, heating up the air and thus creating the inversion that is the prerequisite for the stratosphere. The division between the stratosphere and the troposphere, the tropopause, acts as a stabilising agent on the weather and a barrier against upwards transport of pollution, but the most important direct feature of ozone is that it shields life on the Earth's surface from harmful ultraviolet solar radiation (figure 19.1).

Figure 19.1 UV-B radiation (280-315 nm) is partly absorbed by the ozone layer. Human exposure to UV-B increases the risk of skin cancer and a suppressed immune system. UV-B exposure can also damage terrestrial plant life, single-cell organisms and aquatic ecosystems. (After WMO, 2003).

19.1 Ozone depletion

Depletion of the stratospheric ozone layer permits penetration of additional damaging UV-B radiation (280-320 nm) to the Earth's surface. This will cause increase in human skin cancer and suppression of the human immune system. There can also be serious effects on a large number of other biological species.

It is important to distinguish between stratospheric ozone and tropospheric ozone. In the stratosphere ozone is "good" because it filters the damaging UV radiation from the sun. In the troposphere ozone is "bad" because it has harmful effects on humans, animals and plants.

CFC compounds and their impact

Chlorofluorocarbons (CFC) are compounds containing only carbon, fluorine and chlorine. They were synthesised in the 1920s under the trade name "Freon" and were unknown in nature. Extensive research led to many different CFC compounds with great technological importance.

CFC and related compounds are numbered as follows: The first digit from the left is the number of C atoms minus one. The next digit is the number of H atoms plus one. The final digit is the number of F atoms.

CFC-12 (CCl_2F_2) and HCFC-22 ($CHClF_2$) are still important refrigerants. CFC-12 replaced undesired "dangerous" compounds like NH_3 and SO_2 as refrigerants. CFC-11 is the key in foaming in so-called open-celled plastics. A mix of CFC-11 and CFC-12 was the primary choice as aerosol propellant. CFC-113 (CCl_2FCClF_2) was used to clean electronics because of its ability to dissolve grease deposited on small components.

The use of CFC-11 and CFC-12 grew rapidly (10% annually for three decades!) in the US after World War II. Almost all the uses of CFCs lead to eventual release to the atmosphere. It was the creative thinking and work of two outstanding individual scientists, which led to the recognition of the adverse effect of the release of CFCs into the atmosphere.

In the early 1970s James E. Lovelock invented the very sensitive electron capture detector for gas chromatography and used it to discover that most of the released CFCs were still present in the atmosphere. F. Sherwood Rowland asked the obvious question: "What is going to happen to the CFCs in the atmosphere?" The answer to this question was published in Nature in 1974 (Molina, 1974). CFCs are completely inert in the lower atmosphere and are transported to the stratosphere. Here ultraviolet light photolyse the CFCs and atomic chlorine is released. As seen below chlorine atoms react with stratospheric ozone to form ClO, which reacts to regenerate Cl. This catalytic cycle can remove 10.000-100.000 molecules of ozone per Cl atom.

Although the paper by Rowland and Molina was based on solid physical chemical data, it was heavily disputed. The use of CFCs as aerosol propellants in spray cans was banned only in the US, Canada, Norway and Sweden. The production decreased slightly for a few years and then increased again. Loss of several percent of ozone around 40-kilometre height was reported (WMO, 1985).

Nothing really happened until 1985. However, in May of 1985 a significant drop in atmospheric ozone was reported from the British Antarctic Survey station at Halley bay. The total ozone column had decreased from 320 to 200 DU in the Antarctic spring (October). (1 Dobson Unit (DU) is a 0.01 mm column of O_3 at STP). This finding was published in 1985 (Farman, 1985). It turned out that this had been recorded earlier in satellite data, but the low values had been deleted as "errors". The combination of unique meteorological conditions and chemistry that leads to the so-called Antarctic ozone hole became fully understood in 1987 (Molina et al., 1987).

Chemistry of the unperturbed stratosphere

Ozone is formed in the stratosphere, when radiation of wavelengths less that 242 nm dissociates molecular oxygen:

$$O_2 + h\nu \rightarrow O + O \qquad (\lambda < 242 \text{ nm}) \tag{19.1}$$

The oxygen atoms react with O_2 to form O_3. M is a so-called third body molecule (mediator) that takes up excess kinetic energy. It is usually N_2 or O_2:

$$O + O_2 + M \rightarrow O_3 + M \tag{19.2}$$

This reaction is, for all practical purposes, the only one that produces ozone in the atmosphere. The ozone formed in the stratosphere is removed again in two ways. It can be photolysed back to an oxygen molecule and an oxygen atom:

$$O_3 + h\nu \rightarrow O_2 + O \qquad (\lambda < 320 \text{ nm}) \tag{19.3}$$

Or it can react with oxygen atoms:

$$O_3 + O \rightarrow O_2 + O_2 \tag{19.4}$$

This first photochemical theory of stratospheric ozone, named after the inventor (Chapman, 1930) can yield excited fragments, but collision deactivation to $O(^3P)$ is the exclusive fate of any $O(^1D)$ formed. Therefore it can be useful to think of the sum of O and O_3 as a single species called "odd oxygen" and denoted O_x. O_x is produced in reaction (19.2) and lost in reaction (19.4).

The steady state analysis shows that rate of reaction (19.1) is equal to rate of reaction (19.4) or $j_{O2}[O_2] = k_4[O_3][O] = k_4[O_x]^2$, which gives:

$$[O_x] = (j_{O2}[O_2]/k_4)^{1/2}$$

Ozone is only formed and destroyed under influence of sunlight. Diurnal variations in stratospheric ozone concentrations are therefore expected, but they are found to be small.

A layer-like structure is expected for a species such as O_3 whose concentration depends on the photochemical production rate in a stable atmosphere of varying optical density. Until the mid 1960s the Chapman mechanism was thought to be the principal set of reactions governing ozone formation and destruction in the stratosphere. With more advanced measurements and improved rate data for ozone reactions (Figure 19.2) it became obvious that the Chapman mechanism over-predicted the concentration. The scientific community therefore looked for additional or faster loss processes for O_x.

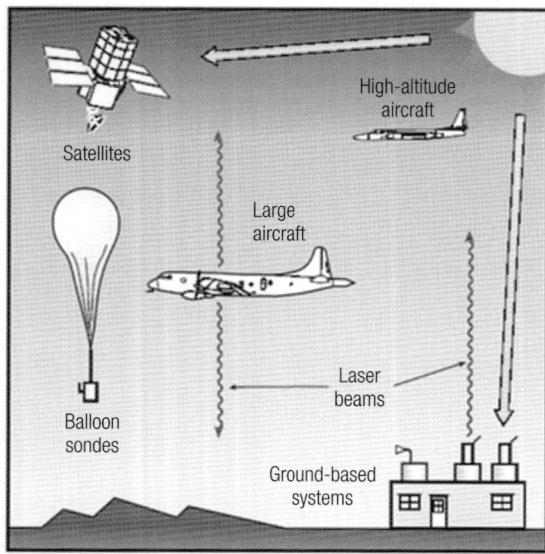

Figure 19.2 Ozone is measured throughout the atmosphere by a variety of instruments both on the ground and platforms moving in the atmosphere (balloons, aircrafts, and satellites). Measurements at many locations around the globe are made weekly to monitor total ozone amounts. (After WMO, 2003).

19.2 HO_x, NO_x and CIO_x cycles

The breakthrough in understanding stratospheric ozone chemistry came in the early 1970s pioneered by the work of Crutzen, Johnston, Storlarski and Cicerone. Finally Rowland and Molina (Molina, 1974) proved the ozone depleting effect of chlorine containing compounds. The catalytic cycle processes that must be added to the Chapman mechanism have the general form:

$$X + O_3 \rightarrow XO + O_2 \tag{19.5}$$

$$XO + O \rightarrow X + O_2 \tag{19.6}$$

Net:

$$O_3 + O \rightarrow O_2 + O_2$$

where X can be free radical catalysts as e.g. H, OH, NO, NO_2, F, Cl, Br or I.

Catalytic atmospheric loss processes were first described by Bates and Nicolet (1950). The idea was that H and OH would originate from photolysis of water vapour high in the atmosphere, the mesosphere. There are so-called HO_x (x = 0, 1, 2) cycles that have been numbered differently in different textbooks:

$$H + O_3 \rightarrow OH + O_2 \tag{19.7}$$

$$OH + O \rightarrow H + O_2 \tag{19.8}$$

Net:

$$O_3 + O \rightarrow O_2 + O_2$$

$$OH + O_3 \rightarrow HO_2 + O_2 \tag{19.9}$$

$$HO_2 + O \rightarrow OH + O_2 \tag{19.10}$$

Net:

$$O_3 + O \rightarrow O_2 + O_2$$

$$OH + O \rightarrow H + O_2 \tag{19.8}$$

$$H + O_2 + M \rightarrow HO_2 + M \tag{19.11}$$

$$HO_2 + O \rightarrow OH + O_2 \tag{19.12}$$

Net:

$$O + O + M \rightarrow O_2 + M$$

$$OH + O_3 \rightarrow HO_2 + O_2 \tag{19.9}$$

$$HO_2 + O_3 \rightarrow OH + O_2 + O_2 \tag{19.13}$$

Net:

$$O_3 + O_3 \rightarrow O_2 + O_2 + O_2$$

All these cycles convert O_x to O_2, but since the concentrations of O, OH and HO_2 vary with height, they are important at different altitudes.

The most important source of NO_x in the stratosphere is the degradation of N_2O where 90% of N_2O is photolysed:

$$N_2O + hv \rightarrow N_2 + O(^1D) \tag{19.14}$$

$$O(^1D) + N_2O \rightarrow 2NO \tag{19.15a}$$

$$O(^1D) + N_2O \rightarrow N_2 + O_2 \tag{19.15b}$$

50% of reaction 9.14 proceeds through channel 15a. As in the case of HO_x there are also NO_x cycles:

$$NO + O_3 \rightarrow NO_2 + O_2 \tag{19.16}$$

$$NO_2 + O \rightarrow NO + O_2 \tag{19.17}$$

Net:

$$O_3 + O \rightarrow O_2 + O_2$$

$$NO + O_3 \rightarrow NO_2 + O_2 \tag{19.16}$$

$$NO_2 + O_3 \rightarrow NO_3 + O_2 \tag{19.18}$$

$$NO_3 + h\nu \rightarrow NO + O_2 \tag{19.19}$$

Net:

$$2O_3 + h\nu \rightarrow 3\,O_2$$

The second NO_x cycle is more important in the lower stratosphere, where the ozone concentration is relatively higher.

$$NO_3 + NO_2 \leftrightarrow N_2O_5 \tag{19.20}$$

Since reaction (20) is not a permanent sink for NO_x, it forms a so-called reservoir species, N_2O_5.

In 1974 the importance of halogens for stratospheric ozone was recognised (Molina 1974). The most important source was chlorofluorocarbons (CFC) that had, because of their great stability, a variety of technological applications. Besides these man-made compounds the only natural source of chlorine to the stratosphere is CH_3Cl. Similar bromine compounds play a somewhat minor role. The primary source gases are shown in Figure 9.3. Having a long atmospheric lifetime, they can be transported to the stratosphere where they can be photolysed by the strong UV-radiation, e.g.:

$$CFCl_3 + h\nu \rightarrow CFCl_2 + Cl \tag{19.21}$$

$$CF_2Cl_2 + h\nu \rightarrow CF_2Cl + Cl \tag{19.22}$$

The chlorine atoms attack ozone and form ClO that can react with oxygen atoms and reform chlorine:

$$Cl + O_3 \rightarrow ClO + O_2 \tag{19.23}$$

$$ClO + O \rightarrow Cl + O_2 \tag{19.24}$$

Net:

$$O_3 + O \rightarrow O_2 + O_2$$

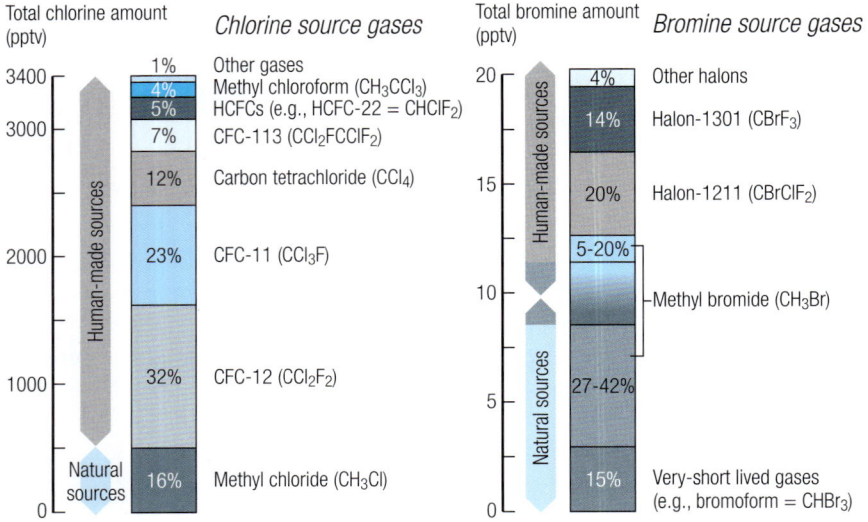

Figure 19.3 Primary sources of chlorine and bromine to the stratosphere. Methyl bromide has both man-made and natural sources. The HCFC, which are substitutes for the CFCs are also regulated under the new versions of the Montreal Protocol. Note the difference in abundance scale for chlorine and bromine. (After WMO, 2003).

The catalytic cycle continues until the catalytic species is removed by other reactions. One chlorine atom can thus remove maybe 100,000 O_3 molecules before it reacts with methane:

$$Cl + CH_4 \rightarrow HCl + CH_3 \tag{19.25}$$

Another example of a sink reaction for other "catalytic" species is:

$$OH + NO_2 + M \rightarrow HNO_3 + M \tag{19.26}$$

N_2O_5 has already been mentioned as a reservoir species. Chlorine nitrate is another important reservoir species:

$$ClO + NO_2 + M \rightarrow ClONO_2 + M \tag{19.27}$$

By photolysis chlorine nitrate can be converted back to active catalyst species:

$$ClONO_2 + h\nu \rightarrow ClO + NO_2 \tag{19.28a}$$

$$ClONO_2 + hv \rightarrow Cl + NO_3 \qquad (19.28b)$$

Reactions 9.27 and 9.28 couple the NO_x and ClO_x cycles. Hence, simultaneous emission of CFCs and N_2O (the source of NO) will have an effect less than the sum of the effects because of the formation of $ClONO_2$.

All the different HO_x, NO_x and ClO_x cycles are coupled and this coupling determines the fate of ozone in the stratosphere. The coupling is complicated and it is hard to get an overview without a computer model.

There is a BrO_x cycle analogous to the ClO_x cycle. Bromine atoms enter the stratosphere via photolysis of the flame-retardants Halon-1301 ($CBrF_3$) and Halon-1211 ($CBrClF_2$). BrO_x and ClO_x also couples:

$$BrO + ClO \rightarrow Br + Cl + O_2 \qquad (19.29a)$$

$$BrO + ClO \rightarrow Br + OClO \qquad (19.29b)$$

$$BrO + ClO \rightarrow BrCl + O_2 \qquad (19.29c)$$

$$Br + O_3 \rightarrow BrO + O_2 \qquad (19.30)$$

$$BrCl + hv \rightarrow Br + Cl \qquad (19.31)$$

$$Cl + O_3 \rightarrow ClO + O_2 \qquad (19.23)$$

Net:

$$O_3 + O_3 \rightarrow 3O_2$$

The bromine reservoirs HBr and $BrONO_2$ have concentrations much smaller than the chlorine analogue and the reaction of Br atoms with CH_4 is very slow. All in all Br-atoms are approximately 40 times less efficient in removing ozone than Cl-atoms.

19.3 The ozone hole

The severe depletion of stratospheric ozone over Antarctica in the polar spring was soon termed the "Ozone Hole" (Farman 1985). It may appear strange that it occurred in a region without sources, but Antarctica has special important weather conditions not found anywhere else on the globe. It is the coldest place, where temperatures below -78 °C allow formation of special ice clouds called Polar Stratospheric Clouds (PSCs). The temperatures in the Arctic are not low enough to allow PSCs to be formed on a regular basis and to the same extent (Figure 9.4). PSCs act as traps for the critical chlorine compounds that are later released in the Antarctic spring.

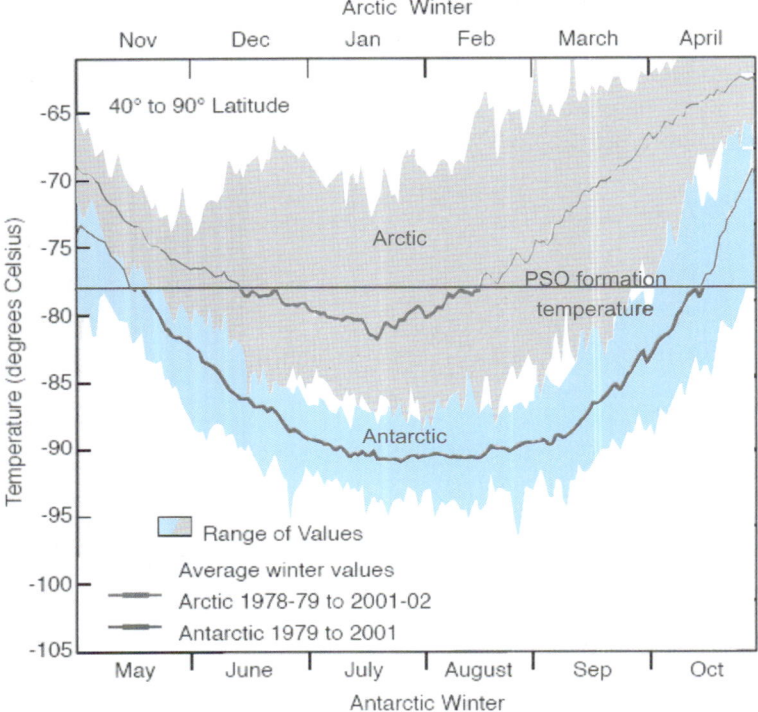

Figure 19.4 Minimum air temperatures in the polar lower stratosphere (After WMO 2003).

PSCs promote activation of chlorine from the absorbed reservoir species. The primary step in this process is the efficient absorption of gaseous HCl on PSCs. Then follows the heterogeneous reaction of $ClONO_2$ (from Eq 27) with HCl on the particle and the photolysis of the liberated Cl_2:

$$ClONO_2(g) + HCl(s) \rightarrow Cl_2 + HNO_3 \tag{19.32}$$

$$Cl_2 + h\nu \rightarrow Cl + Cl \tag{19.33}$$

$$2[Cl + O_3 \rightarrow ClO + O_2] \tag{19.23}$$

$$ClO + NO_2 + M \rightarrow ClONO_2 + M \qquad (19.27)$$

Net:

$$HCl(s) + NO_2 + 2O_3 \rightarrow ClO + HNO_3(s) + 2O_2$$

Another important heterogeneous reaction is:

$$ClONO_2 + H_2O(s) \rightarrow HOCl + HNO_3 \qquad (19.34)$$

HOCl can photolyse to give a Cl atom and OH. Furthermore HOCl can react heterogeneously with HCl:

$$HOCl + HCl(s) \rightarrow Cl_2 + H_2O \qquad (19.35)$$

Yet another heterogeneous reaction is possible:

$$N_2O_5 + HCl(s) \rightarrow ClNO_2 + HNO_3(s) \qquad (19.36)$$

where the formed $ClNO_2$ photolyses rapidly to give active chlorine.

The O atoms needed in reaction 19.4 to complete the cycle are absent in the polar stratosphere. Hence, ClO accumulates until the self-reaction becomes active at large ClO concentrations, and the self-reaction initiates yet a new ozone depleting cycle:

$$ClO + ClO + M \rightarrow ClOOCl + M \qquad (19.37)$$

$$ClOOCl + h\nu \rightarrow Cl + ClOO \quad (\lambda \sim 350 \text{ nm}) \qquad (19.38)$$

$$ClOO + M \rightarrow Cl + O_2 + M \qquad (19.39)$$

$$2[Cl + O_3 \rightarrow ClO + O_2] \qquad (19.23)$$

Net:

$$2O_3 + h\nu \rightarrow 3O_2$$

The above mechanism is the main reason for the destruction that began in the 1980s (Figure 19.5) when the sun rises over Antarctica. An example of the well-known images of the Ozone Hole from space-based measurements is shown in Figure 19.6.

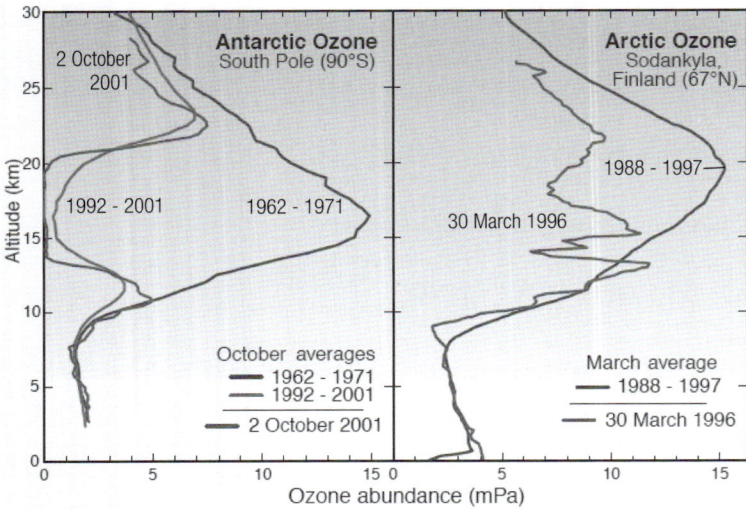

Figure 19.5 In the Antarctic the destruction of ozone began in the 1980s. Observations from balloon measurements are shown in the figure. March Arctic ozone values are often below average values as shown on the right hand side of the figure. (After WMO, 2003).

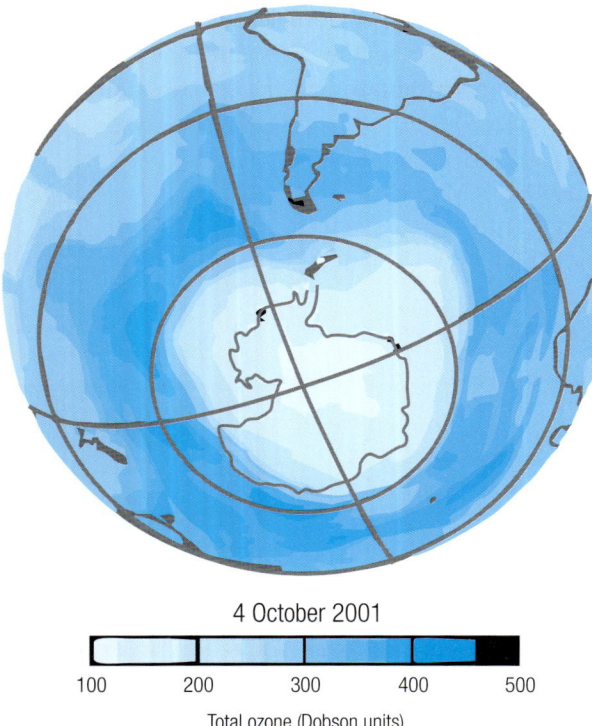

4 October 2001

Total ozone (Dobson units)

Figure 19.6: Total ozone values over Antarctica measured med a satellite instrument. Inside the hole total ozone is 100 Dobson Units (DU) as compared to previous values of 300 DU. (After WMO, 2003).

Stratospheric ozone has decreased globally since the 1980s, and the global depletion is clearly larger than natural variations (Figure 19.7). The depletion varies significantly with latitude, and there is little or no depletion in the tropics.

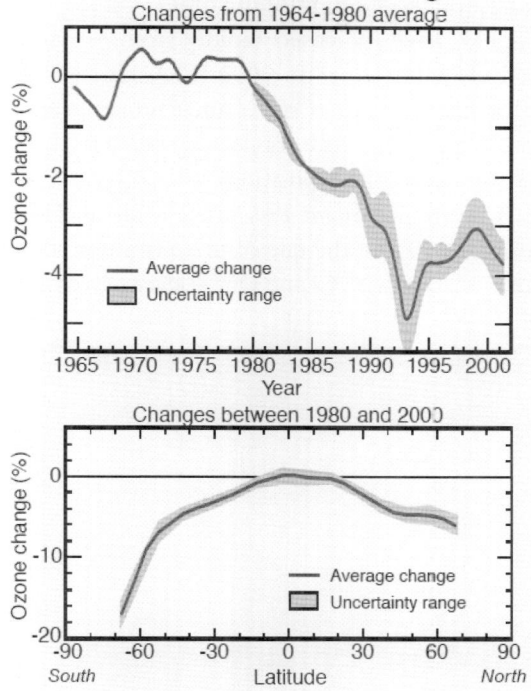

Figure 19.7 Global total ozone changes. The loss in the Southern Hemisphere is larger than in the Northern Hemisphere because of the greater losses every year in the Antarctic. (After WMO, 2003).

19.4 CFC alternatives

Recognition of the adverse impact of chlorofluorocarbons (CFCs) on stratospheric ozone (Molina, 1974) has prompted an international effort to replace CFCs with environmentally acceptable alternatives (WMO 1989). Hydrofluorocarbons (HFCs), hydrochlorofluorocarbons (HCFCs) and hydrofluoroethers (HFEs) are three classes of CFC replacements.

In contrast to CFCs the HFCs, HCFCs and HFEs contain one or more C-H bonds and are therefore susceptible to attack by OH radicals in the troposphere.

As the result of a substantial research effort to define the environmental impact of HCFCs, HFCs, and HFEs there is a good understanding of the atmospheric degradation mechanism of such compounds (Wallington 1994a). From a commercial point of view, the most important HFC is CF_3CFH_2 (HFC-134a), which has found widespread use as a replacement for CFC-12 (CF_2Cl_2) in domestic refrigeration and automobile air conditioning units.

Annual emissions of CF_3CFH_2 have increased rapidly from essentially zero in 1990 to approximately 85,000 tonnes in 2000 (McCulloch 2003). HFEs are used in much smaller quantities than HFCs and find applications as cleaning agents for electronic equipment, heat transfer agents in refrigeration systems, and carrier fluids for lubricant deposition.

Reaction with OH radicals is the dominant loss process for all HFCs, HCFCs and HFEs, accounting for >90% of the fate of these compounds. In the stratosphere photolysis and reaction with Cl and O(^{1}D) atoms give minor

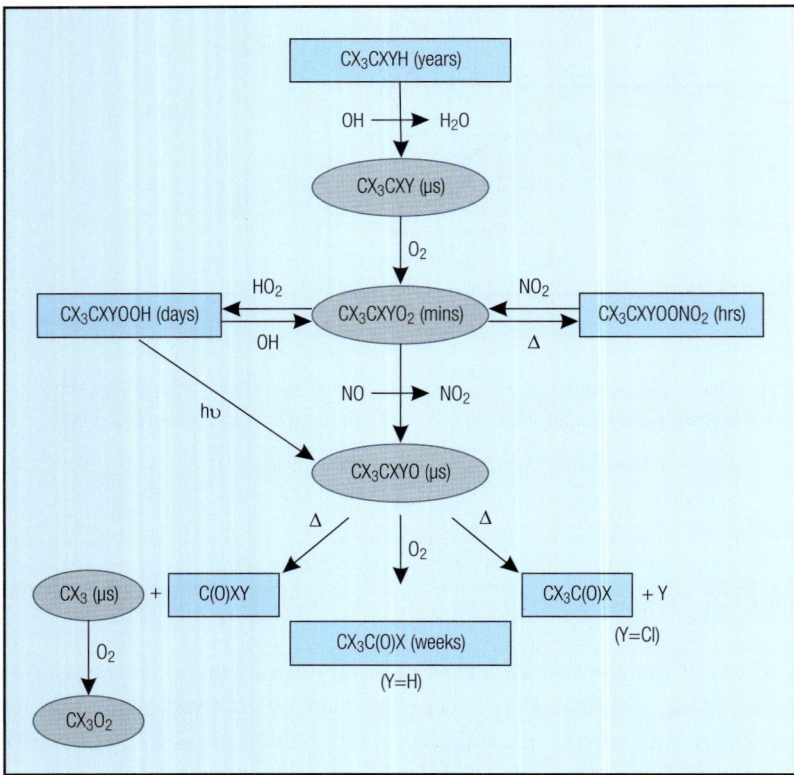

Figure 19.8 Generic scheme for the oxidation of a C_2 halocarbon, X and Y represent Cl and/or F. Values in parentheses are order of magnitude estimates for the lifetimes of the various species. Closed shell species are enclosed in boxes. Radical species are denoted by ellipses.

contributions to the overall loss. A substantial kinetic database exists concerning the reaction of OH radicals with HFCs, HCFCs and HFEs (DeMore 1994). From these data atmospheric lifetimes can be calculated. Lifetimes range from 1 to 270 years. A generic scheme for the atmospheric oxidation of a C_2 haloalkane is given in Figure 19.3.

Reaction with OH radicals gives a halogenated alkyl radical which reacts with O_2 to give the corresponding peroxy radical (RO_2). As discussed in previous sections, peroxy radicals can react with three important trace species in the atmosphere: NO, NO_2, and HO_2 radicals.

Peroxy radicals react rapidly with NO_2 to give alkyl peroxynitrates (RO_2NO_2), by analogy to the measured rate of reaction of CF_2ClO_2 and $CF_3CH_2O_2$ radicals with NO_2. The lifetime of RO_2 radicals with respect to reaction with NO_2 is approximately 10 minutes. Alkyl peroxynitrates are thermally unstable and decompose to regenerate RO_2 radicals and NO_2. At room temperature in an atmosphere of air the peroxynitrates derived from HCFC-22 and HFC-134a have average lifetimes of 24 seconds and < 90 seconds, respectively. Thermal decomposition dominates the atmospheric chemistry of halogenated alkyl peroxynitrates.

The lifetime of CX_3CXYO_2 radicals with respect to reaction with HO_2 has been estimated to be 2-8 minutes. The reaction of peroxy radicals with HO_2 radicals gives hydroperoxides and, in some cases, carbonyl products. Product data are available for two haloperoxy radicals: CH_2FO_2 and CF_3CFHO_2. Reaction of CH_2FO_2 radicals with HO_2 gives 30% yield of the hydroperoxide, CH_2FOOH, and 70% yield of the carbonyl product, HC(O)F (Wallington, 1994b). In the reaction of CF_3CFHO_2 with HO_2 radicals less than 5% of the products appear as the carbonyl $CF_3C(O)$ F and, by inference, >95% of the reaction proceeds to give the hydroperoxide $CF_3CFHOOH$ or the alkoxy radical CF_3CFHO (Maricq, 1994).

The factors, which determine the relative importance of the hydroperoxide and carbonyl forming channels, are unknown. The hydroperoxide $CX_3CXYOOH$ is expected to be returned to the CX_3CXYO_x radical pool via reaction with OH and photolysis. The fate of the carbonyl product $CX_3C(O)$ X produced in the CX_3CXYO_2 + HO_2 reaction is discussed later.

The peroxy radicals derived from HFCs, HCFCs and HFEs react rapidly with NO to give NO_2 and an alkoxy radical RO. The lifetime of peroxy radicals with respect to reaction with NO is approximately 3-7 minutes. Numerous product studies of halocarbon oxidation have shown that the atmospheric fate of the alkoxy radical, CX_3CXYO, is either decomposition or reaction with O_2. Decomposition can occur either by C-C bond fission or Cl-atom elimination. Reaction with O_2 is only possible when an α-H atom is available (e.g. in CF_3CFHO). In the case of the alkoxy radicals derived from HFC-

32, HFC-125, and HCFC-22, only one reaction pathway is available. Hence, CHF_2O radicals react with O_2 to give $C(O)F_2$, CF_3CF_2O radicals decompose to give CF_3 radicals and $C(O)F_2$, and CF_2ClO radicals eliminate a Cl-atom to give $C(O)F_2$.

The alkoxy radicals derived from HFC-143a, HCFC-123, HCFC-124, HCFC-141b and HCFC-142b all have two or more possible fates, but one loss mechanism dominates in the atmosphere. For HCFCs 123 and 124 the dominant process is elimination of a Cl atom to give $CF_3C(O)Cl$ and $CF_3C(O)F$, respectively. For HFC-143a, HCFC-141b, and HCFC-142b reaction with O_2 dominates, giving CF_3CHO, $CFCl_2CHO$, and CF_2ClCHO respectively. The case of HFC-134a is the most complex. Under atmospheric conditions, the alkoxy radical derived from HFC-134a, CF_3CFHO, decomposes (to give CF_3 radicals and $HC(O)F$) and reacts with O_2 (to give $CF_3C(O)F$ and HO_2 radicals) at comparable rates. In the atmosphere 7-20% of the CF_3CFHO radicals formed in the CF_3CFHO_2 + NO reaction react with O_2 to form $CF_3C(O)$ F while the remainder decompose to give CF_3 radicals and $HC(O)F$ (Wallington, 1996).

Before moving on to consider the fate of the carbonyl products, it is appropriate to discuss the atmospheric fate of CF_3O radicals. The usual modes of alkoxy radical loss are not possible for CF_3O radicals. Reaction with O_2 and decomposition via F atom elimination are both thermodynamically impossible under atmospheric conditions. Instead, CF_3O radicals react with NO and hydrocarbons:

$$CF_3O + NO \rightarrow C(O)F_2 + FNO \qquad (19.40)$$

$$CF_3O + CH_4 \rightarrow CF_3OH + CH_3 \qquad (19.41)$$

Reaction with NO yields $C(O)F_2$ that does not react with any gas phase trace atmospheric species and whose photolysis is slow (Nölle, 1992). $C(O)F_2$ is removed from the atmosphere by incorporation into water droplets followed by hydrolysis to give CO_2 and HF or by photolysis in the upper stratosphere to give FCO radicals and F atoms. FNO is photolysed to give NO and an F atom (Wallington, 1995). F atoms reversibly form FO_2 radicals by combining with O_2, and also react with CH_4 and H_2O to give HF, which is rained out of the atmosphere. The reaction of CF_3O radicals with hydrocarbons such as CH_4 produces CF_3OH. The CF_3O-H bond is unusually strong (120 kcal mole^{-1}). Therefore CF_3OH is not attacked by any trace atmospheric radical (Schneider, 1993) and is not photolysed (Schneider, 1995). CF_3OH undergoes heterogeneous decomposition to give $C(O)F_2$ and HF and reaction with atmospheric water droplets to give CO_2 and HF (Wallington, 1994c).

Incorporation into water droplets

This far the oxidation of the halocarbons into halogenated carbonyl products has been discussed. The sequences of gas phase reactions that follow from the initial attack of OH radicals on the parent halocarbon are sufficiently rapid that heterogeneous and aqueous processes play no role. In contrast, the lifetimes of the carbonyl products (e.g., $HC(O)F$, $C(O)F_2$, $CF_3C(O)F$) are relatively long. As discussed below, incorporation into water droplets followed by hydrolysis plays an important role in the removal of halogenated carbonyl compounds (DeBruyn, 1992). For $HC(O)F$, $C(O)F_2$, $FC(O)Cl$ and $CF_3C(O)$ F reaction with OH radicals (Wallington, 1993) and photolysis (Nölle, 1992) are too slow to be of any significance. These compounds are removed entirely by incorporation into water droplets.

For $CX_3C(O)H$ species reaction with OH radicals is important (Scollard 1993). The lifetimes of $CF_3C(O)H$, $CF_2ClC(O)H$, and $CFCl_2C(O)H$ with respect to OH attack have been estimated to be 24, 19, and 11 days, respectively (Scollard, 1993). Photolysis is probably also an important sink for $CF_3C(O)H$, $CF_2ClC(O)H$, and $CFCl_2C(O)H$ (Scollard, 1993). Although the absorption spectra for these compounds are known, their photolysis quantum yields are unknown; therefore it is only possible to establish upper limits for their rates of photolysis. Finally, scavenging by water droplets probably also plays a role in the atmospheric fate of these halogenated aldehydes and needs to be investigated. For $CF_3C(O)Cl$, reaction with OH is not feasible. Photolysis of $CF_3C(O)Cl$ is important (Rattigan, 1993) and competes with incorporation of $CF_3C(O)Cl$ into water droplets.

Photolysis of $CF_3C(O)Cl$ gives CF_3, CO, and Cl. In addition, trace amounts (<1% yield) of CF_3Cl were reported. CF_3Cl is a long-lived compound that efficiently transports chlorine from the lower atmosphere to the stratosphere. However, the low yield of CF_3Cl from $CF_3C(O)Cl$ photolysis renders this pathway of negligible environmental significance. Following reaction with OH radicals, $CF_3C(O)$, $CF_2ClC(O)$, and $CFCl_2C(O)$ radicals can either react with O_2, or decompose to give CO and a halogenated methyl radical. Reaction with O_2 is essentially the sole atmospheric fate of $CF_3C(O)$ radicals (Wallington, 1994d) and possibly $CF_2ClC(O)$ and $CFCl_2C(O)$ radicals. The resulting $CX_3C(O)O_2$ radical can react with NO or NO_2. Reaction with NO_2 gives a halogenated acetyl peroxynitrate, which undergoes thermal decomposition (Wallington, 1994d) to regenerate $CX_3C(O)O_2$. Reaction with NO gives a $CX_3C(O)O$ radical which rapidly dissociates to give CX_3 radicals and CO_2 (Wallington, 1994d).

The final step in removal of any species from the atmosphere involves heterogeneous deposition to the Earth's surface. Removal processes include wet deposition via rain-out (following uptake into tropospheric clouds) and dry deposition to the Earth's surface, principally to the oceans. The rates of these processes are largely determined by the species' chemistries in aqueous solution. Due to their low aqueous solubility and reactivity heterogeneous lifetimes of the parent HFCs, HCFCs and HFEs are of the order of hundreds of years.

A substantial body of data concerning the atmospheric degradation of HFCs, HCFCs and HFEs is available (Wallington, 1994a). While some uncertainties exist, the current understanding of the atmospheric degradation of the commercially important HFCs, HCFCs and HFEs is well estab-

Halocarbon	Updated Model- Derived	Updated Semi- empirical	Semi- empirical Schauffler et al (2002)	WMO (1999) Model	WMO (1999) Semi- empirical	Montreal Protocol
CFC-11		1				
CFC-12		1.0	0.96	0.82	0.9	1.0
CFC-113		1.0	0.90	0.90	0.9	0.8
CFC-114	0.94		1.0	0.85		1.0
CFC-115	0.44			0.40		0.6
Halon-1301	12			12	13	10.0
Halon-1211	6.0	6.0		5.1	5	3.0
Halon-2402	<8.6					6.0
Halon-1202	1.3					
CCl_4	0.73	0.78		1.20		1.1
CH_3CCl_3	0.12	0.15		0.11	0.12	0.1
HCFC-22	0.05	0.041		0.034	0.05	0.055
HCFC-123	0.02			0.012	0.02	0.02
HCFC-124	0.02			0.026		0.022
HCFC-141b	0.12	0.037		0.086	0.1	0.11
HCFC-142b	0.07	0.014		0.043	0.066	0.065
HCFC-225ca	0.02			0.017	0.025	0.025
HCFC-225cb	0.03			0.017	0.03	0.033
CH_3Cl	0.02					
CH_3Br	0.38			0.37	0.37	0.6
Upper limits for selected hydroflourocarbons						
HFC-134a			$< 1.5 \times 10^{-5}$			
HFC-23			$< 4 \times 10^{-4}$			
HFC-125			$> 3 \times 10^{-5}$			

Table 19.1 Ozone Depletion Potentials for long-lived halocarbons

lished. HFCs have no impact on stratospheric ozone. HCFCs have small but non-negligible ozone depletion potentials (ODPs). The ODP of a compound is defined as the total steady-state ozone destruction that results per unit of mass of the compound emitted per year relative to that of a unit mass emission of CFC-11. The ODP is a relative measure. The ODP value for a compound depends on two factors: Its atmospheric lifetime and its ability to release Cl and/or Br atoms.

All the ODP compounds mentioned in this chapter are also potent greenhouse gases. The direct global warming potentials (GWPs) of HFCs, HCFCs and HFEs are approximately an order of magnitude less than those of the CFCs they replace. GWP is defined similar to ODP but relative to CO_2. However, as the use of some of these alternatives becomes more important they may make a very significant positive contribution to radiative forcing. Finally, HFCs, HCFCs and HFEs are sufficiently un-reactive and are released in such small quantities that they do not contribute to urban smog formation (Hayman, 1997).

Ozone depletion potentials for long-lived halocarbons are shown in Table 19.1.

19.5 Ozone depletion and climate change

Ozone change and climate change is linked in several important ways. An IPCC special report on this issue has recently been published (IPCC 2005). Ozone itself is a greenhouse gas (GHG) like CO_2, CH_4, N_2O and the halocarbons. As shown in Figure 20.4 (p. 421) the depletion of stratospheric ozone has caused a negative radiative forcing of -0.15±0.10 W m^{-2}, corresponding to a cooling. The ozone depleting substances (ODSs) have produced a positive forcing of 0.34±0.03 W m^{-2}. Warming due to ODSs and cooling due to ozone depletion does not offset each other. The seasonal and spatial distribution of the cooling effect differs from those of the warming.

The Earth's surface is warming and the stratosphere is cooling. A cooler stratosphere will affect the time period and spatial distribution of PSCs. The PSCs will be present for a longer time and over a larger area and will probably delay the recovery of the ozone layer.

Ozone, the solar cycle and volcanic eruptions

The UV radiation from the sun produces stratospheric ozone, and increase in this radiation will increase the formation. The Sun´s output varies over the

11-year solar sunspot cycle as seen in Figure 19.9, but as it is seen from the figure top and middle panel the changes in solar radiation cannot account for the long-term decrease in ozone.

During volcanic eruptions HCl and SO_2 can be emitted directly high into the atmosphere. HCl is removed by rainout whereas SO_2 is converted into H_2SO_4, which will form new particles in the stratosphere and increase ozone depletion. This is also clearly seen in connection with the two large volcanic eruptions, El Chinchon (1982) and Mt. Pinatubo (1991). However, these particles only remained in the stratosphere for a few years. Comparing the middle and the bottom panel of Figure 19.9 it is seen that volcanic eruption can not account for the long-term trend in global ozone.

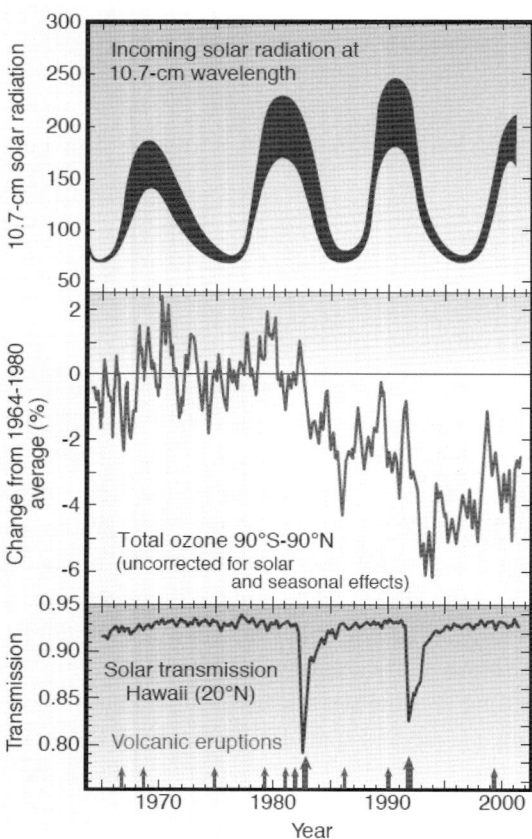

Figure 19.9 Global ozone, volcanic eruptions and the solar cycle.

19.6 The Montreal Protocol

In the 1980s the stratospheric ozone depletion received political attention. It was clear that mankind had used its power to alter the Earth's atmosphere on a global scale. In Vienna in 1985 the first step towards CFC regulation was taken with the United Nations Convention for the Protection of the Stratosphere. In September 1987 the Montreal protocol was signed posing restrictions to cut global emissions of CFC to 50% of the 1986 level. CFC substitutes of more environmentally acceptable compounds were derived and used (Wallington, 1994a). Due to better understanding of the serious effects on atmospheric ozone the Montreal protocol has had amendments in London in 1990, in Copenhagen in 1992, in Montreal in 1997 and in Beijing in 1997. Each amendment has posed tighter regulations on CFCs, HCFCs and bromine containing compounds.

Did the Montreal Protocol work?

In Figure 19.10 is shown a truly well understood emission-concentration relationship for CFC-11. The Southern Hemisphere lagged behind the Northern

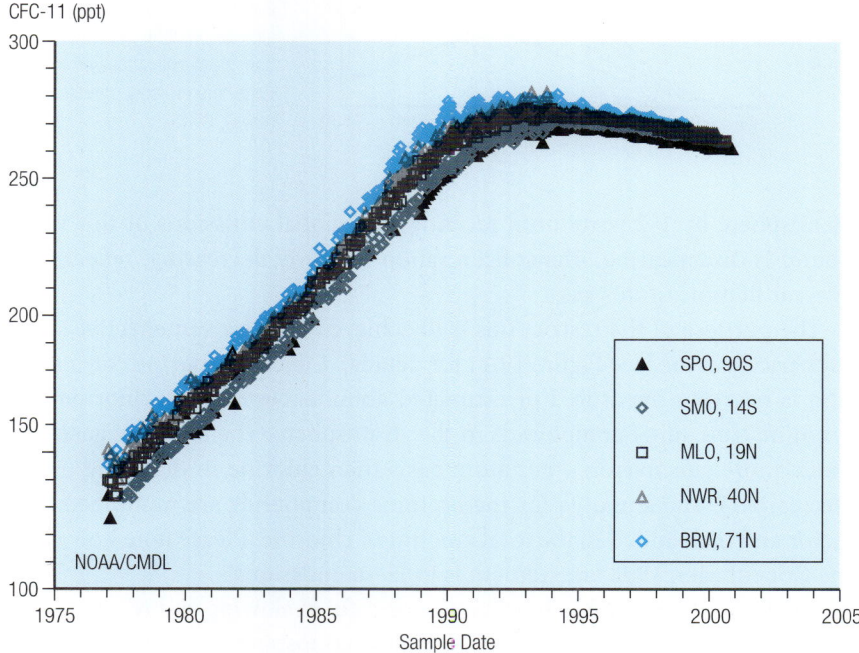

Figure 19.10 CFC-11 concentrations from a global network of measuring stations. Figure from Stephen Montzka, NOAA.

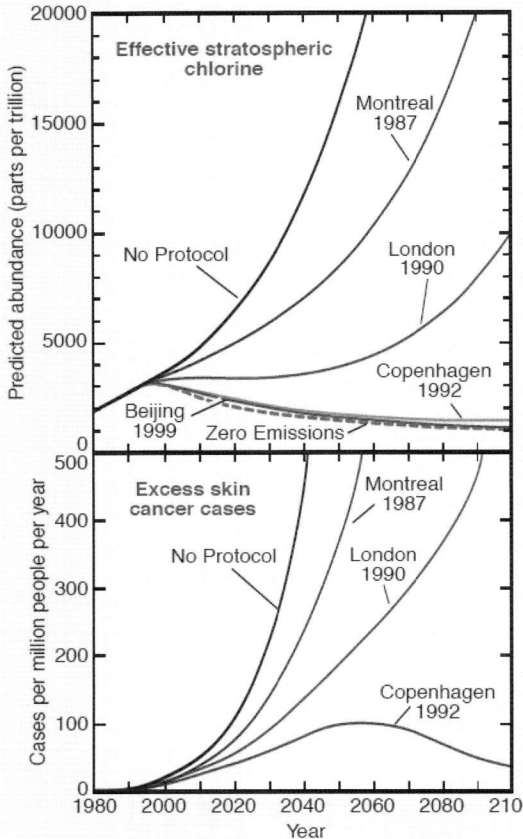

Figure 19.11. The top panel show the predictions for the future abundance of effective stratospheric chlorine assuming: no protocol and the different protocols. The "zero emissions" line shows the stratospheric abundances if all emissions were reduced to zero in 2003. The lower panel shows how excess skin cancer would increase with no regulations and how the development would be under the protocol provisions. (WMO, 2003).

Hemisphere by 1-2 years until recently. The global emissions of this compound is disappearing. The concentration is slowly decreasing, reflecting a 50-year lifetime of this gas.

The purpose of the restrictions is to achieve reductions in effective stratospheric chlorine (see Figure 19.11 for details). Effective chlorine concentration is based on measured or estimated abundances of both chlorine and bromine containing compounds in the stratosphere. The bromine gases are much more effective on a per atom basis than chlorine in depleting ozone (see earlier). The amounts for the bromine compounds are multiplied by a factor and then added to the total amount of chlorine. Please note how close the Copenhagen 1992 amendment is to "zero emissions".

The Montreal protocol has been very successful in slowing and reversing the increase of ozone depleting substances in the atmosphere. This is clearly seen in Figure 19.12.

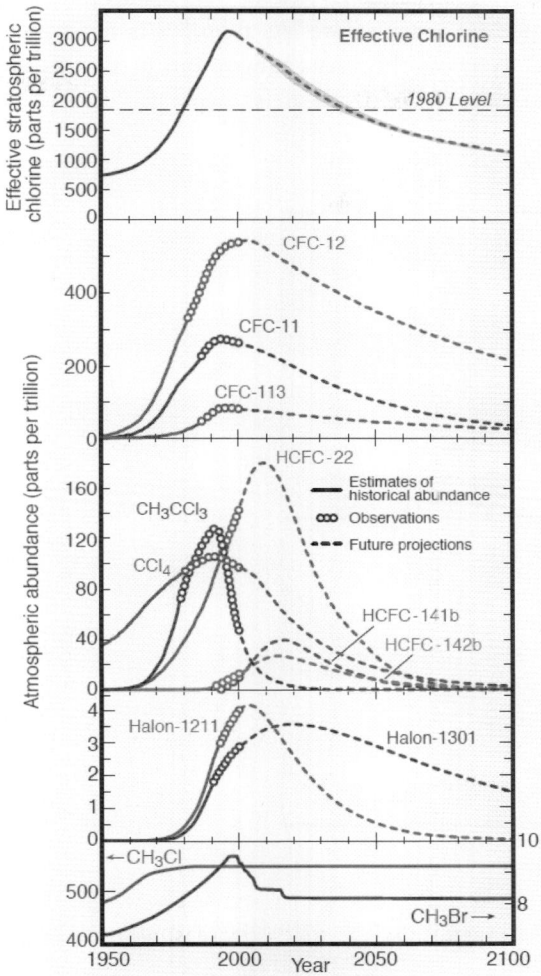

Figure 19.12: The rise in effective stratospheric chlorine has slowed and reversed in the last decade (top panel). This decrease is a result of reductions in emissions of individual compounds (bottom four panels). The largest reduction occurred for CH_3CCl_3 as a result of a protocol that reduced the global production to near zero. Open symbols are measurements and solid lines are estimates. (WMO, 2003).

Recovery

Man-made chlorine and bromine containing compounds are expected to gradually disappear by the middle of the 21st century. The emission reductions are based on the assumption of full compliance by all nations in the world. Natural chemical and transport processes determine the rate at which halogen can be removed from the stratosphere. Compounds with long life-time will require several hundred years before less than 5% is remaining. Computer models are used to assess the past changes and predict future changes. In Figure 19.13 two measures of ozone are shown.

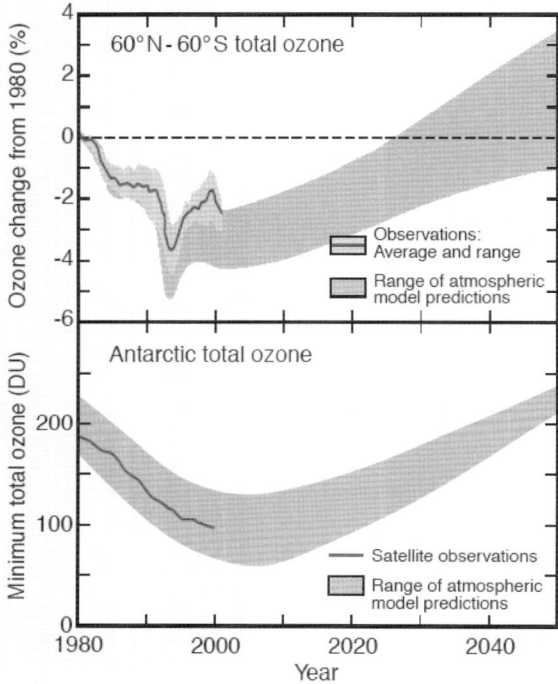

Figure 19.13: Observed and predicted global ozone between 60ºN and 60ºS (top panel) and minimum total ozone over Antarctica (bottom panel).

The ozone layer is expected to recover during the 21st century. However, volcanic eruptions could delay ozone recovery and the influence of climate change could accelerate or delay ozone recovery. Finally, slow release of the ODP substances enclosed in appliances and in discarded material in landfills may also delay the recovery.

19.7 Litterature

Chapman, S. (1930): *A theory of upper-atmosphere ozone,* Mem. R. Meteorol. Soc., *3,* 103.

DeBruyn, W., et al. (1992): *Tropospheric heterogeneous chemistry of haloacetyl and carbonyl halides.* Geophys. Res. Lett., *19,*1939-1942.

DeMore, W.B. et al. (1994): JPL Publication 94-26.

Farman, J.D. et al. (1985): *Large losses of total ozone in Antarctica reveal seasonal Clox/NOx interaction.* Nature, *315,* 207-210.

Hayman, G.D. and Derwent, R.G. (1997): *Atmospheric chemical reactivity and ozone-forming potentials of potential CFC replacements* .Environ. Sci. Tech., *31,* 327-336.

IPCC 2005: Special report on *"Safeguarding the ozone layer and the global climate system".* Cambridge University Press, ISBN 0-521-68206-1.

Maricq, M.M. et al. (1994): *Atmospheric chemistry of HFC-134a: kinetic and mechanistic study of the CF_3CFHO_2 + HO_2 reaction.* J. Phys. Chem., *98,* 8962-8970.

McCulloch, A. et al. (2003): *Releases of refrigerant gases (CFC-12, HCFC-22 and HFC-134a) to the atmosphere.* Atmos. Environ., *37,* 889-902..

Molina, M.J. and Rowland, F.S. (1974): *Stratospheric sink for chlorofluoromethanes – clorine atomic- catalysed destruction of ozone.* Nature, *249,* 810-812.

Molina, M.J. et al. (1987): *Antarctic stratospheric chemistry of chlorine nitrate, hydrogen-chloride and ice – Release of active chlorine.* Science, *238,* 1253-1257.

Molina, L.T and Molina, M.J. (1996): *Ultraviolet spectrum of CF_3OH: Upper limits to the absorption cross sections.* Geophys. Res. Lett., *23,* 563-565..

Nölle, A. et al. (1992): *UV absorption-spectrum and absorption cross-section of COF_2 at 396K in the range 200-230 nm.* Geophys. Res. Lett., *19,* 281-284..

Rattigan O.V. et al. (1993): *Temperature dependent absorption cross-sections of CF_3COCl, CF_3COF, CH_3COF, CCl_3CHO and CF_3COOH.* Photochem. Photobiol. A: Chemistry, *73,* 1-9.

Schneider W.F. and Wallington T.J. (1993): *Ab Initio investigation of the heats of formation of several trifluromethyl compounds.* J Phys Chem , *97,* 12783-12788.

Schneider W.F. et al. (1995): *Atmospheric Chemistry of CF_3OH: is photolysis important?* Environ Sci. Tech., *29,* 247-250.

Scollard D.J. et al. (1993): Rate constants for the rections of hydroxyl radicals and chlorine atoms with halogenated aldehydes, J Phys Chem., *97,* 4683-4688.

Wallington T.J. and Hurley M.D. (1993): *Atmospheric Chemistry of HC(O)F: reaction with OH radicals.* Environ. Sci. Tech., *27,* 1448-1452.

Wallington T.J. et al. (1994a): *Atmospheric chemistry and environmental impact of CFC replacements: HFCs and HCFCs.* Environ. Sci. Tech., *28,* 320A-325A.

Wallington T.J. et al. (1994b): *Atmospheric chemistry of $CF_3C(O)O_2$ radicals: Kinetics of their reaction with NO_2 and kinetics of the thermal decomposition of the product $CF_3C(O)O_2NO_2$.* Chem. Phys. Lett., *226,* 563-569.

Wallington T.J. and Schneider W.F. (1994c): *The stratospheric fate of CF_3OH.* Environ. Sci. Tech., *28,* 1198-1200.

Wallington T.J. et al. (1994d): *Mechanistic study of the gas phase reaction of CH_2FO_2 radicals with HO_2.* Chem Phys Lett , *218,* 34-42.

Wallington T.J. et al. (1995): *Atmospheric chemistry of FNO" and FNO_2: Reactions of FNO with O_3, $O(^3P)$, HO_2, and HCl and the reaction of FNO_2 with O_3.* J. Phys. Chem., *99,* 984-989.

Wallington T.J. et al. (1996): *Role of excited CF_3CFHO radicals in the atmospheric chemistry of HFC-134a.* J Phys Chem ,*100,* 18116-18122.

WMO (World Meteorological Organization) **(1985):** Scientific Assessment of ozone depletion Report No. 16

WMO (World Meteorological Organization) **(1989):** Scientific Assessment of Stratospheric Ozone Report No. 20.

WMO (World Meteorological Organization) **(2003):** Scientific Assessment of ozone depletion Report No. 47.

20 The increasing greenhouse effect

Jes Fenger

The Earth is nearly 5 billion years old, and in its history it has been subjected to drastic climate changes, the first related to various impacts during the formation. There are indications that the axis of rotation has been different, and it appears that the Earth has been nearly completely frozen down about 700 million years ago (Hofman and Schrag, 2000). The climate, however, is only known in more detail at a later date, where it can be reconstructed from measurements in deep-sea sediments, ice caps, moors etc. Also in this period the Earth has undergone a series of marked – and sometimes quite rapid – climatic shifts. The reasons are not fully understood, but continental drift may be part of the explanation.

20.1 The climate of the Earth up to present

A short history

In the carboniferous age, about 300 million years ago, the climate on Earth was warm and humid. Marshes and tropical forests covered large parts of the globe, and the coal deposits were laid down. In the Permian, about 50 million years later, the climate was still warm, but very dry. It resulted in evaporation from the oceans and formation of some of the Earth's large salt deposits.

In the Cretaceous period 100 million years ago it was still relatively warm. The dinosaurs lived then, until they disappeared rather abruptly and gave room for mammals. A possible explanation is drastic short-term climatic changes as a consequence of a large meteor impact on the Yucatan peninsula about 65 million years ago. Another possibility is gigantic volcanic eruptions in Northern India. In both cases large amounts of dust in the atmosphere in shorter or longer time must have shielded the Earth from sunlight and thus resulted in a drastic cooling. Also a pronounced formation of nitrogen oxides that depleted the ozone layer is possible.

Five million years ago the climate was mild with an average temperature a couple of degrees above the present and with much smaller differences between the tropical and the Polar Regions. About two million years ago the Earth went into the present Ice-age cycle. In this period the Earth has seen at least 8 larger or minor cold periods. The length of these so-called ice ages is about 100,000 years and the intermediate warm intervals are much shorter – possibly only 10,000 – 20,000 years. These shifts in climate are not fully understood, but part of the explanation must be the so-called Milankovitch cycles, that arise from variations in Earth's movement around the Sun and thus in the amount and pattern of radiation energy deposited on the Earth.

We have relatively good knowledge of the interglacial period proceeding the last ice age. In this so-called Eem period, which started about 125,000 years ago, the climate was warm in Europe with water buffaloes and hippopotamus living in Central Europe and Southern England.

The Eem period only lasted about 10,000 years before a new and so far last ice age set in. 20,000 years ago the northern part of Scandinavia was covered by a nearly 1 km thick layer of ice, which began to melt 15,000 years ago. After a short interval, when the temperature fell about 6 °C in large parts of Europe, the warming continued and the global average temperature 6000 years ago was 2 °C higher than today. It resulted in a rapid melting of the polar ice and a rising of the sea level of 120 m above the minimum during the ice age – maybe at a speed of up to 5 cm per year.

The recent years

The last 4000 years have been characterised by various climatic shifts, but about 1000 years ago the climate on the Northern Hemisphere went into a short, but relatively stable warm period, that e.g. resulted in colonisation of Greenland. This was abandoned around 1400 due to a new cooling (the little ice age), which plagued the North European societies until around 1700. In the beginning of the nineteenth century there was a short, but marked cooling, which may have been connected to volcanic activity e.g. the Tambora eruption in 1850.

From the middle of the nineteenth century and simultaneously with the industrial revolution the global mean temperature has generally been increasing. It is now about 0.7 °C higher than 150 years ago (Figure 20.1). This increase has already had its impact in terms of prolonging the growth seasons and redistribution of natural species. Also melting of glaciers and reduction of snow covers have been observed. The sea level has always changed – sometimes quite drastically – but in the last 150 years there has been a continuous increase that has been attributed to rising water temperatures and melting of glaciers.

These changes are so far modest compared to what the Earth has experienced in its history, but previous climate changes would now have resulted

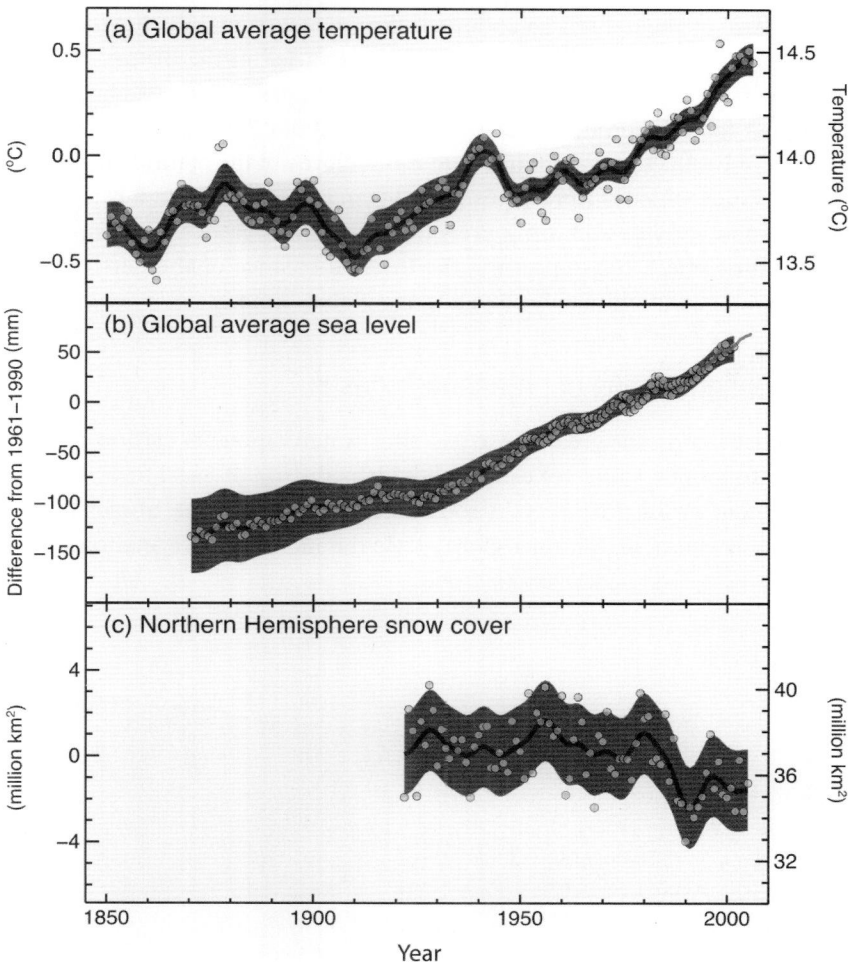

Figure 20.1 The increasing average global temperature since year 1850, the increasing global average sea level, and the recent Northern hemisphere snow cover in March and April (IPCC, 2007).

in impacts that we would consider quite unacceptable, both with respect to nature, human living conditions and societal infrastructures.

20.2 The greenhouse effect and the greenhouse gases

Essential to the question of Earth's heat balance and climate is the so-called greenhouse effect (Section 1.3).

The size of the natural greenhouse effect, which is mainly due to atmospheric water vapour and carbon dioxide, has varied in the history of Earth and has been connected to previous climate changes. It may also have compensated for the output of the Sun that was about 25% lower in the early history of Earth. The present problem is therefore not the natural greenhouse effect as such which is a prerequisite for life in its present form. It is that various human activities emitting greenhouse effective gases that *enhance* the greenhouse effect, thus upsetting the thermal balance of the Earth-atmosphere system and ultimately changing the global climate.

The concentrations of greenhouse gases

The most important greenhouse gas after water vapour is carbon dioxide. The main human source of carbon dioxide is the use of fossil fuels, but also e.g. cement production plays a role. Forest clearing in the tropics gives a significant emission, but on a global scale the increase in vegetation on the

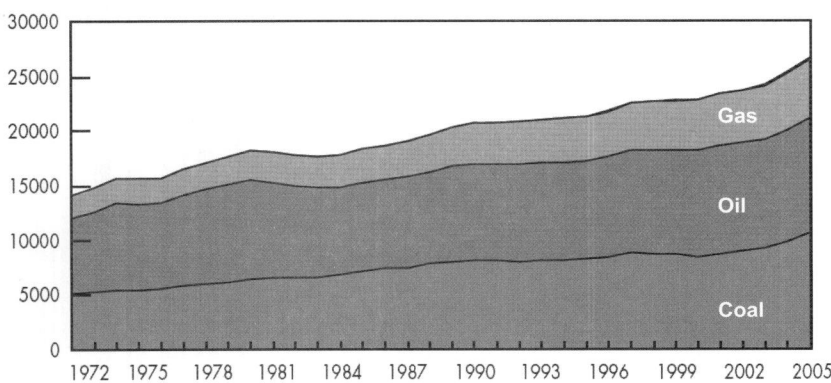

Figure 20.2 Global emission of carbon dioxide in (Mt y^{-1}) from use of fossil fuels since 1972. (IEA, 2006).

Northern Hemisphere more than compensate this and globally vegetation is at present a drain.

The lifetime of carbon dioxide in the atmosphere is determined by a series of processes of varying rate and involves quantities much larger than the human contribution (Section 1.2), but overall it is approximately 100 years. This means that emitted carbon dioxide is mixed into the atmosphere over the entire globe. Thus the location of emission is insignificant.

The global human emission of carbon dioxide has nearly doubled in the last 40 years (Figure 20.2), and it appears (Figure 20.3) that the atmospheric concentration has increased correspondingly. Roughly half of this emitted carbon dioxide is still in the atmosphere, and the atmospheric concentration has increased about 30% since the industrialisation began.

Methane is mainly formed by anoxic fermentation. It partly arises from the digesting systems of ruminants and the use of natural fertilisers (manure). There are also large emissions from wet rice paddies. More than half of the present emission is of human origin and 70% of this relates to food production. The lifetime in the atmosphere is about 10 years, but it depends e.g. upon the concentration of hydroxyl radicals and thus again upon the

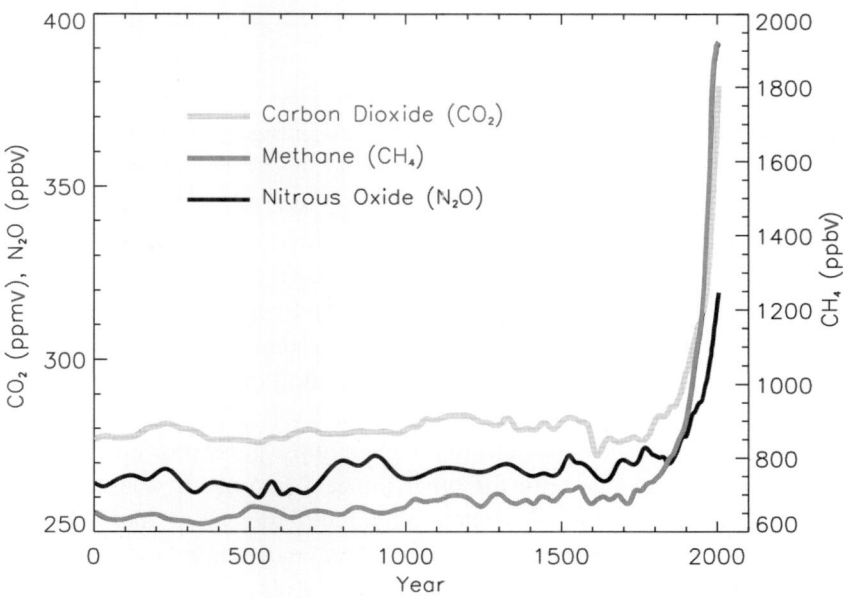

Figure 20.3 Concentrations of greenhouse gases the last 2000 years. Up to about 1950 the values are largely based on measurements of ice cores. More recent results are direct measurements. It appears that the concentration of carbon dioxide has increased more than 30%, methane concentration has more than doubled, and nitrous oxide has increased by about 16%. (IPCC, 2007).

concentration of other pollutants. The concentration has more than doubled since the middle of the 19th century (Figure 20.3).

Nitrous oxide is formed in various organic processes and human activities as use of fertilisers, burning of biomass and industrial processes. The lifetime is about 114 years and the concentration has increased 16% (Figure 20.3).

Halocarbons (CFC and related compounds) are purely of human origin and have had a series of technical applications. Their concentrations in the atmosphere are small, but their GWPs are large and thus their contribution to the greenhouse effect is important.

The "greenhouse impact" of the different gases depends upon their optical properties and upon to what extent the radiation bands, where the compound absorbs, are saturated. But it also depends upon their lifetime in the atmosphere. Thus methane as such is very radiation efficient, but it has a smaller impact, because it has a relatively short lifetime (about 10 years) in the atmosphere. The greenhouse impact of the different compounds is expressed as GWP (Global Warming Potential) for a defined time horizon – typically 100 years – and expressed relative to carbon dioxide (Table 20.1).

Compound	20 years	100 years	500 years
Carbon dioxide, CO_2	1	1	1
Methane, CH_4	72	25	7.6
Nitrous oxide, N_2O	289	298	153

Table 20.1 Global warming potentials for the most important green house gases. CFC's and related compounds can have GWPs of several thousands, but due to low concentrations they are less important. (IPCC, 2007).

In addition various other factors of human origin influence the heat balance e.g. contrails from high-going aircrafts and particles. The concentration of tropospheric ozone has about doubled in some regions since the 19th century. The scientific understanding of these additional impacts is low, e.g. because the impact in the atmosphere is local.

The increasing concentrations of the greenhouse gases thus give an extra heating, which together with the other impacts amounts to about 2 Wm^{-2} (Figure 20.4). It may not seem much compared to the natural thermal input of 342 Wm^{-2}, but it is a constant change that is enough to upset the global heat balance – and to cause a global warming.

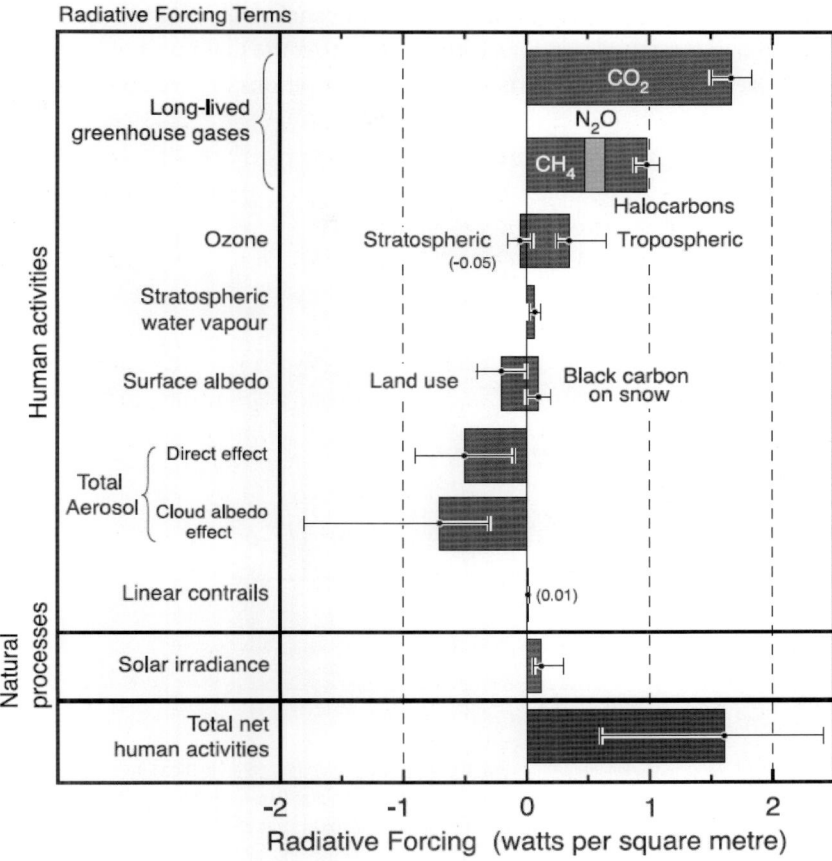

Figure 20.4 Estimated contributions to global heating in terms of radiative forcing from various factors. Year 2005 compared to 1750 (IPCC, 2007).

Global temperature increases

The 0.7 °C, which the average temperature has raised since the beginning of the industrialisation in the 19th century, may seem modest, but the temperature has increased faster than seen before – especially in the recent decades. The increase has not been even, thus there was a temporary cooling around 1950, which could be caused by emission of particles from coal burning. It has also been attempted to explain the change as a result of natural processes, e.g. variations in the Sun's magnetic field demonstrated in the periodicities of sunspots. This should influence the amount of cosmic particles reaching the Earth and thus influence the formation of clouds and the general heat balance of Earth. IPCC (The Intergovernmental Panel of Climate Change) finds, however, in model calculations (Figure 20.5) that the main part of heating in recent decades is due to the increasing concentrations of greenhouse gases.

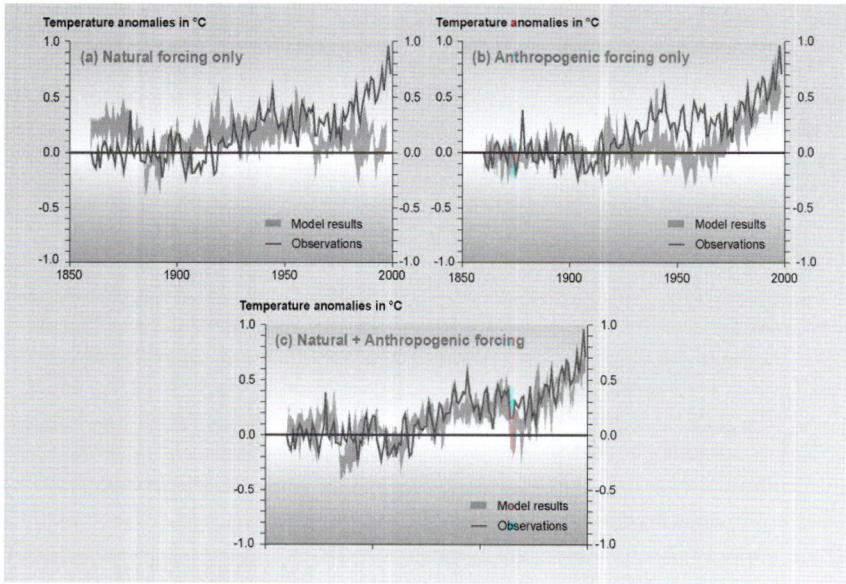

Figure 20.5 Model calculation of the average global temperature compared to measurements. The black curves are the same in all three diagrams and show the actually measured temperature. The light grey curves are the calculated temperature development due to respectively: A: Natural causes B: Human impact C: Natural *and* human impacts. It appears that both human and natural causes are necessary to explain the observations, but that the human influence has dominated in recent decades. (IPCC, 2001).

Computer models for climate

The climate system comprises the atmosphere, oceans, land cover with ice and snow, and the entire biosphere. Climate models are mathematical equations that describe the systems components and interactions. Climate models are processed on large computers, where the atmosphere, the land surface and the oceans are divided into a three dimensional grid. In each grid cell the changes in temperature, humidity, wind etc. over time is calculated. Typically, in a model the atmosphere is divided into 20-30 layers and the oceans in 10-20 layers. In the horizontal plane the resolution can be a few hundred kilometres. The models are under constant development and various types of feedback are included, e.g. the reaction of vegetation to climate change.

The spatial resolution in global climate models of a few hundred kilometres is not enough for more detailed studies of climate impacts. Therefore regional climate models are in various ways imbedded in global models to give a resolution of down to 10 km. As a sensitivity study such calculations are interesting, but how useful such accuracy is from a political point of view remains a question, considering the uncertainty of the underlying emission scenarios.

The future climate

It may be possible to explain the past climate with models that can be tuned to the observations, but it is much more complicated to predict the future climate. No one knows how the world will develop. It is therefore only possible to describe various more or less probable scenarios. Normally they do not reach further than about 100 years into the future. This is partly because of the uncertainty in the preconditions, partly because of the limited computer power.

IPCC (2001) has developed a series of scenarios for the global economical, technological and demographic development in the 21st century. They are divided into four families according to their focus on either economy or environment (A and B) and global or regional development (1 and 2). Together they cover 40 different combinations of increases in world population (range 7-15 billion), increase in global gross domestic product (range 11-26 times), distribution of different fossil and non-fossil energy sources etc. For each family is given a typical "marking" scenario. For A1, however, three different scenarios are presented: A1FI has emphasis on fossil fuels, A1T emphasizes non-fossil fuels, and A1B a balanced energy production (Figure 20.6a). The IPCC does not indicate any probability for the various scenarios, but in the nature of things at most one of them will be realised. Some criticism has been raised against the assumptions on development in the third world, but this does not influence the basic uncertainty in the prediction of the global development.

On the basis of the emission scenarios and global cycles are then calculated how the atmospheric concentration will develop (Figure 20.6b). Even though some optimistic scenarios (B1) describe a reduction in the emission of carbon dioxide, they all show an increase in the concentration from the present 390 ppmv (2008) from less than 500 and to over 1000 ppm in year 2100. Simultaneously changes in the concentration of other greenhouse gases are expected – roughly along the same lines. Although the carbon dioxide concentration is not stabilised directly in any of the scenarios, it can be done by a global technological effort. Of course the price depends upon the level of ambition of the original scenario.

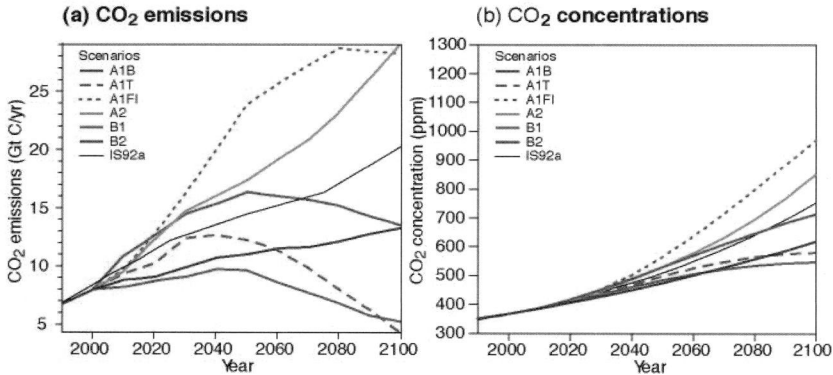

Figure 20.6a Carbon dioxide, CO_2, emissions in the IPCC "marking" emission scenarios. (IPCC, 2001).

Figure 20.6b The atmospheric CO_2 concentrations corresponding to the emission scenarios in Figure 20.6 a. Note that the concentrations increase in all scenarios. (IPCC, 2001).

With global climate models the corresponding changes in global average temperatures up to 2100 are eventually calculated (Figure 20.7).

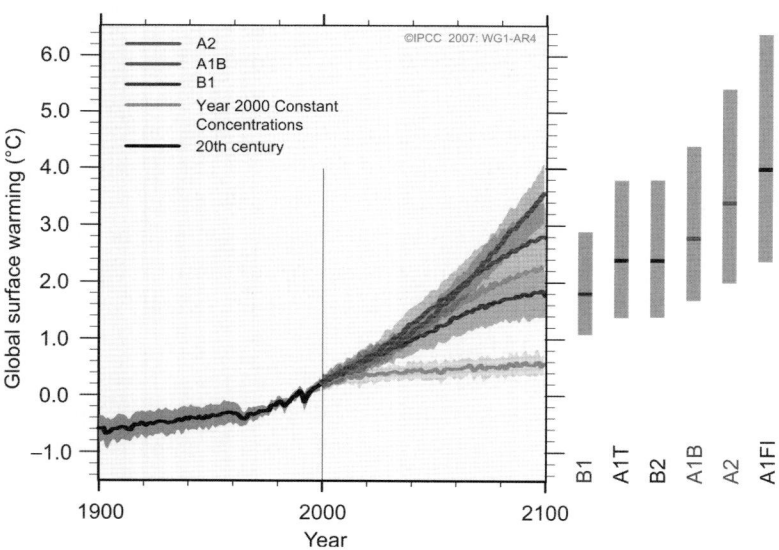

Figure 20.7 The global temperature developments corresponding to the atmospheric CO_2 concentrations in figure 20.6.b. (IPCC, 2007).

The temperature increases will, however, not be the same in different regions of the Earth. Figure 20.8 shows as an example the B1, A1B and A2 scenario. It appears that the largest increases in temperature are seen in the northern part of Europe. Other calculations show e.g. that in the Mediterranean there may be a serious decrease in precipitation.

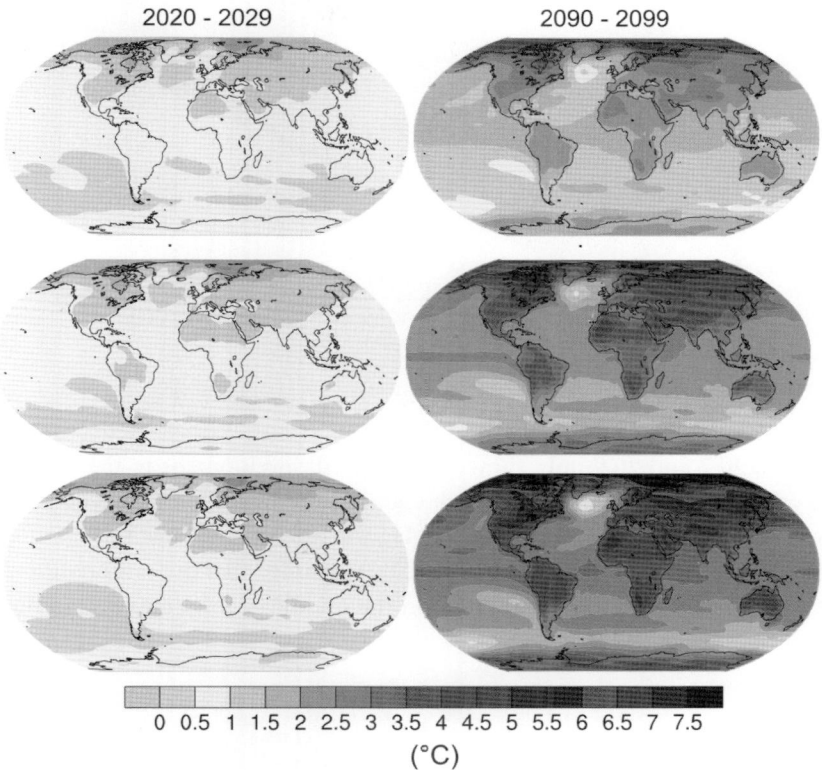

Figure 20.8 Annual temperature changes in Europe for the B1, A1B and A2 scenarios (from above) from 1980-99 to 2090-2099. Averages for different models (IPCC, 2007).

Simultaneously with the increase in temperature the global sea level will raise. Mostly because water expands during heating, but also because of earthbound glaciers melt. The calculated global sea level increase for the next 100 years is calculated to be up to about half a meter and is shown in Figure 20.9. In practice the effective sea level change will be influenced by changes in wind pattern and by vertical movements of the land. It is therefore not the same all over the world. Thus in Finland the land has been rebounding (rising) since the last ice age, and the effective sea level change may even in the beginning be negative.

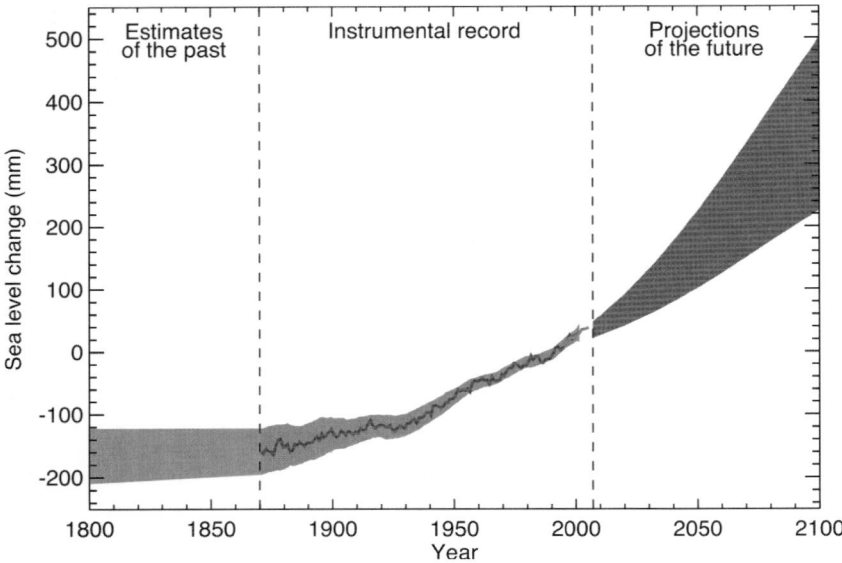

Figure 20.9 Global sea level rises in the past and in spectrum of futures (IPCC, 2007).

Figures 20.7 and 20.9 only show the expected development in the 21-century, but the situation will not reach equilibrium during the next 100 years, even if the emissions were stabilised. Figure 20.10 shows how the situation may develop in the following centuries. It is assumed that the carbon dioxide emissions pass through a maximum in the middle of this century and then

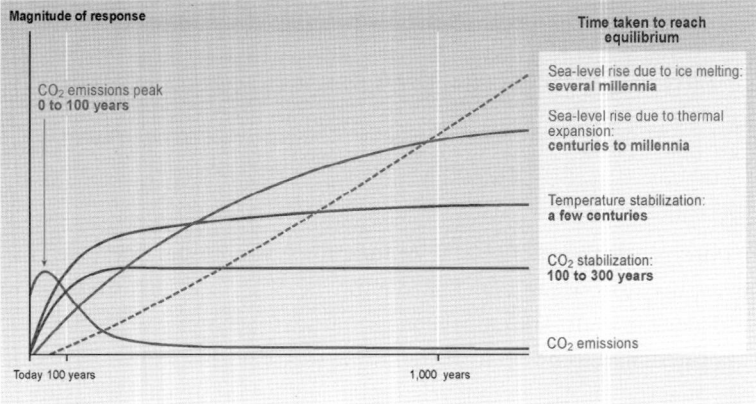

Figure 20.10 General developments of emissions, concentrations, temperature and sea level. No units are shown, but it could mimic a stabilisation of the concentration at 550 ppm, a temperature increase of 2.2 °C in 2100 and a final temperature increase of 3.5 °C. The sea level will rise for several hundred years and will not stabilise in a foreseeable future. (IPCC, 2001).

decrease to practically zero. The carbon concentration is stabilised, but the temperature continues to rise for some time. The sea level will rise for several centuries due to a slow heat exchange and slow melting of glaciers. This may not end for a foreseeable future and may eventually end several meters higher than today.

Surprises in the greenhouse

The Earth has previously seen significant, rapid changes in climate. It has therefore in various connections been discussed whether something similar could be triggered by an increase in the greenhouse effect e.g. by a change in the pattern of large ocean currents. In the long run and without limitation of emissions it cannot be excluded. It has especially been discussed whether the input of melt water from the Greenland glaciers could result in a collapse of the North Atlantic Current (a part of the Gulf Stream). It could result in a colder climate in North Europe in an otherwise warmer world. Such changes in ocean currents would also result in changes in the local sea level. Most model calculations, however, do not show more than a weakening of the current in the first 100 years. It must be stressed, though, that the mechanisms governing the North Atlantic Current is not fully understood, and that the available computer capacity is not sufficient for more detailed studies.

Another equally important phenomenon is the impending thawing of the perma-frozen tundra where large amounts of methane (CH_4) are tied up. A slow release of the gas to the air spaces in the soil at low temperatures may prevent microbiological oxidation to CO_2 and lead to escape of a large proportion of the methane right into the atmosphere. As methane is a powerful GHG and the bound quantities are vast the result may be a rapid temperature increase. Again the predictions are rather uncertain especially as the process is self-reinforcing.

The latest example of still insufficient knowledge for predicting the future is the active dynamics involved in melting of the ice caps in the Arctic, and the rapid disappearance of the winter sea ice around Greenland. The appearance of barren ground and open dark sea will inevitably accumulate more solar energy than white snow and ice and thereby increase the global warming.

20.3 Stabilisation scenarios

As it is shown it is possible to calculate the most probable development in concentration of greenhouse gases for a given scenario. But it is also possible to calculate "backwards": To describe a development in concentrations and then determine the corresponding permissible emissions. It has been done by the IPCC in so-called stabilisation scenarios (Figure 20.11a). If it e.g. is assumed that the concentration of CO_2 must be stabilised at 550 ppm, it will give problems that are not negligible, but can be managed. But even this will require that the present growth in emissions is stopped before 2050, then reduced to the present value in year 2100 and then further reduced in the coming centuries (Figure 20.11b). This should be accomplished simultaneously with an increase in the global population to maybe 10 billion by the end of the century and a substantial increase in the material standard of living in the developing countries. It may not be technically impossible, but it is certainly a challenge – and the calculations do not describe any particular useful strategy for emission reduction!

It is therefore obvious, that further climate changes cannot be completely avoided for any realistic development in CO_2-emissions. They can only be limited!

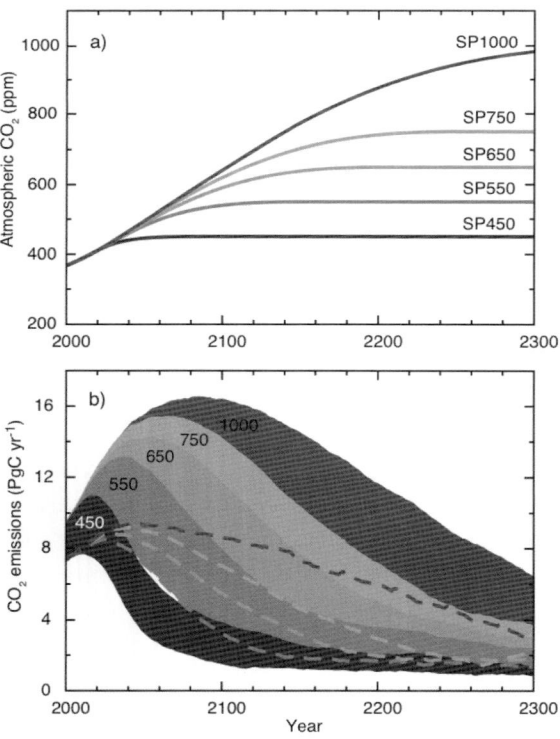

Figure 20.11 Stabilisation scenarios. The upper curves (a) show prescribed developments in CO_2-concentrations. The lower curves (b) show the corresponding permitted emissions. (IPCC, 2007).

20.4 Climatic impacts

The climate determines how nature and human enterprises develop. In historical time natural, more or less local, changes have had important impact and sometimes have resulted in collapse of cultures (e.g. Lamb, 1982; deMenocal, 2001). It appears that drought (lack of water) is more important than temperature increases.

A global temperature increase of a few degrees does not make Earth, taken as a whole, less habitable, but uncontrolled climate changes may be impractical or even disastrous, if they happen too quickly. It may be especially dangerous, when certain areas cannot any longer support the population. Already now the world has seen problems with dwindling water resources and millions of "environmental refugees".

A logical chain

The increasing emission of greenhouse gases, the ensuing climate changes and their impacts can be perceived as a chain of decisions and phenomena

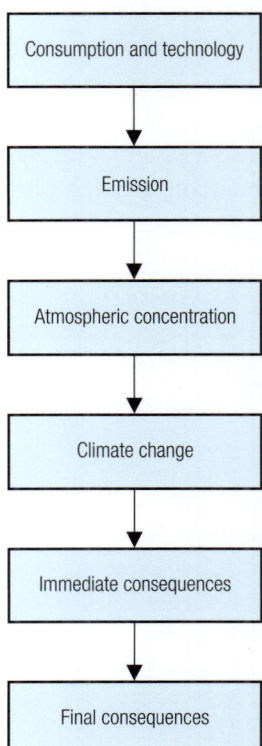

Figure 20.12 Simplified presentation of the course of events in human induced climate changes. In practice there may be a series of feedbacks between the different stages.

(Figure 20.12). It starts with a societal development, where the choice of consumption and technology determine the emissions of greenhouse gases and the resulting atmospheric concentrations. The climate changes have direct, regional impacts on agriculture, forestry, water resources, health etc. Also natural ecosystems are affected. Finally, these direct impacts have consequences for international relations and trade. Societal reactions can happen at any stage, and it is possible to model them, but the uncertainty is added, when one calculates from stage to stage. In addition the time horizon is 50 years or more, and the changes will take place in a world that may be completely different from the one we know today. Therefore any adaptation must be planned with great flexibility.

It is important to note that phenomena in the real world are not described by a one-dimensional chain of relations from emissions to final impacts. There are feedbacks. Sometimes they are natural (e.g. changes in the Earth's albedo or the concentration of water vapour in the atmosphere), sometimes they are human (e.g. changes in energy consumption or various types of adaptation). The situation is complicated by interactions between climate change and other environmental impacts. If e.g. natural systems are weakened by climatic changes, they may be more vulnerable to other environmental stresses as e.g. acidification. The cause-effect relation may therefore not be obvious, and climate changes must be evaluated in connection with a series of technical and societal conditions that may differ from country to country.

It is – both from a technical-scientific and a political viewpoint – important that the entire climate question contains an interaction between phe-

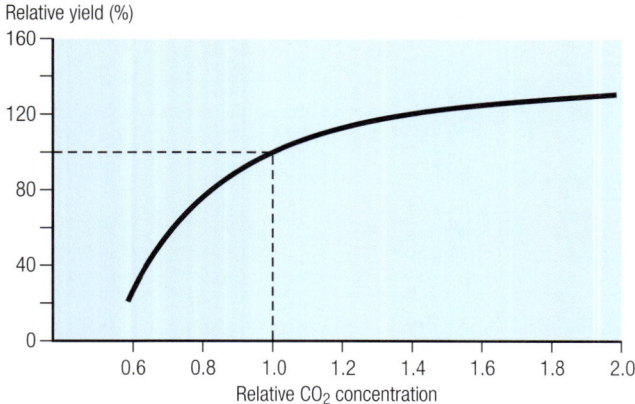

Figure 20.13 Relative effects of carbon dioxide concentration on wheat grain yield. Present concentration is set to 1. (Downing et al. 2000).

nomena with widely varying time horizons. If the basic physical-chemical processes are taken into account the time span is from under a day to more than a thousand years.

Agriculture

Most important for humanity is the production of foodstuffs for an increasing population. An evaluation of the global potential for growing of grains suggests that it will keep pace with the population at least until 2050. These projections are, however, based on assumptions for the degree of agricultural development and the liberalisation of international trade. They are therefore very uncertain, but they suggest that distribution will be more important than production.

In Europe a moderate warming may have some advantages for agriculture, as a longer growth season in combination with the fertilising effect of increasing concentration of carbon dioxide may increase the agricultural yields (Figure 20.13). Chiefly the northern part of Europe will benefit more than the southern, and a shift towards north is anticipated. A drawback will be the need for increased use of pesticides and fungicides, which may have impacts on water quality.

Water resources

Water resources are crucial for the habitability of a particular area, and the catastrophic consequences of drought are e.g. known from the Sahel-desert in 1968-73, when 250,000 died from hunger. At present globally about 2/3 of the available water is used for irrigating agricultural crops. Only ¼ is used in industry and 1/10 in households.

When it has been possible to increase the agricultural production in the last part of the 20th century it is due to a more than doubling of the extent of irrigation. This development is expected to continue in large parts of the United States, China and the Mediterranean. This may cause problems, especially in the Mediterranean, where a reduced precipitation is expected and may offset the advantage of temperature change.

Although a series of model calculations presents a somewhat confusing picture it is obvious that the water requirements will increase with increasing temperature. IPCC (2001) estimated that 1,7 billion people, one third of the worlds population, presently live in countries that are water sressed (using more that 20% of the renewable water supply). This number is projected to increase to 5 billion by 2025.

The available amount of water depends primarily on precipitation and evaporation and thus also on temperature and vegetation. But also the water

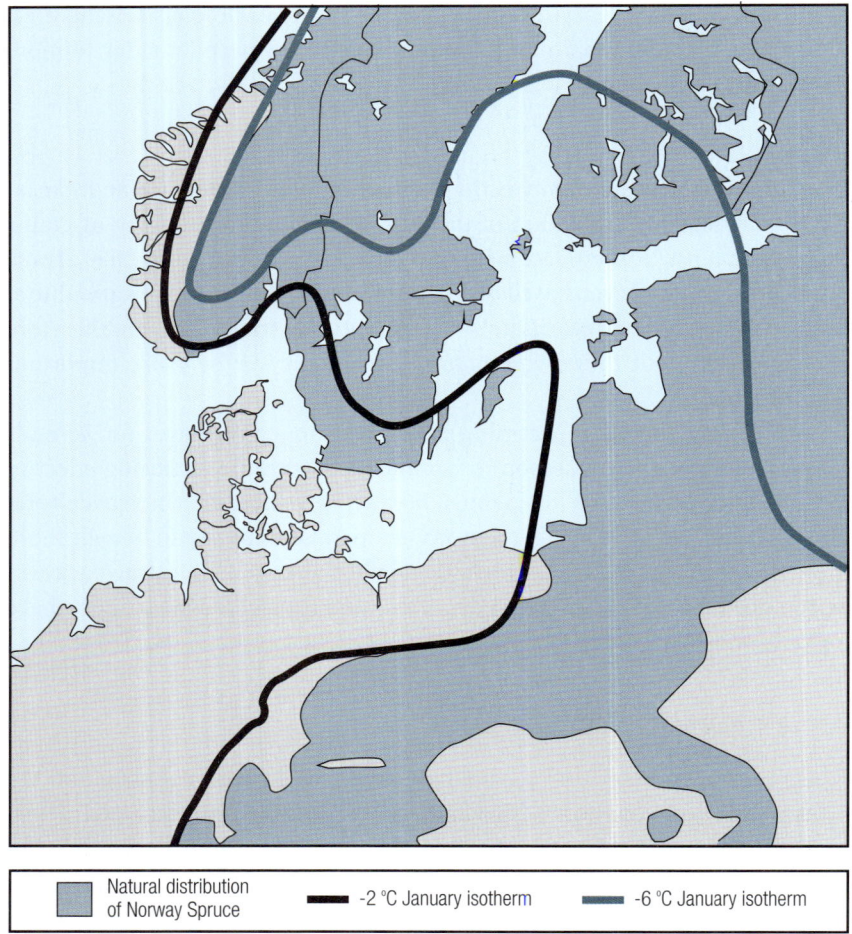

Natural distribution of Norway Spruce	▬▬ -2 °C January isotherm	▬▬ -6 °C January isotherm

Figure 20.14 Norway Spruce needs low temperatures during winter, and its natural growth limit closely follows an isotherm of −2 °C in January. That means that if the temperature increase e.g. 4 °C the distribution can change to the present −6 °C isotherm. Note that Denmark is already situated outside the natural growing zone.

quality will depend on climate and will generally be reduced with increasing temperature due to oxygen consuming biological activity. In coastal areas the increasing sea level may lead to intrusion of salt water into the ground water.

Forestry

Forests cover about 30% of the global land area, if the Polar Regions is disregarded. 57% is situated in the developing countries, where the greatest threat so far has been deforestation for agriculture or ruthless exploitation with loss of bio-diversity and degradation of soil.

Deforestation has previously been a large source of carbon dioxide. At present an increase in temporal forests more than compensates the loss, but in the future the increasing temperature will displace the forests towards the pole at a rate larger than natural migration. Many forests in industrialised countries are, however, heavily managed and may be adjusted to a changing climate. This is e.g. the case in Denmark, where the Norway Spruce, that at present constitute 40% of the forest vegetation, is gradually being substituted with various deciduous trees in a so-called "natural forest management" (Figure 20.14).

Natural ecosystems

Natural ecosystems are more at risk i.a. because urban areas, motorways and other infrastructure fracture the landscape and prevent natural migration. A loss of bio-diversity is probable. Another difficulty in migration may be that the different partners in an ecosystem do not migrate with the same velocity. Special problems may appear in mountainous regions, where the vegetation will be forced upwards and is caught without possibility of further movement.

Infrastructure

All types of infrastructure are at risk under changing climatic conditions. Sewage systems may be inadequate, houses too weakly constructed, coast-near buildings risk flooding etc. In the industrialised countries an adaptation will normally be possible. In developing countries – especially in estuaries or at coral reef islands – the sea level rise may lead to that the area must be given up. In the arctic a melting of permafrost has already had deleterious impact on roads and buildings.

Fishery

The oceans cover 70% of the global surface and yield 15% of the foodstuffs. Variations in the marine environment, as seen e.g. with the "El Nino", can have large local or regional societal significance. At a lesser scale in The North Sea there is a distinct climate effect on the catch of cod, which varies with the North Atlantic Oscillation (NAO). Globally over-fishing and destruction of breeding places are, however, at present considered more important than climate change.

Health and welfare

Human health and welfare depend directly and indirectly on the climate. In general the impact is considered negative with increased risk of heat strokes and extension of the vector borne diseases as dengue and malaria. In colder climates warming may be advantageous. With the increasing globalisation travel all over the world will become more frequent and will increase the risk of infection. A more effective vaccination strategy and higher degree of hygiene may counteract this.

Economic impacts

Recent years have seen a series of natural disasters, e.g. flooding of the European river systems. It is questioned whether such phenomena have become more frequent and are related to climate change, or they are a result of changed land use in the particular areas. Anyway the insurance sector is alert and increases in premiums are relevant.

A general increase in temperature and a change in precipitation will probably change the pattern of tourism, and it is speculated that Northern Europe will be more popular than the Mediterranean. Many ski resorts, especially in low-lying areas, may suffer from lack of snow cover.

20.5 Means of mitigation

It appears that climate changes cannot be avoided, and adaptation will be necessary. This is, however not a reason for giving up emission reductions, as any slowing down of the global warming is an advantage. It will in many cases also limit other types of air pollution. The mitigation takes place in many ways and by many different methods. For carbon dioxide the principal goal is an out-phasing of the use of fossil fuels and introduction of renewable energy sources and possibly fission and fusion power.

Reductions in energy consumption

The immediate mitigation is energy savings in the form of better insulation of buildings, more effective cars and other means of transport, and a transfer to collective transport. Most utensils like radios, TV, refrigerators etc. are becoming much more effective, but unfortunately the technological development has been overtaken by increased use, and generally the consumption has gone up.

Reductions in power production

Power plants are more efficient especially when driven in combined electricity and heat production for district heating. Natural gas emits less carbon dioxide per produced energy unit and is an option as long as the gas is available (Table 12.3 p. 262). It is a question though whether the gas is not better used a raw material in chemical production or as fuel for transportation.

Renewable energy sources

Use of all types of renewable energy should be developed: wind power, hydropower, solar power, wave energy etc. A remaining problem is the storing of energy from the sometimes irregular production of electricity, which may partly be overcome in diversified system with several of these options included. Especially hydropower may act a good buffer. Formation of hydrogen as a carrier may be a possibility but requires an extensive infrastructure at high costs.

It should be remembered though, that renewable energy sources also have environmental impact. Thus strong local opposition to wind power and hydropower has been observed. The same applies to nuclear power.

The opposite effects of climate impact on renewable energy sources has so far received little attention. Changes in precipitation, wind pattern, biomass production and solar irradiation may, however, be important in the future.

Carbon Capture and Storage

To the extent that formation of carbon dioxide cannot be avoided there are various types of depositing in geological formations, in forestation etc.

There is at present considerable world wide interest in developing the Carbon Capture and Storage, where CO_2 is captured at the power plants (many technical solutions) and subsequently suggested stored at the sea bottom or in suitable geological formations.

Reduction of the emission of other greenhouse gases

Methane from ruminants and animal manure can be limited by different feeds, and methane from rice paddies can be reduced by a change from wet to dry fields. Mitigation is, however, difficult because the emission is related to food production and arises from many small sources.

Methane from refuse can be collected and used as fuel. It both gives an energy profit and transfers the effective methane into the less dangerous carbon dioxide. The technique is not fully developed. Nitrous oxide comes mainly from fertilisation and a more effective management reduces the emission. Tropospheric ozone is reduced via reduction of the primary pollutants. CFCs are being phased out in connection with the protection of the ozone layer.

20.6 International negotiations

The report of the Brundtland Commission, "Our Common Future" (1987) was the first political recognition of the seriousness of the problem. The report saw the conflict between on the one hand rising global population and material standard of living, and on the other protection of environment and resources. But it did also see growth as a necessary tool in the fight against poverty and environmental degradation. Only the growth should be "sustainable". The report concluded that the industrialised countries within the next 40 years (i.e. before about 2030) should halve the energy use per capita, thus giving space for an increase in the developing countries of 25% per capita. Already now this ambition appears unrealistic. The industrialised countries can (or will) hardly reduce their energy consumption sufficiently, and the developing countries will not be satisfied with a 25% increase.

The already mentioned IPCC was established partly as a result of the Brundtland report, and after a series of preparatory meetings the UN organised in 1992 in Rio de Janeiro a world conference on environment and development. Here 155 parties signed a framework convention on climate change. The ultimate objective of the UNFCCC (The United Nations Framework Convention on Climate Change) is:

> "-- a stabilisation of greenhouse gas concentrations in the atmosphere at a level that would prevent dangerous anthropogenic interference with the climate system. Such a level should be achieved within a time frame sufficient to allow ecosystems to adapt naturally to climate change, to ensure that food production is not threatened and to enable economic development to proceed in a sustainable manner".

This convention is expressed in very general terms, but the main point is, that it acknowledges the fact that climate change can only be limited and not avoided. And that a certain measure of adaptation is necessary.

Precisely what is dangerous is to a certain degree a political question, but technically and scientifically it has been proposed that a temperature increase of 0.2 °C and a sea level rise of 2 cm per decade may be acceptable. Comparing this with the IPCC calculations it appears that we – without an extraordinary effort – risk the double of this. In addition it must be realised that these figures are average values, both with respect to climate change and sensitivity. Locally the impact can be greater, especially outside the temperate regions.

The Kyoto protocol

The climate convention has been followed by a series of international meetings, the most important being an international conference in Kyoto in 1997. Here it was agreed (the so-called Kyoto Protocol) that the industrialised countries, including Russia and Eastern Europe, as a whole should reduce their emission of greenhouse gases by at least 5.2% before 2008-12 compared to 1990. The protocol has legally binding commitments for six GHG's: CO_2, CH_4, N_2O, SF_6, HFC's and Perfluorocarbons. The GWPs for these gases are counted in the equivalent $CO_{2,eq}$ regulated by the Protocol.

There are different goals for the individual countries. Thus EU as a whole must reduce by at least 8%, whereas Iceland may increase the emission by 10%, Australia by 8% and Norway by 1%. Within EU there have subsequently been negotiated large differences between the different countries; Denmark e.g. has promised to reduce by 21%.

The goals for the individual countries need not be fulfilled directly; there is a series of mechanisms, by which obligations can be transformed into co-operation or technological development and support:

- Countries with emission obligations can in collaboration agree on a common total reduction – so-called "joint implementation". They can also buy emission permits from each other, so-called "emission trading" via negotiable permits or be credited for reduction projects in countries without reduction obligations
- Since it is only the effective emission that matters established sinks in a country, e.g. forestation can be used in the calculation of the total emission. Such sinks need not be located in the country proper, but can i.a. via economical instruments be established in another country – typically a developing country

These so-called "flexibility mechanisms" are all in principle rational tools that can ensure a cost-effective effort. In practice, however, they give complications, because they are difficult to control, and because they open up for a misuse of emission control in developed countries. Finally, some find it immoral that one can buy a right to pollution.

Another weakness with the Kyoto-protocol is that former Soviet countries have kept their emission rights from 1990. In fact these countries have seen a marked reduction in emissions since the collapse of the Soviet Union. That leaves an unused possibility of emissions – so-called "hot air" – that can be sold to other countries, permitting them to avoid otherwise necessary reductions. After the conference the former chairman of IPCC, Bert Bolin wrote that:

> "The inertia of the climate system was not appreciated fully by the delegates in Kyoto".

and that

> "The Kyoto conference did not achieve much with regard to limiting the build up of greenhouse gases in the atmosphere. If no further steps are taken during the next 10 years, CO_2 will increase in the atmosphere during the first decade of the next century essentially as it has done during the past few decades".

The development so far

The Kyoto protocol could only come into force if 55% of the partners representing at least 55% of the emission ratified it. That happened in 2004 when Russia ratified, but still without the United States and Australia. It thus only covers a part of the industrialised world – and not the developing world at all. This is major drawback, since the developing countries with about 80% of the world population are now already responsible for about half of the emissions. Therefore the protocol covers little more than one fourth of the emissions.

It may seem unfair to put the responsibility on the developing countries after the industrialised countries have had their material standard of living and they still have a large per capita emission. In a classical book the Indian scientists Argawal and Narain (1991) ask whether you can compare the carbon dioxide emissions from petrol consuming cars in Europe and North America, or for that sake anywhere in the third world, with methane emissions from the poor peasant's oxen or rice paddies in West Bengal or Thailand?

The greenhouse effect has, however, no feelings. It is only the emissions that count. In the long run the main objective for the industrialised countries will therefore not so much be a limitation of their own emissions as it will be a transfer of technology to the developing countries that allow them to obtain a reasonable standard of living without the pollution.

A necessary balance

Some look at the climate projections with scepticism or find that the costs of limiting the impacts are too high compared to the profits. Much of what has been written against "the climate threat" is however typical for a short-sighted attitude that puts the humanity in the centre. But even if one conceive the world as the property of mankind, one must realise that too rapid changes in global climate can increase international wealth differences and stresses flora and fauna by an impact on the natural and managed ecosystems.

It is therefore not a choice between the two possibilities: To prevent climate change or to adapt to them. It is a problem to find a balance between them. Sometimes discussions on adaptation is conceived as giving up the basic goal of preventing climate change – and thus "politically incorrect". This attitude is totally outdated. There is a growing awareness in the scientific, and also political, world that the entire climate problem must be seen from a comprehensive viewpoint and not only as a problem of reducing the emission of greenhouse gases.

Human induced climate change is, however, only a part of our overload of the Earth. Even if we succeed in controlling the isolated climate problem, there will still be problems with the growing world population and the related ethnic and territorial conflicts: Pollution, reduction of natural areas, overloaded water resources and rising living costs due to diminished food production.

There is no magical figure for the number of people that Earth can accommodate. It depends upon the wanted standard of living, and how much nature should be sacrificed. But the goal should be obvious: We must limit our influence on nature (including climate) in the best possible way. And when we do not succeed, we must be willing to pay the price in the form of adaptation to changes we brought about ourselves.

20.7 Literature

Agarwal, A. and Narain, S. (1991): *Global warming in an unequal world: a case of environmental colonialism.* Fuente, New Delhi

Cotton, W.R. and Pielke, R.A. (1995): *Human impacts on weather and climate.* Cambridge University Press, Cambridge, UK

deMenocal, P.B. (2001): *Cultural Responses to Climate Change During the Late Holocene. Science, 292, 667-673*

Downing, T.E. et al. (2000): *Quantification of uncertainty in climate change impact assessment.* In: Downing, T.E. Climate Change, Climate Variability and Agriculture in Europe. Environmemtal Change Unit. University of Oxford pp. 415-434

Greadel, T.E. and Crutzen, P.J. (1995): *Atmosphere, Climate, and Change.* Scientific American Library, New York

Hoffman, P.F., Schrag, D.P. (2000): *Snowball Earth.* Scientific American, January, 50-57

IEA (International Energy Agency) **(2006):** *Key World Energy Statistics*, IEA, Paris

IPCC (Intergovernmental Panel on Climate Change) has published four (Assessments) in respectively 1992, 1996, 2001 and 2007 plus a series of technical reports. These reports contain short "Summaries for Policymakers". May be obtained at: www.ipcc.ch

IPCC (2001): Climate Change 2001. Cambridge University Press, Cambridge. I. The Scientific Basis,. II. Impacts, Adaptations, and Vulnerability,. III. Mitigation, . They can be found at the internet: www.IPCC.ch

IPCC (2007): Climate Change 2007. Cambridge University Press, Cambridge. I. The Scientific Basis,. II. Impacts, Adaptations, and Vulnerability, III. Mitigation, . They can be found at the internet: www.IPCC.ch

Jørgensen, A.M.K. et al. (eds.). (2001): *Climate Change Research – Danish Contributions.* Danish Climate Centre, Copenhagen, Denmark.

Lamb, H.H. (1982): *Climate, history and the modern world.* Methuen, London.

Lutz, W. (ed.) (1994): *The future population of the world: what can we assume today?* Earthscan Publications, London.

Mackenzie, F.T. (1998): *Our Changing Planet.* Prentice Hall, Upper Saddle River, New Jersey.

Meyers, N. (1993): *Environmetal Refugees in a Globally Warmed World.* BioScience, *43,* 752-761.

Parry, M. (ed.), 2000: *Assessment of Potential Effects and adaptations for Climate Change in Europe.* The Europe Acacia Project. University of East Anglia, Norwich.

The Brundtland Commission (1987): *Our common Future.* Oxford University Press, Oxford.

Worldwatch Institute in Washington, USA, publishes every year a report: "State of the World" with articles about energy consumption, agriculture, population increase, climate change etc.

VI

Economy and legislation

The extent and composition of air pollution depend upon the overall socio-economic development. That again depends upon the production and demand, the technological development and the institutional conditions in the form of various national and international legislations and conventions.

To some extent the chosen level of pollution is determined by a tacit or explicit cost-benefit consideration, which implies the evaluation of external costs of impacts of air pollution. Most controversial in this context is the estimation of the cost of a human life. This is discussed in Chapter 21.

Finally Chapter 22 gives a more detailed and comprehensive treatment of the EU legislation and administration than appears from the previous more specialised chapters.

21 Economics of air pollution control

Mikael Skou Andersen

Smoking chimneys have traditionally been associated with economic growth and regarded as a more or less inevitable nuisance to be tolerated in order to improve welfare. However, at least since the famous London killer fog of 1952, it has been clear that air pollution can have lethal impacts and, in developed as well as in developing countries perceptions are changing. Although the myth of an inevitable trade-off between economy and ecology continues to be nurtured, perhaps especially by those that thrive on free access to the commons, a philosophy of ecological modernisation which emphasizes pollution control as a means to improve overall welfare by minimising unwanted residuals and economising on energy resources appears to underlie several recent policy initiatives.

Three basic reasons for the changing perspectives on air pollution

First of all due to extensive scientific research over recent decades the negative impacts of air pollution on health, buildings and vegetation are now better understood and documented. Secondly, a variety of air pollution abatement techniques are available on the market and their costs are far from excessive, especially not those that address sulphur and other conventional pollutants. An expanding market for air pollution control helps drive technology costs down and supports fuel shifts. Thirdly, methods for cost-benefit analysis have been improved in recent years, so that the costs of measures for controlling pollutants can be directly compared with monetary estimates of the benefits that pollution control will bring, especially with respect to expected health improvements. As air pollution is trans-boundary in nature, policymakers are often interested in the extent to which costs accepted under a regional agreement involving a number of countries will be matched by benefits in a way that is equitable and fair to all involved parties. Building on state-of-the-art in atmospheric modelling economic analysis can be extended to address such questions.

The Kuznet's curve

Under the conceptual label of the Kuznet's curve (Ekins, 1997) it has been hypothesised that countries will have to grow rich before they can afford to become cleaner by reducing pollution (Figure 2.10, p. 59). It can indeed be shown that countries with high levels of GDP per capita have greater accomplishments in air pollution control than have developing countries. While Kuznet's curve has captured a tendency valid for recent decades with regard to SO_2 reduction, there are indications that recent improvements in cost-benefit analysis are changing the patterns under which abatement is introduced. China is one country that, informed by economic analysis, is introducing stricter air pollution control measures.

How to promote implementation

Once air pollution abatement has been identified as economically desirable, questions arise as to how to promote implementation. Concern has traditionally focused on avoidance of daily peak concentrations and regulations detail thresholds for concentrations that should not be exceeded. In recent years and under the aegis of regional conventions, focus has shifted to approaches that aim to reduce overall emissions in order to reduce long-term impacts on sensitive ecosystems and health conditions, based on annual emissions ceilings and national reduction targets for pollutants. While stationary sources are regulated under permit systems with guide values, mobile sources are usually controlled by mandatory technological requirements which are based on technology-specific solutions, sometimes derived from a best-available-technology approach. Economists are uneasy with such detailed prescriptions, as mandatory technological requirements may lead to a loss of economic efficiency, as will be explained in more detail below. The regulatory tool-kit economists favour use of economic policy instruments, such as pollution fees, taxes or emissions trading systems which can lead to the economically most effective abatement allocation. In addition, economic policy instruments have proven to be very helpful tools for providing the right incentives to attain the desired emission reductions within a reasonably short time, without extensive bargaining between regulators and emitters.

This chapter provides nontechnical introduction to and overview of methods and principles for economic analysis of air pollution abatement, in order to make the reader more familiar with the fundamentals of this topic. The polluter-pays principle as formulated by the OECD in 1975 suggests that control measures should not be subsidised by tax-payers, in order to avoid distortion of trade and maintain a level playing field for competition. Use of economic policy instruments reflecting this principle extends economic principles into the legislative framework and the chapter explains how this can be seen as a useful complement to the framework of limit values and allowable daily/hourly means which is traditionally applied for regulatory purposes.

21.1 A basic concept: externalities

The famous economist from the University of Cambridge, A.C. Pigou, in his 'Economics of Welfare' (1920) cited observations that on an annual basis no more than 12 per cent of the sunlight astronomically available was reaching the city of London, and he hypothesised that the notorius fogs were in fact exacerbated by the smoke and air pollution from burning of coal. Air pollution occurs as side-effects of otherwise legal economic activities between economic parties in the marketplace. Pigou proposed to conceptualise air pollution as an un-priced 'externality' – external cost – a concept which in recent decades has risen to prominence.

By definition external costs are un-priced costs that arise as a side-effect of transactions in the marketplace between two parties and affecting third parties. If in order to market goods or services and sell it to a consumer a producer causes pollution inflicting other parties, the damages incurred by this transaction are regarded as external costs. Externalities need not always be negative, effects on third parties can also be positive, for instance if pollution acts as fertiliser for crops or, taking an example from the natural world, where a bee pollinates a flower. The point is that as long as these external effects are not assigned a price, they do not form part of the calculi in the marketplace and they constitute a form of what is known as 'market failure'.

Prices reflecting the actual costs

Ideally prices should take account of all relevant factor costs. Consumers that face alternative products and services are misguided as to the relative use and value of society's resources if external costs are not factored in. Power generated from coal can, at face value, be brought to the market at a lower cost than wind-based electricity; as long as external costs are not factored in, the consumer has no information as to the real relative costs. In consequence of the distorted market transaction, demand for coal-based electricity dominates, whereas there will be little or no demand for wind energy, apart from activities due to possible government support. As demand fails to reflect the underlying 'true' costs market failure occurs, as the appropriate signals concerning the economically best use of factors of production and society's resources are not provided. Economists favour 'efficiency' as the best solution, which is a situation where allocation of factors of production takes place in an optimal way, considering respective values and costs.

Evidently, arriving at relative external costs remains an empirical question and only few transactions are completely free of external costs. Whereas fossil fuels inflict external costs on health, buildings and vegetation, wind energy may inflict external costs by means of noise, light reflection and lost amenity values. Being unable at the time to account for the specific costs imposed by air pollution on Londoners, Pigou indicated its relevance by quoting surveys identifying monetary damages to buildings and extra costs for laundry and artificial light (1920). However, as we shall see in the next section relatively advanced methods are now available for providing good assessments of externalities.

The use of externality estimates

Once reliable estimates for externalities have been attained they can be useful in a number of ways. Many government agencies use such estimates for the purpose of economic analysis in which various alternatives are compared, e.g. cost-benefit analysis. The difficulty with cost-benefit analysis is that civil servants try to arrive at the optimal solution. This may work reasonably at project level, but once projects affect wider market dynamics the ability of government agencies to predict the wider macroeconomic effects reliably is not always convincing.

Pigou's favoured idea was to introduce externality adders directly to the market; that is, to introduce a correction of the prices distorted by the absence of externality costs. The principal method is to introduce a levy, known as a Pigouvian tax. Such taxes, for the external effects in question, allow the marketplace itself to clear demand and supply without additional government interference and are hence regarded as the first-best solution. Where positive externalities arise, the estimates can also be used to justify a subsidy.

Many countries have introduced Pigouvian taxes and an overview of experience is provided in EEA (2006). As it can be difficult to introduce taxes for political reasons, negative externalities are sometimes used to justify subsidies for the preferred alternatives. The negative cost of CO_2 is used to justify a positive premium on renewable energy, for instance. From an economics perspective this approach is less attractive, as it fails to bring about an optimal outcome. Subsidies have to be financed by other taxes, for instance on labour, which may lower economic activity and employment. Only taxes that directly internalise un-priced external costs can be regarded as non-distorting and as a way to avoid market failure and improve the functioning of the market.

It follows from this logic that a price tag should be attached to air pollution. The price or tax will not lead to full elimination of such pollution, but to a level of pollution that can be regarded as economically efficient. Some amount of pollution will need to be tolerated, because the costs of completely eliminating all negative effects most likely exceed the benefits of production and consumption. When the marketplace is informed about the costs of pollution, it will adjust its use of resources to take better account of the losses inflicted on society, and the marketplace, at least in theory, will ensure that supply and demand are balanced to achieve an optimum.

The key question which the above brief summary of the theoretical insights of environmental economics leaves us with is therefore whether there are sound and scientifically based methods available to account for external costs. In the context of this book it is fortunate that some of the pioneering work on calculating external costs has been carried out with particular emphasis on air pollution.

21.2 The impact pathway approach

Economics is based on the assumption that individuals are utility maximising and have clear preferences. In order to ascertain the price tag that could be attached to air pollution, economists need to unravel the preferences for avoiding air pollution and to have these preferences expressed in monetary terms. For this purpose environmental economists apply a variety of methodologies. Sometimes observations in the market relating to the demand for other goods can offer insights to the willingness-to-pay to avoid air pollution, for instance where property prices are affected by changes in environmental quality. Calculating the precise signal from changes in environmental quality requires advanced econometric methods, and for this reason a simpler approach is sometimes employed, whereby willingness-to-pay questions are posed in surveys of individuals. In order to determine external costs, individuals are asked how much they are willing to pay to obtain a change in environmental quality – the price arrived at can be used to represent the damage caused by pollution. By aggregating the results from such surveys insights can be gained on the monetary values of environmental goods and damages as appreciated by human beings.

Valuation challenges

These methods are not however uncontested and especially the second method – hypothetical valuation, in which individuals are required to express a price for the value of an environmental good – has been deplored for its inability to generate consistent results (Diamond and Haussmann, 1994). Hypothetical, or contingent, valuation remains a relatively young discipline, however, and methodological rigour in survey and aggregation techniques is important for reliability of results. One needs always to check the original survey design and to reflect whether monetary values have been attained with a methodology that can pass the test of validity and reliability.

With respect to air pollution, a relatively advanced methodology has been developed which aggregates information from valuation surveys together with atmospheric modelling and epidemiology to provide for damage estimates for individual pollutants (Friedrich and Bickel, 2002). This method, which is named the 'impact pathway' method, traces the transport and transformation of pollutants to allow damage estimates that are in principle site-specific. This is because the economic impact of air pollution will depend critically on exposure; in cities more people will be affected whereas in rural areas fewer, and the numbers exposed will affect the economic damage that e.g. one ton of sulphur will cause when emitted at different sites.

Marginal change: concentration increments

A main methodological challenge in accounting for external costs is that it is necessary to know the marginal contribution of a particular economic activity to the state of the environment. In the case of air pollution one needs to know the contribution of, say, a particular power plant's emissions to the air pollutant concentrations in a particular area. Such changes in air pollution concentrations, and in subsequent exposures, can be calculated by applying state-of-the-art atmospheric models. This is the first step in the impact pathway methodology.

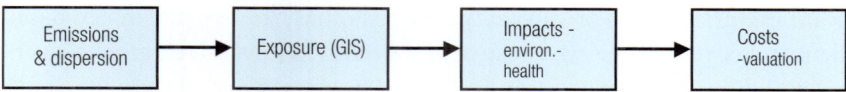

Figure 21.1 The impact pathway chain.

The modelling defines the concentration increments that arise as a result of an addition to a source, and is performed by combining conventional local-scale air pollution models with comprehensive regional-scale air pollution models. A baseline scenario provides the background concentrations and chemistry, and a scenario for the emitter shows the new concentrations. Then, by deducting the former from the latter the marginal concentrations are arrived at.

Once the marginal changes in concentrations are known, the second step in the impact pathway is exposure. Geographical information systems (GIS) are used here to keep record of land-use and population densities so that, for instance, the number of people exposed to air pollution is matched directly with changes in air pollution exposure in specific parts of the geographical grid. In this way the analysis can capture potentially significant exposures in areas with high population densities. The method is highly site-specific in its accounting for external effects, as both the emissions and the exposures are calculated according to the specific site in question.

The third step in the impact pathway assessment involves exposure-response functions for the effects of air pollutants. Exposure-response functions describe how various entities will react to changes in exposures. Vegetation cover, for instance, reacts to changes in air pollutant concentrations, as do humans. The medical literature is especially rich on exposure-response functions for impacts on human health; impacts relate both to morbidity and mortality – as statistically some individuals, especially among elderly, die from exposure to air pollution. These exposure-response functions are, with current knowledge, often treated as linear, but there can be important thresholds to consider, below which it is possible that no effects occur.

The fourth and final step involves attaching unit values to the endpoints of the impact pathway analysis. With regard to air pollution, important endpoints comprise both mortality and morbidity effects for humans; the latter comprise work-loss days, cases of bronchitis, intake of asthma medication, hospital admittances, etc. The endpoints can be valued with preference-based values based on contingent valuation surveys or alternatively with cost-of-illness figures that reflect actual costs. Because of the uncertainties related to hypothetical valuation, authorities in some countries prefer to use cost-of-illness data. As cost-of-illness data only include the direct costs for hospitalisation, medicine and sick-days, etc. they are often scaled by a cost-of-illness factor, usually of 2 or 3, to take account of the welfare loss associated with pain and grievance. For mortality it is customary to use preference-based figures. Because of the significance of mortality for overall external costs of air pollution a separate section is devoted to present and discuss this issue. Before proceeding to this, the following section provides examples of how external costs have been calculated for air pollution.

21.3 Modelling of external costs from power plants

In the following it is shown how external costs of air pollution have been modelled using the impact pathway methodology and taking advantage of atmospheric modelling for three different combined heat and power (CHP) producers situated in the greater Copenhagen area. Two of the CHPs are mainly coal-based, whereas the third is a large waste incinerator that produces heat and power. Of the coal utilities only one has complete scrubber facilities with desulphurisation and de-NO_x. It is situated in the centre of Copenhagen, whereas the other is situated outside central Copenhagen and is without de-NO_x.

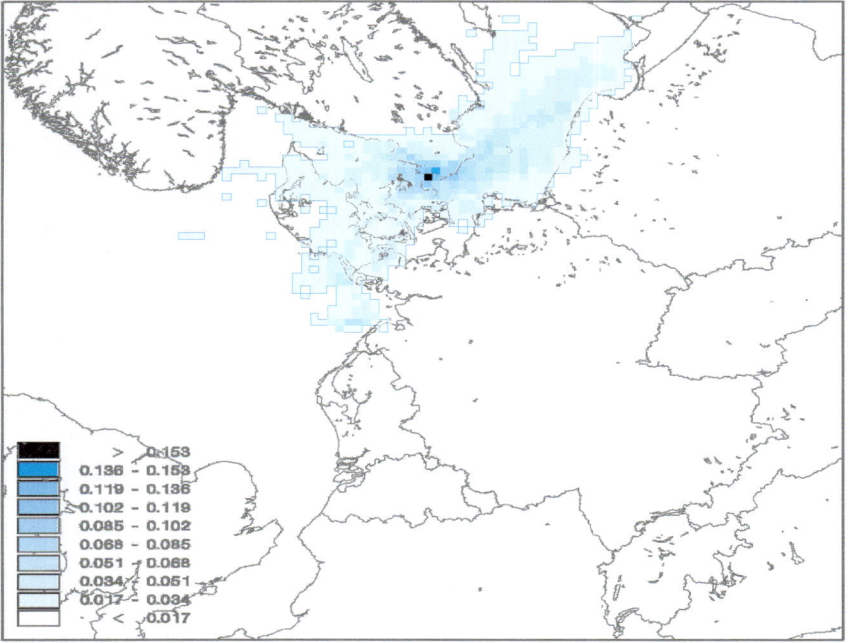

Figure 21.2 Changes in annual delta-concentrations of O_3 (mg m^{-3}) in year 2000 from emissions of Amagerværket according to the regional-scale model of EVA (Lise Frohn, cf Andersen et. al., 2008).

Modelling principles

Based on local-scale and regional-scale atmospheric models (OLS and DEHM respectively), transport and transformation of the primary pollutants, SO_2, NO and $PM_{2.5}$, have been modelled in two scenarios. Secondary pollutants that are formed in atmospheric-chemical processes, notably SO_4^{-2}, NO_3^- as well as O_3 are accounted for too. The modelling arrives at delta-concentrations that represent the difference between the two scenarios and which are regarded as the marginal concentrations attributed to the plume from the power plant in question.

Modelling is performed in a grid of cells in conformity with the EMEP

Health effect endpoint	Exposure-response function per µg m^{-3} y^{-1}	Valuation Euros (2004-prices)
MORBIDITY		
Chronic bronchitis	3.2×10^{-5} (adults)	50,360 per case
Restricted activity days Hospital admissions	8.4×10^{-4} ÷ hosp. adm. (adults)	116 per day
- respiratory	3.46×10^{-6}	7,409 per case
- cerebrovascular	8.42×10^{-6}	9,387 per case
- congestive heart failure	3.09×10^{-5} (>65years)	15,450 per case
- lung cancer	1.26×10^{-5} (adults)	20,150 per case
Asthma children (7,6%<15years)		
- bronchodilator use	1.29×10^{-1}	20 per case
- cough	4.46×10^{-1}	54 per case
- lower resp. symptoms	1.72×10^{-1}	14 per case
Asthma adults (5,9%>15years)		
- bronchodilator use	2.72×10^{-1}	20 per case
- cough	2.8×10^{-1}	54 per case
- lower resp. symptoms	1.01×10^{-1}	14 per case
MORTALITY		
Acute mortality, SO_2	7.85×10^{-6}	1,941,000 per case
Chronic mortality, PM	1.138×10^{-3} (>30 years)	71,000 per yoll
Infant mortality, PM	4.68×10^{-5} (<9 months)	2,911,700 per case
Acute mortality, O_3	3.27×10^{-6} *SOMO35	1,941,000 per case

Table 21.1. Exposure-response functions and unit values applied for assessment of the damage costs of air pollution cf. Andersen et. al. (2008). Exposure-response functions are without scaling and adapted to the age distribution and mortality rate of the Danish population. For mortality effects the monetary values follow Alberini et. al. (2006) adjusted with purchasing-power parities (yoll is years of life lost). *SOMO35 denotes the number of days where the diurnal max. 8-hour mean concentration of 35 ppb O_3 is exceeded.

methodology and the delta-concentrations are modelled separately for each cell in the atmospheric grid, each 16x16 km. With the help of this approach and in combination with detailed data for population density drawn from the CPR (Central Person Registry) it is possible to further model changes in human exposure. In the local-scale modelling a higher grid-resolution has been allowed and CPR data which is available at address level has been aggregated in 1x1 km grid cells (cf. Figure 21.2).

Exposure-response functions are drawn mainly from the European Commission's Clean Air for Europe project, where they were verified through a consultation procedure with medical experts in committees of WHO, the World Health Organization. The most significant exposure-response functions refer to mortality (acute and chronic death) as well as morbidity (in particular work-loss days). Table 21.1 provides an overview of the applied exposure-response functions for the various health-effects as well as the applicable population sub-groups.

In identifying unit values for health effects, reference for morbidity is primarily to costs-of-illness as reported by health authorities, whereas for mortality the chosen value is informed both by international and Danish studies (Table 21.2).

	EVA (m_2000)	EVA (m_2001)	EVA (m_2002)	EVA Average
$PM_{2.5}$-primary	9.95	13.42	11.45	11.61
SO_4^{-2}/SO_2	8.84	8.84	8.86	10.07
SO_2/SO_2	1.11	1.46	1.12	
SO_2-sum	9.95	10.30	9.97	
NO_3^-/NO_x	5.91	7.45	7.10	6.85
O_3/NO_x	-0.04	0.03	0.05	
NO_x-sum	5.92	7.48	7.15	

Table 21.2 External costs of air pollution emissions in Euro per kilogramme pollutant (2004-prices) for three consecutive meteorological years (power plant located in rural area near Copenhagen).

Meteorological variability

The external costs modelled (see Table 21.3) are an average of three meteorological years from 2000-2002, whereas the background meteorology is the 1998 EMEP dataset. NO_3^- and H_2SO_4 are secondary particulates which tend to be transported over longer distances, while $PM_{2.5}$ and SO_2 are emissions that produce damage predominantly in the local area (within the 50x50 km area). Results are specified for primary and secondary components of sulphur and also include the net effect of ozone. Ground level ozone is created mainly in more southern areas of Europe, and transported regionally to affect Denmark.

Here, a positive externality is apparent, as ozone is removed in the immediate vicinity of the smoke gas plume due to reactions with NO_x. However, in the summer period photochemical reactions will take place which lead to formation of ozone along the smoke plume transport path. Ozone also reacts with NO to form NO_2. Although it is widely acknowledged that NO_2 has negative health effects, a separate exposure-response function that disentangles the specific NO_2 impacts on mortality and morbidity cannot be specified. Once NO_2 is properly accounted for, the net contribution to the damage estimates of ozone formation and depletion may be subject to adjustment.

Three power plants

Table 21.3 provides the results in summary format for three Copenhagen power plants. It should be noted that the damage costs are site specific, in that for the predominantly local-scale pollutants, $PM_{2.5}$ and SO_2, the results in urban locations differ markedly from those in rural locations, whereas for the regional-scale pollutant of NO_x, with transport over a much greater area, differences are averaged out. Fluctuations are limited and would allow regulators to ascribe NO_x reductions a standard value of about €7 kg^{-1}, whereas for $PM_{2.5}$ and SO_2 separate values would apply for urban and rural emissions, respectively.

€ kg^{-1}	Rural CHP	Urban CHP (coal)	Urban CHP (waste)
$PM_{2.5}$-primary	11.61	17.84	19.40
SO_2-sum	10.07	15.48	15.60
NO_x-sum	6.85	6.16	6.16

Table 21.3 Site-specific external costs of air pollution emissions in Euro per kilogramme pollutant (2004-prices) for three CHPs located in the Greater Copenhagen area (cf. Andersen et. al., 2008).

Copenhagen is located on an island and a considerable portion of the pollutants emitted affects annual concentrations over the marine territory only. Although the model captures these changes it should be emphasised that no population is assumed to be exposed offshore. Similar emissions in a location in continental Europe, with greater population exposures and no marine areas over which to dilute the emission, would be appreciated differently in terms of damage costs.

21.4 Uncertainty: valuation of statistical lives

While there are many implications of air pollution, the most significant in monetary terms appears to be related to mortality. The exposure-response function indicated in Table 21.1 above for the relationship between three air pollutants and mortality relies in the main on insights gained in cohort studies carried out in the United States between 1979 and 2002 (Pope, 2002). A cohort of more than 500,000 individuals was established by the American Cancer Society and on the basis of death certificates collected from the deceased, independent researchers have established statistically significant relationships for what is termed chronic death, indicating increased mortality in particular for cardiovascular death and lung cancer, of which some is believed to be attributable to mechanisms relating to air pollution.

Compared with earlier and smaller studies the Pope exposure-response function appears to be in the lower end of the continuum. In a follow-up study it has also proven to be fairly robust. On this basis all-cause mortality is assumed to increase by 6% in response to an increase in air pollution with $PM_{2.5}$ of 10 µg m^{-3}. In the terminology chronic death is distinguished from the acute deaths observed in simple pollution events such as London's killer fog. Chronic death is mortality attributed to longer-term exposure, equivalent to smoking-related mortality (where no single cigarette-smoking event is responsible). All deaths are of course to some extent acute, but acute death is taken to refer to mortality related to an acute episode of both cause and effect.

An ethically sensitive issue

Evidently, valuation of the loss of human lives is an ethically sensitive issue. No single human life can or should be accorded a monetary value as such. It is the value of preventing a statistical fatality which is in question. In order to express the benefits of measures that will reduce mortality as externalities in economic analysis it becomes necessary to be able to assign a monetary value to the prevention of statistical fatalities.

Figures for the 'Value of a Statistical Life (VSL)' have been developed and applied for many years in transport economics. Various methods have been applied to try and figure out the appropriate value of a statistical life. The 'human capital' approach assumes that the value of a statistical life should be based merely on what the individual represents in terms of capital to society, e.g. the productive contribution. Under this perspective there is some ambiguity as to how to treat retired people, who no longer represent a productive value, and thereby may sometimes not be attributed any value at all. With its productivist perspective, the human capital approach also does not accord any monetary value to life itself, nor to the grief and pain of losing it, either for relatives or for the deceased themselves.

A welfare economic approach

In contrast to the reductionist human capital approach OECD's cost-benefit guidelines (2006) recommend using a welfare economic approach under which the welfare loss is attributed a monetary value alongside the productive value; this approach conceptually represents a more comprehensive estimate for the value of preventing a statistical fatality.

Some decades ago there was a tendency to identify welfare economic VSL on the basis of wage-risk studies, i.e. VSLs based on careful studies of the wage premiums that workers would be able to negotiate in jobs more prone to mortality risks (Viscusi and Aldy, 2003). This approach in general led to fairly high and, to some extent, unrealistic estimates. In recent years the tendency has turned more towards surveying people directly about their willingness-to-pay for risk reductions, in some cases complemented by what people actually have been willing to pay, e.g. for airbags in cars. Although these methods have led to somewhat lower estimates than wage-risk studies, many administrators and policymakers have remained sceptical. In a comprehensive study for Denmark's Road Directorate a VSL of more than €2 million was found, but the result was not accepted by the Danish Ministry of Transport who had sponsored the study and who continued to apply a value in the region of €1 million. This experience paralleled that of Jones-Lee (1985); in his famous study for the UK Ministry of Transport he found a value of £1 million, but the authorities required him to present a 'lowest plausible value'. The value of £0.5 million suggested was applied by the UK authorities for many years, and eventually transferred to the European Union, adjusted for inflation, where it continues to be applied!

Difficulties in obtaining reliable and valid expressions of preferences in contingent valuation studies based on willingness-to-pay questions have already been indicated above. In the case of questions of risk avoidance, difficulties rest not only upon obtaining reliable estimates for payments but also whether individuals have an appropriate understanding of small risk changes. Some individuals are very risk-adverse and may indicate unrealistically high estimates of the payments they are willing to make, while neglecting the constraints posed by their personal budgets and consideration of the many other risks which they face and for which a premium also would have to be paid. Survey methods are continually improving, however, and uncertainties are appearing to diminish.

Lost life-years

In the case of air pollution the belief is that it is mainly the elderly who are prone to be vulnerable to air pollution, and hence the number of life years lost is somewhat lower than in the transport sector. Many elderly people do die in traffic incidents, but the average transport victim is nevertheless believed more likely to be more middle-aged (45-50) than air pollution victims. For this reason and to avoid assigning unrealistically high VSL values to air pollution reductions it has become customary to use an alternative method – the 'Value-of-Life-Year' (VOLY) – whereby a value is assigned to each life-year that statistically can be saved. This method seems to have gained ground in the EU Environment Directorate, whereas the US EPA continues to apply the VSL approach for all fatalities. Whereas many studies provide support for VSL estimates in the range of €1-2 million, there are only a handful of original studies that report VOLY estimates. Earlier in this chapter are used the median-value reported by Alberini et al. (2006), which resulted from the EU-funded NewExt study. One of the difficulties with the VOLY approach is that it becomes a requirement to develop an estimate for the actual number of life-years lost as a result of the risk in question, an estimate which is often not at hand and needs to be modelled separately based on population life-tables and risk factors (the interested reader is referred to Friedrich and Bickel, 2002).

21.5 Use of externality estimates in cost-benefit analysis

The above derived externality estimates can be applied in project-based cost-benefit analyses as well as in more comprehensive macroeconomic modelling of the economy-energy-environment interfaces. The European Union requires for example that projects supported by EU structural funds undergo a project-based cost-benefit analysis. Although externalities of course represent a cost to society, in cost-benefit analysis they constitute the benefits. This is because the pollution costs foregone as a result of pollution control represent the monetary benefits of the policy.

Site-specific externalities

It follows from the atmospheric modelling integrated in the impact-pathway approach that the externalities are to a great extent location specific and hence that they are most suitably modelled with reference to individual projects. There are also ambiguities related to the use of cost-of-illness figures, which will differ between countries with different health-care systems. Finally the valuation of mortality is based on willingness-to-pay for risk reductions, which will differ between countries with different income levels. For this reason the above figures for Copenhagen cannot easily be transferred to other countries; on one hand the offshore area off Copenhagen offers a dilution opportunity which does not exist everywhere, on the other hand the level of income is high compared with the EU average.

In the Danish context the figures have been used to assess implications for the marginal social costs of electricity production. Denmark's power sector provides more than 70 per cent of electricity based on coal, but Denmark also has a thriving wind energy industry. At present coal has a competitive advantage, but this has narrowed after the introduction of the EU's emissions trading scheme which requires that carbon costs are factored in. Even with a conservative assessment of air pollution costs, wind energy gains significantly in appraisal.

Modelling in RAINS and E3ME

While use of externality estimates for assessment of electricity generation options represents a project-based appraisal, such figures have also been integrated in much more comprehensive macroeconomic modelling merely as one set among many other economic figures to take into account when considering the implications of different policy scenarios. In the context of climate policy the reduction of conventional air pollutants such as SO_2 and NO_x are usually termed 'ancillary benefits' as they constitute an additional advantage of the carbon reduction achieved when fossil fuels are phased out by means of energy savings or shifts towards renewable energy. For this reason several models of the European economy try to factor in this gain by using the externality estimates. One such model is the E3ME model from Cambridge, which has been the preferred macroeconomic tool for the UK government for appraisal of climate and energy policy over a number of years.

At a multilateral level, within the UN's Economic Commission for Europe (UNECE), where negotiations take place on international agreements to reduce in particular trans-boundary air pollution, a different and very comprehensive modelling framework has been established to produce cost-benefit analysis of proposed measures. The name of this modelling framework, which guides all the international negotiations, is the RAINS-GAINS model developed by IIASA in Vienna.

The original RAINS model is an atmospheric model which can account for transport and deposition of emissions on a pan-European scale. The RAINS model can show the implications of reducing emissions by x % in one country, say Germany, for all the remaining countries. The RAINS model is important for obtaining a sense of fairness in international negotiations but also for tuning reduction targets in to what makes sense from an ecology and health point of view. Some ecosystems are more sensitive to acidification than others and, in a similar line of thought, large population centres will derive more benefit from air pollution control than more peripheral regions.

The RAINS-GAINS model in its present version makes an ambitious attempt to integrate atmospheric modelling with an extended cost-benefit model that captures the externalities as well as the costs related to measures in different sectors and countries.

On the benefit side the basic methodology is outlined above; via the impact-pathway methodology the physical implications of changes in emissions and exposure can be modelled. Besides health impacts, the RAINS-GAINS model includes impacts on vegetation. It does so, however, with averaged figures for health benefits, disregarding possible differences in cost-of-illness and mortality valuation across Europe. Decision-makers, especially in the European Union, seem to be uneasy with assigning different values to reductions in mortality risks in different member states, hence seeking to avoid the implication that human lives are worth more in rich countries than in poor.

On the cost side the RAINS-GAINS model includes a comprehensive database of technological measures available for reduction of various pollutants and their implementation. Considerable effort has been put into assuring that the cost module accurately reflects national price levels and so the economic accuracy here is greater than on the benefit side. One caveat is that the model only includes specified technological measures; it fails to take account of more structural changes in economies, such as new transport modes or fuel changes in the power and heating sectors. As such it includes mainly conventional end-of-pipe measures.

Great strides have been made in recent years in improving the basis for decision-making and cost-effectiveness in the field of air pollution control. Above three examples of available methods are mentioned; project-level, macroeconomic and integrated assessment. No doubt these models and methods will undergo further development and improvement in the years to come.

21.6 Policy instruments

Once decision-makers have agreed on the reduction targets to aim for in air pollution control the question arises on which policy instruments that should be put in place to facilitate implementation. This is by no means a simple technical issue; choice and design of policy instruments can be fairly decisive for whether or not policies will succeed at all. A law is only a piece of paper, how far it will be implemented depends on the institutional framework for its administration, the resources available and whether the regulated have incentives to cooperate. These circumstances are unfortunately not always given due attention. Sometimes odd compromises are reached, where ambitious targets are outlined, but policy instruments are lacking in capacity – symbolic decision-making can for some policymakers be a convenient way to balance opposing interests.

Command-and-control schemes

Economists are usually critical of the command-and-control schemes that engineers tend to recommend for pollution control. Command-and-control approaches, where polluters are required to adopt pre-specified best-available technologies do work in some countries with a strong legal and regulatory culture, such as Germany, whereas in other countries where business interests are more prone to challenge state regulations, such as the USA, they are more or less doomed to some degree of failure. There is an extensive literature on the failure of the 1971 US Clean Air Act, which on paper had the strictest requirements globally, but failed to produce the right incentives to car manufacturers and industry. While US car manufacturers, in the culture of US litigation, fought the requirements for catalytic converters in court, Swedish and Japanese producers brought the technology to the market (Lundqvist, 1980; Tsuru and Weidner, 1989).

Economists have provided some support to the case of business interests by showing that command-and-control measures often lack the flexibility which will lead to adoption of the most efficient responses to control requirements. If all emitters are required to reduce emissions by the same percentage it may *look* fair, but as they will often in reality have very different costs of abatement it will neither be fair nor economically efficient for society as a whole. From an economic perspective the preferred solution is that pollution is reduced at those sources where the lowest abatement costs can be identified. If targeting the cheapest reduction opportunities introduces a fairness issue it is better subsequently to transfer money from actors with high abatement costs to actors with low abatement costs to level out the burden.

The Pigouvian pollution tax

From an economic perspective there are policy instruments which are economic in nature, in that they provide the right incentives. A Pigouvian tax will provide a continuous incentive to polluters to consider whether some amount of control and abatement is more efficient than paying the tax – and it is likely to promote abatement among all liable to the tax up to the level of the tax rate. Coase (1960) lambasted the Pigouvian pollution tax scheme for the complexity potential in its management and claimed that transactions costs of administration would by far outweigh possible gains. It was only by coincidence that the world learned the potential of using a market-based instrument to address pollution and came to appreciate its significance; Japan pioneered such an approach, but due to geographic and cultural distances, it took some time before experiences became known outside the narrow circles of environmental economists.

Japan's SO_2-tax

Japan introduced a levy on SO_2 from point sources from 1974 – and emissions quickly began to drop. A few years later Japan with SO_2 emissions at less than 10 kg cap^{-1} y^{-1} was the world champion in emissions control! Compared with USA, where industry took air pollution control regulations through the courts and delayed implementation, Japan had taken an effective lead in emissions abatement. When European Union directives were issued in 1988, after lengthy negotiations and a Treaty mandate, it was the Japanese companies who were in control of the desired technologies. The first scrubbers arrived in Germany from Japanese suppliers via the Trans-Siberian railway and Japanese licences prevail in the manufacture of this equipment across the globe (Sprenger, 1998). The emissions trading system with tradable quotas for SO_2 introduced in the USA in the 1990's was effective too, but largely on the background of relatively high emission levels compared with Japan and Europe.

Introduction of Japan's SO_2 levy was largely uninformed by economic theory and was in fact an outcome of the famous 'pollution trials' in which the compensation claims of victims of Japan's rapid post-war industrialisation were upheld. The Japanese courts decided that the financial claims should be addressed by imposing a 'penalty fee' on air pollution emissions, but the scheme's motive of compensation was largely unimportant in relation to its subsequent environmental and economic effectiveness in curbing emissions. As noted by Matsuno and Ueta (2000) the economic stimuli of the levy triggered action in the form of a race to avoid payments (to the compensation fund for victims). At the same time, a number of large emitters took emission control even further than action based on pure optimality criteria would suggest, reflecting a desire to act in a socially responsible manner under the new paradigm. Pigou's externality principle requires that emissions are priced according to their social costs. To the extent that compensation reflects social costs, the Japanese scheme closely echoed this principle.

21.7 The present situation

In current times, air pollution taxes or levies are in place in a range of countries, but differ widely in coverage and rates (see EEA, 2006, for a detailed inventory). In Japan, SO_2 emissions declined to a level so low that a different scheme had to be introduced to guarantee the financial basis for pensions to pollution victims. One country that in recent years has introduced an SO_2 tax is Denmark and here emissions, which already were at a low level, declined by about 80 per cent, making Denmark the reigning champion with regard to SO_2 emissions per unit of GDP. The success of the relatively modest tax rate for SO_2 emission of about €1.3 kg^{-1} shows that control technologies are available at an even lower rate. In virtually all countries where taxes or levies on air pollutants are applied, they form just part of a comprehensive regulation package involving specific limit values as well as thresholds for peak values on a daily or a per hour basis. Economic instruments are to an extent regarded as blunt in that they cannot always guarantee that specific quality criteria are met. In the interest of safeguarding quality criteria, environmental inspectors are often not willing to sacrifice detailed control for more flexible economic instruments; however what they often tend to neglect is that the tax inspector administers from a code that is far more rigorous than that surrounding many environmental guidelines.

21.8 Literature

Alberini, A. et al. (2006): *Willingness to pay to reduce mortality risks: evidence from a three-country contingent valuation study*. Environmental & Resource Economics, 33, 251-264

Andersen, M.S. et al. (2008): *EVA – a non-linear Eulerian approach for assessment of health-cost externalities of air pollution*, paper presented at EAERE conference in Gothenburg, Sweden, June 2008. http://www.dmu.dk/samfund/miljoeoekonomi/EVA

Coase, R. (1960): *The problem of social cost,* Journal of Law and Economics, Oct., 1-44.

Diamond, P.A. and Hausman, J.A. (1994): *Contingent valuation – is some number better than no number.* Journal of Economic Perspectives, *8,4,* 45-64.

Ekins, P. (1997): *The Kuznet's curve for the environment and economic growth: examining the evidence,* Environment and Planning, *A 29,5* 805-830.

EEA, European Environment Agency (2006): *Using the market for cost-effective environmental policy: Market-based instruments in Europe,* Report 1, Copenhagen, Denmark.

Friedrich, R. and Bickel, P. (2001): *Environmental External Costs of Transport.* Springer. München, Germany.

Jones-Lee, M.W. et al(1985): *The value of safety: results of a national sample study,* Economic Journal, *95,* 49-72.

Lundqvist, L.J. (1980): *The hare and the tortoise: Clean air policies in the US and Sweden,* The University of Michigan Press. Ann Arbor, USA.

Matsuno, Y. and Ueta, K. (2000): *A socio-economic evaluation of the SOx charge in Japan,* pp. 194-214 in Andersen, M.S. and Sprenger, R.U. (eds.), Market-based instruments for environmental management, Edward Elgar, UK.

OECD (2006): *Cost-benefit analysis and the environment: recent developments,* Paris, France.

Pigou, A.C. (1920): *The Economics of Welfare.* Macmillan, London, UK.

Pope, C.A. et al. (2002): *Lung cancer, cardiopulmonary mortality and long-term exposure to fine particulate air pollution.* Journal of American Medical Association, *28,:9,* 1132-1141.

Sprenger, R.U. (1998): *Environmental policy and competitiveness: the case of Germany,* pp. 197-240, in Barker, T. and Köhler, J. (eds.), International competitiveness and environmental policies, Edward Elgar, UK.

Tsuru, S. and Weidner, H. (1989): *Environmental policy in Japan,* Edition Sigma, , Berlin, Germany.

Viscusi, W.K. and Aldy, J.E. (2003): *The value of a statistical life: a critical review of market estimates throughout the world,* Journal of Risk and Uncertainty, *27:1,* 5-76.

22 Legislation and administration

Finn Palmgren

The European legislation and other types of agreements affecting acidification and eutrophication are mainly the UNECE (United Nations Economic Commission for Europe) Convention on Long-range Transboundary Air Pollution (CLRTAP) and its Protocols, the EU directives and national laws.

22.1 The Convention on Long-range Transboundary Air Pollution

Sulphur emissions in Europe started to increase after the Second World War. Acid precipitation, acidification and the subsequent serious damages to life in lakes and rivers were observed around 1970 in the Scandinavian countries. This was reported at the UN Stockholm Conference in 1972. The problem of transboundary air pollution – not only for sulphur, but also other pollutants – was put on the political agenda.

With reference to the declaration of the 1972 UN Conference on the Human Environment in Stockholm, to the effect that states have an obligation to ensure that activities carried out in one country do not give rise to environmental damage in others, the Scandinavian countries jointly presented a draft for a convention. The then 35 members of the UNECE, including the European Community signed the Convention on Long-range Transboundary Air Pollution (CLRTAP) in Geneva in 1979. After ratification by 24 of the signatories, it came into force in March 1983. 49 countries signed in 2003. The Convention does not in itself call for any binding commitments, but countries shall "endeavour to limit and, as far as possible, gradually reduce and prevent air pollution," and they shall use "the best available technology which is economically feasible."

The Convention sets up an institutional framework bringing together research and policy. For example, since 1977 the monitoring of transboundary air pollution has been carried out under the European Monitoring and Evaluation Programme (EMEP). The EMEP collates data on the national emissions of sulphur and nitrogen (ammonia and nitrogen oxides), as well as data on transformation and transport in the atmosphere and deposition. The parties of the Convention took over the long-term financing of EMEP in 1984.

The Convention has been extended by specific protocols, five of which are significant for addressing acidification and eutrophication:

- The 1985 Protocol on the Reduction of Sulphur Emissions or their Transboundary Fluxes by at least 30% (entered into force 1987)
- The 1988 Protocol concerning the Control of Nitrogen Oxides or their Transboundary Fluxes (entered into force 1991)
- The 1991 Protocol concerning the Control of Emissions of Volatile Organic Compounds or their Transboundary Fluxes (entered into force 1997)
- The 1994 Protocol on Further Reduction of Sulphur Emissions (entered into force 1998)
- The 1999 Protocol to Abate Acidification, Eutrophication and Ground-level Ozone, the socalled Gothenburg Protocol, entered into force 2005 (UNECE, 1999)

The Convention has helped generate data. It has moreover promoted the exchange of knowledge and experience and influenced the decisions of various countries with regard to their measures for reduction of emissions. The process has put pressure from public opinion to get a protocol signed and respected.

The First Sulphur Protocol

In the spring of 1983 the Scandinavian countries put forward a proposal for limiting the emissions of sulphur. After two years of negotiating, a protocol was signed in Helsinki, Finland, in 1985, and it came into force in September 1987. It requires the signatories to reduce their national yearly emissions of sulphur, or its transboundary fluxes, by at least 30% by 1993 at the latest, from their 1980 levels. The 30-percent criterion was to be regarded as the first step in a long-term project for reducing emissions. Some of the greatest polluters, such as Poland, Britain, and Spain, did not sign the protocol. Between 1980 and 1993, the 20 European countries which ratified the protocol reduced their annual emissions by 55 per cent, while total European emissions of sulphur dropped by 43 per cent.

The critical loads approach

In 1988 the Convention appointed a new working group to develop a common critical-loads approach and to evolve abatement strategies based on that approach. The essence of the critical loads approach is that reductions of emissions are to be negotiated with a view to the effects of air pollutants, rather than by setting an equal percentage of reduction for all countries. The aim is to reduce, in a cost-effective manner, the emissions of air pollutants to levels where the critical loads will not be exceeded. This concept provided an acceptable, effects-based scientific approach to strategies for the abatement of air pollution. Each country was to make maps, showing the critical loads and levels for various areas, receptors, and pollutants in its own territory. The resulting data was assembled into Europe-wide maps showing exceedances of the critical loads and level. Computer models for integrated assessment enabled comparisons to be made of the cost-effectiveness of various strategies for achieving specified interim targets for environmental quality and the protection of health. Agreements were then reached on the reduction of emissions (interim targets) strategies for the abatement of emissions, and the reductions to be allocated among the various countries in the form of national ceilings for emissions.

The Second Sulphur Protocol

The first result of the critical loads approach was the 1994 Second Sulphur Protocol, which came into force in 1998. It sets differing requirements for each country – the aim being to attain the greatest possible effect for the environment at the least overall cost. It also includes some specific requirements for large combustion plants. The text for basic obligations says that "parties shall control and reduce their sulphur emissions in order to protect human health and the environment from adverse effects and ensure that sulphur depositions do not, in the long term, exceed critical loads". The scientific analysis of the protocol showed that in order to comply with the long-term goal; the emissions of sulphur should be reduced by at least 90%. The countries are commited under the protocol to reduce total European emissions of sulphur by 50% by 2000, and 58% by 2010, in relation to the level in 1980.

The NO_x Protocol

In the meantime eutrophication was observed on sensitive ecosystems, e.g. raised bogs, moors, lakes and coasts-near seas. In addition, ground-level ozone was now realised as an environmental problem in relation to health as well as damages on vegetation. NO_x plays an important role in both cases.

The 1988 Protocol on the control of nitrogen oxides, which came into force in 1991, provides that emissions after 1994 should not exceed the 1987 level. It does not call for reduction, but defines the basis for a next step involving measures to reduce emissions, taking into account internationally accepted critical loads. Twelve signatories pointed at the weakness of this protocol by proposing separately, in a joint declaration, to reduce their NO_x emissions by 30% by 1998 at the latest. By 1994 the European emissions were reduced by about 16% in relation to the 1987 levels. From the reported emission data it appeared however that three countries that had ratified the Protocol – Greece, Luxembourg, and Spain – had not managed even to freeze emissions. And of the 12 countries that were aiming at a 30% reduction, only four or five had succeeded.

The Gothenburg protocol

The 1999 Gothenburg Protocol aims at significant reduction of acidification, eutrophication, and the formation of ground-level ozone by setting national ceilings for emissions of the four pollutants that give rise to these effects, namely SO_2, NO_x, VOCs, and ammonia. Starting from the critical loads approach and by attacking several environmental problems and several pollutants simultaneously in a co-ordinated manner, the overall level of cost-effectiveness could be improved even further. The Protocol also contains binding requirements in the form of emission limit values both for stationary and mobile sources, as well as fuel standards. The European emissions of SO_2, NO_x, VOCs, and NH_3 are expected to decrease by respectively 63, 40, 40, and 17% between 1990 and 2010. In order to attain the internationally agreed long-term aim of no more exceeding of the critical loads, a stepwise approach involving reviews of this protocol is foreseen. The Protocol is not yet in force.

22.2 The EU directives

The EU directives, which directly affect emissions and concentrations of air pollutants are briefly considered in the following. In addition, a number of directives and other actions at EU level can have indirect effect, e.g. those aimed at reducing the emissions of greenhouse gases and others capable of influencing developments in the energy, transportation, and the agricultural sectors.

EU directives on emissions of air pollutants

The key legislation dealing with industrial emissions to air was the Air Framework Directive 84/360/EEC on Industrial Plants (EEC 1984). This framework Directive requires that certain types of industrial plants should only operate if they have been authorised. The authorisation must include emission limits for important pollutants, including SO_2, NO_x, CO and heavy metals. An authorisation may only be issued if, amongst other things, the competent authority is satisfied that best available technology not entailing excessive costs has been applied to reduce pollution and that applicable air quality limit values have been taken into account. The directive "Control of Emissions from Large Combustion Plants" (2001/80/EC) covers plants with a rated thermal capacity of at least 50 MW and replaces the former daughter directive of 1988 (88/609/EC). It contains emission limits for sulphur dioxide, nitrogen oxides, and dust. It tightens up the requirements for new plants, and introduces for the first time emission limits for existing ones. In 1996 The European Council adopted Directive 96/61/EC on "Integrated Pollution Prevention and Control" (IPPC) (EC 1996b). Its purpose is to reduce pollution to the environment as a whole, avoiding transfer from one medium to another. As far as air pollution is concerned, the IPPC Directive will replace Directive 84/360/EEC. Industrial operators will be required to employ Best Available Techniques, taking into account economic considerations.

The directive on "Sulphur Content of Certain Liquid Fuels" (99/32/EC) sets the maximum permitted concentration for sulphur in heavy fuel oil used in the EU at 1% from 2003, and for gas oils at 0.2%, to be reduced to 0.1% from 2008.

Emissions of air pollutants from road vehicles

EC legislation on vehicle emissions and fuel quality standards has evolved greatly since the first Directive setting emission limits for petrol vehicles in 1970. The early legislation, whilst aiming to reduce pollution, was designed to avoid different standards in different Member States which might act as a barrier to trade. Increasingly, legislation on vehicle emissions is designed with a view to meeting air quality targets.

The directive "Quality of Petrol and Diesel Fuels" (98/70/EC) prescribes among other things 350 and 150 ppm as maximum sulphur content for diesel and petrol respectively. From 2005 the figure was lowered in both cases to 50 ppm (0.005 per cent).

Three directives address mainly the emissions of nitrogen oxides, non-methane volatile organic compounds, and small particles. The directive for passenger cars and light commercial vehicles (98/69/EC) specifies emission standards to be introduced in two steps – the first in 2000 and the second in 2005. Directive 99/96/EC takes a similar stepwise approach for heavy vehicles, but with the inclusion of a third step (for 2008). Directive 97/24/EC sets emission standards for two and three-wheeled vehicles, mopeds and motorcycles. A proposal for an amendment, with stricter standards for motorcycles, which was presented in 2000, was agreed in 2002.

General emission directives

The directive on "National Emission Ceilings for Acidifying and Ozone-Forming Air Pollutants" (2001/81/EC) sets binding ceilings to be attained by each member state by 2010 and covers four air pollutants: SO_2, NO_x, VOCs and NH_3. The member countries' total emissions of these four pollutants are to be reduced by 77, 51, 54, and 14% respectively between 1990 and 2010. The directive includes environmental objectives on which the ceilings are based. A new directive update in connection with the "Thematic Strategy on Air Pollution" was presented in 2005.

Air quality directives

The first EC Directive on ambient air quality was adopted by The European Council in 1980. Council Directive 80/779/EEC (EEC 1980) on Air Quality Limit.

'limit value' shall mean a level fixed on the basis of scientific knowledge, with the aim of avoiding, preventing or reducing harmful effects on human health and/or the environment as a whole, to be attained within a given period and not to be exceeded once attained.

Values and Guide Values for Sulphur Dioxide and Suspended Particulates were adopted to protect human health and the environment against adverse effects from SO_2 and Suspended Particulates.

The Directive lays down limit values for SO_2 and Suspended Particulates which are mandatory throughout the territory of Member States. The Directive also sets long term guide values. Member States are required to measure SO_2 and particulate matter, to ensure that the limit values are met, and to inform the Commission of any breaches of the limit values and to undertake any necessary abatement measures. The Directive was followed by Directives setting air quality limit values for lead (Pb) and NO_2 (EEC 1982) (EEC 1995). Table 22.1 shows the main air quality limit values in force in the European Union since the end of 1997.

Pollutant	Directive	Parameter	Limit Value, $\mu g\ m^{-3}$
Sulphur dioxide, SO_2	80/779/EEC	98 percentile of all daily mean values taken throughout the year.	250 (if particles < 150)
		98 percentile of all daily mean values taken throughout the year.	350 (if particles ≥ 150)
Particulate matter (measured as Black Smoke)	80/779/EEC	98 percentile of all daily mean values taken throughout the year.	250
		median of daily mean values throughout the year	80
Lead, Pb	82/884/EEC	Annual mean	2
Nitrogen dioxide, NO_2	85/203/EEC	98 percentile of all daily mean values taken throughout the year.	200

Table 22.1. EC Air Quality Limit Values in force since 1997.

Their *primary aim* was to protect human health, though it was specifically recognised in the case of SO_2 and NO_2 that meeting the limit values would also reduce damage to the environment.

The Directives were based on the best scientific evidence available at the time, and in particular the work of the World Health Organisation, but there has been further research on the effects of air pollution on both human health and the environment, which should be taken into account. In addition, implementation of the existing Directives revealed a number of problems:

- The Directives included more ambitious long-term objectives for SO_2 and NO_2. They are not mandatory and have not been widely adopted as operational targets by Member States
- There were large differences in air quality monitoring and assessment strategies within EU
- The methods and techniques used for monitoring air quality were not always comparable
- The Directives required Member States to report only exceedances of air quality limit values to the Commission, but better information is needed in relation to future legislation and information to the public

It was therefore decided that the European Union should bring air quality limit directives up to date. The goals are set out in the Fifth Programme of Action on Sustainable Development and the Environment. They are:

- Provision of effective protection of all citizens against recognised effects of air pollution
- Establishment of permitted concentration levels of air pollutants which take into account the protection of the environment

The first result is Directive 96/62/EC on Ambient Air Quality Assessment and Management, adopted by Council in September 1996 (the Air Quality Framework Directive) (EC 1996a). The second is the Council Decision establishing a reciprocal exchange of information and data from networks and individual stations measuring ambient air pollution within the Member States (EC 1997a).

The main aims of the Air Quality Framework Directive are to
- Define and establish objectives for ambient air pollution in the Community designed to avoid, prevent and reduce harmful effects on human health and the environment as a whole
- Assess ambient air quality in Member States on the basis of common methods and criteria
- Oobtain adequate information on ambient air quality and ensure that it is made available to the public inter alia by means of alert thresholds
- Maintain ambient air quality where it is good and improve it in other cases

The Directive is a framework directive which provides a basic structure, which must be filled for pollutant by pollutant by means of daughter legislation. Figure 22.1 shows how this basic structure will operate.

The legislation requires the Member States to divide their territory into zones and must assess air quality annually in all of them. The Directive applies throughout the Member States.

Daughter Directives setting limit values will include the date by which the limit values must be attained. The margin of tolerance is a new concept in EC legislation on air quality. It is not a derogation of the limit value, but a trigger for action in the period before the limit value must be met. As Figure 22.1 shows, the *margin of tolerance* is added to the limit value when the legislation setting the limit value comes into force. It is reduced each year to reach zero on the attainment date. The purpose of the margin of tolerance is to identify the zones with the worst air quality. Member States must prepare detailed action plans for these Group 1 in Figure 22.1 showing how the limit value will be met on time.

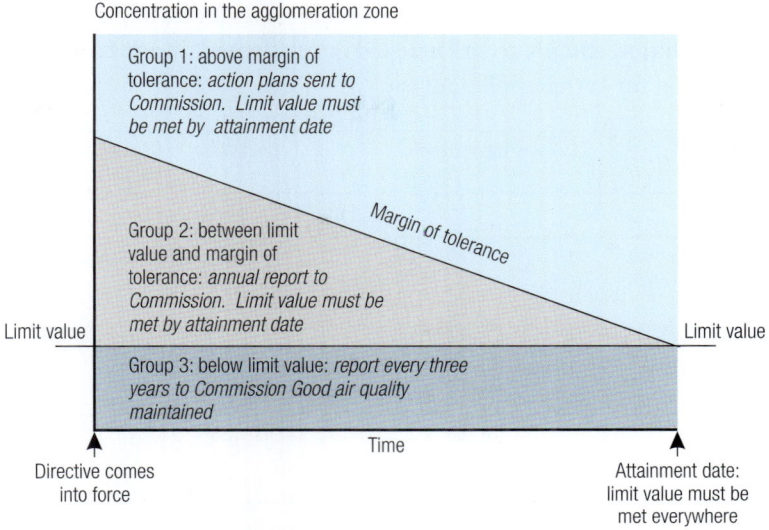

Figure 22.1. Schematic diagram of structure of the framework directive.

Zones where maximum pollution levels are between the limit value and the limit value plus margin of tolerance (Group 2 in Figure 22.1) are not required to forward detailed action plans to the Commission. But they must report concentrations annually to the Commission and must take any necessary steps to ensure that the limit value is met by the attainment date. The Commission will regularly publish data on air quality in the Member States.

Annex I of the Air Quality Framework Directive lists the pollutants for which air quality objectives are to be set (Table 22.2).

Pollutants for which there is existing EC legislation	Other pollutants
Sulphur dioxide	Benzene
Nitrogen dioxide	Carbon monoxide
Lead	Cadmium
Suspended particulate matter	Arsenic
Fine particulate matter such as soot (including PM_{10})	Nickel
Ozone	Mercury

Table 22.2. Pollutants listed in Annex I of Directive 96/62/EC.

The Air Quality Framework Directive envisages two types of air quality objectives. The air quality limit values in the first two daughter directives are listed in Table 22.3 and 22.4. It is defined as a level fixed on the basis of scientific knowledge with the aim of avoiding, preventing or reducing harmful effects on human health and/or the environment as a whole, to be attained within a given period and not to be exceeded once attained. In addition, alert thresholds are set for certain pollutants.

	Averaging period	Limit value	Date by which limit value is to be met
Sulphur dioxide			
1. Hourly limit value for the protection of human health	1 hour	350 µg m^{-3} not to be exceeded more than 24 times per calendar year	1 January 2005
2. Daily limit value for the protection of human health	24 hours	125 µg m^{-3} not to be exceeded more than 3 times per calendar year	1 January 2005
3. Limit value for the protection of ecosystems, to apply away from the immediate vicinity of sources	Calendar year and winter (1 October to 31 March)	20 µg m^{-3}	two years from entry into force of the Directive
Oxides of nitrogen			
1. Hourly limit value for the protection of human health	1 hour	200 µg m^{-3} NO_2 not to be exceeded more than 18 times per calendar year	1 January 2010
2. Annual limit value for the protection of human health	Calendar year	40 µg m^{-3} NO_2	1 January 2010
3. Annual limit value for the protection of vegetation	Calendar year	30 µg m^{-3} $NO + NO_2$	two years from entry into force of the Directive
Particulate matter			
Stage 1			
1. 24-hour limit value for the protection of human health	24 hours	50 µg m^{-3} PM_{10} not to be exceeded more than 35 times per year[1]	1 January 2005
2. Annual limit value for the protection of human health	Calendar year	40 µg m^{-3} PM_{10}	1 January 2005
Stage 2			
1. 24-hour limit value for the protection of human health	24 hours	50 µg m^{-3} PM_{10} not to be exceeded more than 7 times per year	1 January 2010
2. Annual limit value for the protection of human health	Calendar year	20 µg m^{-3} PM_{10}	1 January 2010
Lead			
Limit value for the protection of human health	Calendar year	0.5 µg m^{-3}	1 January 2005

Table 22.3. The limit values for SO_2, NO and NO_2, particulate matter and lead in the first daughter directive (EC 1999).

	Averaging period	Limit value	Date by which limit value is to be met
Benzene			
Limit value for the protection of human health	Calendar year	5 µg m^{-3}	2010
Carbon monoxide			
Limit value for the protection of human health	Maximum daily 8-hour mean	10 mg m^{-3}	

Table 22. 4. The limit values for benzene and CO in the second daughter directive (EC, 2000).

No non-effect level has been defined for ozone, the heavy metals and PAH's. The directives therefore set target values for these pollutants, Table 22.5 and 22.6.

Target values for ozone		
	Parameter	Target value for 2010
Target value for the protection of human health	Maximum daily 8-hour mean	120 µg m^{-3} not to be exceeded on more than 25 days per calendar year averaged over three years
Long-term objectives for ozone		
	Parameter	Long-term objective
1. Long-term objective for the protection of human health	Maximum daily 8-hour mean within a calendar year	120 µg m^{-3}
2. Long-term objective for the protection of vegetation	AOT40, calculated from 1 h values from May to July	6 000 µg m^{-3}·h^{-1}

Table 22.5. Target value and long term objectives for ozone in the fourth daughter directive (EC, 2002).

Pollutant	Target value [1]
Arsenic	6 ng m^{-3}
Cadmium	5 ng m^{-3}
Nickel	20 ng m^{-3}
Benzo(a)pyrene	1 ng m^{-3}

(1) For the total content in the PM$_{10}$ fraction averaged over a calendar year.

Table 22.6. Target values for arsenic, cadmium, nickel and benzo(a)pyrene (EC, 2004) ("target value" means a concentration in the ambient air fixed with the aim of minimising harmful effects on human health and the environment).

Air quality assessment includes measurement, the compilation of emission inventories and air quality modelling. The first Commission proposal for a daughter directive refers to these two levels as the *upper and lower assessment thresholds*. Table 22.7 summarises its requirements.

Area	Assessment regime, from the most demanding (top) to the lowest (bottom) requirements
1. Where levels are above the upper assessment threshold	Based on high quality measurements - may be supplemented by modelling
2. Where levels are between the upper and lower assessment thresholds	Combination of high quality measurement (but less intensive than in case 1) and modelling allowed
3. Where levels are below the lower assessment threshold	At least one high quality measuring site per agglomeration, combined with modelling, objective estimation, indicative measurements
a. In agglomerations for pollutants for which an alert threshold has been set	At least one high quality measuring site per agglomeration, combined with modelling, objective estimation, indicative measurements
b. In all other cases	Modelling, objective estimation, indicative measurements

Table 22.7 Assessment requirements under Directive 96/62/EC.

The aim is:
- To ensure that the most intensive assessment is carried out in those agglomerations and other zones within which there is the highest risk of a limit value being exceeded
- To ensure that the least intensive requirements apply only where pollution levels are sufficiently low that there is virtually no risk of an exceedance. It should be noted however that if an alert threshold has been set for a pollutant measurements must be made within agglomerations no matter how low the level of pollution

Daughter legislation on each pollutant will fill in the framework by including:

- Criteria and techniques for measurement, including the location of sampling points, the minimum number of sampling points and reference measurement and sampling techniques
- Criteria for the use of other techniques for assessing ambient air quality, particularly modelling
- Definition of the upper and lower assessment thresholds

The Member States have the responsibility to decide on the best means to tackle local problems.

22.3 Thematic Strategy on Air Pollution

Despite significant improvements, serious air pollution impacts persist within the EU. The Community's Sixth Environmental Action Programme (6th EAP) therefore called for the development of a "Thematic Strategy on Air Pollution" with the objective to attain "levels of air quality that do not give rise to significant negative impacts on, and risks to human health and the environment" (EC, 2005). Under the Clean Air for Europe programme (CAFE), the Commission has examined whether current legislation is sufficient to achieve the 6th EAP objectives by 2020. This analysis looked at future emissions and impacts on health and the environment and has used the best available scientific and health information. It showed that significant negative impacts will persist even with effective implementation of current legislation. The main problems are particles, NO_x, O_3 and NH_3.

The Thematic Strategy establishes interim objectives for air pollution in the EU and proposes appropriate measures for achieving them. It recommends that current legislation be modernised, be better focused on the most serious pollutants and that more is done to integrate environmental concerns into other policies and programmes, (EC, 2005; EC, 2008).

Part of the strategy will be implemented through a revision of the current ambient air quality legislation comprising two main elements:
- Streamlining of existing provisions and merging the four daughter directives and the exchange of information decision into a single directive
- The introduction of new air quality standards for fine particulate matter ($PM_{2.5}$) in air and the PM_{10} limit value for annual average in stage 2 will be replaced by the limit value in stage 1

The standards for $PM_{2.5}$ include a concentration loft on 25 µg m^{-3} to be met by 2010 at al locations in the Member States. In addition a 20% decrease at urban background sites between 2010 and 2020 in relation to the level in 2008-2010.

The national emission ceilings directive (NECD) will also be revised to ensure reduced emissions of NO_x, SO_2, VOCs, NH_3 and primary particulate matter consistent with the interim objectives proposed for 2020.

22.4 National legislation

The EU legislation shall be implemented in the national legislation. It means that the Member States have to prepare national laws, practical guidelines and monitoring programmes for air quality and emissions.

The EU directives must be implemented, e.g. as regulations, which often include the text of the directives.

The guidelines include tools for preparation of permissions to new installations and advices to control programmes. An example is the Danish guidelines (Danish Environmental Protecion Agency, 2002). As an example the Danish guidelines for control of air pollution from industries are mainly based on the IPPC directive (EC, 1996 b) include:

- The BAT (Best Available Technology) principle
- Application of BAT
- Mass-flow limits
- Emission limit values
- Calculation of outlet height using the OML (Operational Meteorological Air Pollution) model
- Drawing up terms and inspection rules
- Methods of sampling and analysis
- Emission limit values for energy plants
- Requirements for the design of tanks and silos
- Recommended emission limit values etc. for thermal and catalytic oxidation installations for the destruction of organic solvents

The air quality monitoring programmes have to comply with the air quality directives (EC, 1996a; EC, 2008). In addition, the Member States have special national interests and needs for monitoring the air quality, and finally the European countries have to participate actively in the EMEP monitoring programme.

22.5 Application of modelling

Modelling plays a central role in regulation of air pollution in the global, regional, urban and the local scales. The models are often used in a combination with the air quality measurements as a supplement to the measurements, for optimisation of the monitoring programmes and for preparation of forecasts in relation to measures taken or planned to be taken to reduce the air pollution.

The international conventions, e.g. the European CLRTAP, are generally based on large scale model results of air quality and deposition of pollutants, including assessment of the effect on the ecosystems (critical load). This type of models is mainly developed under the EMEP programme. Recently, the Thematic Strategy on Air Pollution is based on model calculations of different scenarios by the EMEP model complex and the corresponding cost-benefit model estimates (EC, 2005).

Similar models are used on a finer scale, e.g. country wide. They calculate air quality and deposition of pollutants responsible for eutrophication, acidification and direct damages on vegetation (e.g. ozone), but also heavy metals, persistent organic pollutants etc. The aims of such model calculations are to assess and control the impact of national air emissions, e.g. ammonia, on the national environment.

Point and area source models are widely applied in the national permit systems for industrial installations, power plants and other local sources (e.g. Danish Environmental Protection Agency, 2002).

Special air pollution models are developed for assessment of the very local impact, e.g. street pollution models. The street pollution models are often combined with urban scale models in order to asses the total air pollution exposure of the population.

The air quality models are until now only an indicative tool for test of compliance with the EU limit values, but acceptance of models for this purpose is expected, when the accuracy and reliability is sufficiently good.

22.6 Literature

Danish Environmental Protection Agency. (2002): *Guidelines for Air Emission Regulation. Limitation of air pollution from installations*, Environmental Guidelines Nr. 1

EMEP: *Co-operative Programme for Monitoring and Evaluation of the Long-range Transmission of Air pollutants in Europe.* http://www.emep.int/index_facts.html

EEC (1980): *Directive 80/779/EEC on air quality limit values and guide values for sulphur dioxide and suspended particulates*, Official Journal L 229, 38

EEC (1982): *Directive 82/884/EEC on a limit value for lead in air*, Official Journal L 372, 15

EEC (1984): *Directive 84/360/EEC on the combating of air pollution from industrial plants*, Official Journal L 188, 20

EEC (1989): *Directive 89/427/EEC amending Directive 80/779/EEC air quality limit values and guide values for sulphur dioxide and suspended particulates*, Official Journal L 201, 53

EEC (1985): *Directive 85/203/EEC on air quality standards for nitrogen dioxide*, Official Journal L 87, 1

EC (1996a): Directive 96/62/EC on Ambient Air Quality Assessment and Management, Official Journal L 296, 55

EC (1996b): *Directive 96/61/EC concerning Integrated Pollution Prevention and Control*, Official Journal L 257, 10

EC (1997a): *Council Decision 97/101/EC establishing a reciprocal exchange of information and data from networks and individual stations measuring ambient air pollution within the Member States*, Official Journal L 35, 14

EC (1999): *Council Directive 1999/30/EC relating to limit values for sulphur dioxide, oxides of nitrogen, particulate matter and lead in ambient air*, Official Journal L 163, 41

EC (2000): *Directive 2000/69/EC relating to limit values for benzene and carbon monoxide in ambient air*, Official Journal L 313, 12

EC (2001): *Directive 2001/81/EC on national emission ceilings for certain atmospheric pollutants*, Official Journal L 309, 22

EC (2002): *Directive 2002/3/EC relating to ozone in ambient air*, Official Journal L 67, 14

EC (2004): *Directive 2004/107/EC relating to arsenic, cadmium, mercury, nickel and polycyclic aromatic hydrocarbons in ambient air*, Official Journal L 23, 3

EC (2005): *Communication From the Commission to the Council and the European Parliament. Thematic Strategy on air pollution.* COM (2005) 446 final.21.9.2005

EC (2008): *Directive 2008/50EC of the European Parliament and of the Council on ambient air quality and cleaner air for Europe. Official Journal L 152/1.*

UNECE (1999): *Protocol to Abate Acidification Eutrophication and Groun-level ozone. http://www.UNECE.org/env/lrtap/multi_h1.htm.*

WHO (1998): *Air Quality Guidelines for Europe.* WHO, Copenhagen, Denmark

Concluding remarks

Jes Fenger and Jens Chr. Tjell

Humanity stresses the Earth. We are too greedy, too sloppy and – in reality – too many. One of the problems is our pollution of the atmosphere. It started in the dawn of civilisation, but has in time changed both in character and scale. The first attempt of mitigation was a simple dispersion, but it only meant that the problems were shifted from indoor, over local to regional scale. The modern technology can in many respects give us a breathing space, but modern awareness of the impacts has now moved us into a new, global scale. Concurrently the coupling between the immediate local emission and the unpleasant impacts is reduced – both geographically and in time.

The problem with air pollution used to be simple. Everybody could see the black smoke that came out of a low chimney – and their number was so small that it was not necessarily a serious problem, if the smoke could be dispersed with the wind. Now this has changed, because air pollution is dispersed at all geographical scales, and the relation between emissions of contaminants, the resulting pollution of the air, and the impact of the pollution are by no means simple.

Effective management therefore requires monitoring and research. We must know the causes of pollution and how we can solve the problems in the cheapest and most efficient way. Interplay of many technical disciplines is required.

It is necessary to follow the development, to detect unfortunate tendencies, and to investigate whether regulatory measures have had the expected impact. And these activities must be carried out by international co-operation. Model calculations on computers of increasing strength are becoming increasingly important.

When the NEC Directive on national emission ceilings in Europe was acceded in the EU in October 2001 and the Gothenburg Protocol in December 1999, binding limits for air pollution from the individual countries were defined for the first time. It has been estimated that if and when the Gothenburg Protocol is fully implemented, the acidified areas in Europe will be reduced by 85%, the *eutrophied* areas by 65%, and areas suffering from excessive ozone pressures by 50%. Also the impact on human health will be significantly reduced. It must however be admitted that the emissions from international shipping and aircrafts are still not under control and will become increasingly important.

In the industrialised countries in Western Europe the problem with major air pollutants is thus reasonably well on its way to a satisfactory solution. However, a series of hazardous compounds still await more careful investigations. Small particles are still a very critical issue.

In the long run the most important problem is the increasing emissions and subsequent concentration of carbon dioxide and other greenhouse gases, leading to a distortion of the global energy balance and global climate changes. These compounds will not be toxic in realistic concentrations and they are therefore not subjected to concentration limit values. In any case limit values would have been of no use, since the greenhouse gases are distributed over the entire Earth. In a foreseeable future the Earth – as such – may not be a worse place to live in than it is today, but nature and we are used to it in the form it has, and changes alone may be inconvenient, certainly expensive or even catastrophic.

In may sound unjust, but the greenhouse issue is primarily becoming a problem for the developing countries. Already now the developing countries emit the same order of magnitude of the amounts of greenhouse gases as the industrialised countries – not because the individual citizen emits much, but because they are so many. If the required growth in material standard of living is not obtained without the corresponding increase in emission of greenhouse gases, any savings in the industrialised countries will be completely overruled. Further the impacts of climate change may be most severe in the tropics, where most of the developing countries are situated. Finally, the developing countries do not – since they are developing – have the infrastructure that permits an adjustment to the changing conditions.

Of course the industrialised countries must reduce their own emissions, but equally important is it therefore to help the developing countries with adequate access to clean technology. The ultimate goal is an out-phasing of the use of fossil fuels as an energy source. This will solve not only the climate change problem, but also many other pollution problems.

Index